Advances in Cognitive Neurodynamics

More information about this series at http://www.springer.com/series/11163

José M. Delgado-García • Xiaochuan Pan
Raudel Sánchez-Campusano • Rubin Wang
Editors

Advances in Cognitive Neurodynamics (VI)

Proceedings of the Sixth International Conference on Cognitive Neurodynamics – 2017

Springer

Editors
José M. Delgado-García
Division of Neuroscience
Pablo de Olavide University
Seville, Spain

Raudel Sánchez-Campusano
Division of Neuroscience
Pablo de Olavide University
Seville, Spain

Xiaochuan Pan
Institute of Cognitive Neurodynamics
East China University of Science
and Technology
Shanghai, China

Rubin Wang
Institute of Cognitive Neurodynamics
East China University of Science
and Technology
Shanghai, China

ISSN 2213-3569 ISSN 2213-3577 (electronic)
Advances in Cognitive Neurodynamics
ISBN 978-981-10-8853-7 ISBN 978-981-10-8854-4 (eBook)
https://doi.org/10.1007/978-981-10-8854-4

Library of Congress Control Number: 2008928127

Printed on acid-free paper

This Springer imprint is published by the registered company Springer Nature Singapore Pte Ltd.
The registered company address is: 152 Beach Road, #21-01/04 Gateway East, Singapore 189721, Singapore

Preface

The 6th International Conference on Cognitive Neurodynamics (ICCN2017) was held in Carmona (Seville), Spain, from August 1–5, 2017. It is one of the series conferences held biennially since 2007, with support from the international journal "Cognitive Neurodynamics" (Springer). The research field of cognitive neurodynamics is the frontier of union where experimental and mathematical/computational neuroscience converge with cognitive neuroscience. Experiments generate a huge amount of neural data that must be treated correctly to obtain the best outcomes and the most accurate interpretation of them. At the same time, mathematical/computational methods and modeling are applied to understand and reveal dynamic principles on brain structure and functions concerning some cognitive processes such as brain oscillations, learning and memory, and neural plasticity among other higher-order brain functions or dysfunctions. Undoubtedly, cognitive neurodynamics is highly interdisciplinary, where researchers from biomedical sciences, neuroscience, cognitive neuroscience, mathematics, physics, computer science, technological science, and engineering contribute together to the advance in this field. The series conferences of ICCN provide very good opportunities for scientists from various fields to review their achievements, to share their ideas, and to promote the development of this field.

ICCN2017 attracted more than 100 participants from 17 countries (Australia, Belgium, China, France, Germany, Italy, Japan, New Zealand, Portugal, Russia, Spain, South Korea, Sweden, Switzerland, The Netherlands, United Kingdom, and United States of America), who made this conference a successful and memorable scientific event. There were 6 plenary lectures by leading scientists in the field of cognitive neurodynamics, 12 symposia (with 60 oral presentations) also by prominent researchers, and 1 poster session (a total of 38 posters) by both researchers and PhD students. Posters were permanently displayed along the whole meeting, allowing a long time for questions and discussions. The plenary speakers were Profs. Drs. Pierre-Paul Vidal (France), Salvador Martínez (Spain), Chris De Zeeuw (The Netherlands), Yoshikazu Isomura (Japan), Guo-Qiang Bi (China), and Wu Li (China). The organizers of the symposia were Drs. Alberto Ferrus (Symposium 1); Jan Lauwereyns (Symposium 2); Laura M. Roa

(Symposium 3); Agnès Gruart (Symposium 4); José L. Cantero (Symposium 5); Yutaka Yamaguti, Akihiro Yamaguchi, and Ichiro Tsuda (Symposium 6); Juan de los Reyes Aguilar (Symposium 7); Yoshikazu Isomura (Symposium 8); Hans Liljeström (Symposium 9); Toshishisa Tanaka and Jianting Cao (Symposium 10); Raudel Sánchez-Campusano and Steven L. Bressler (Symposium 11); and Xu Lei (Symposium 12). In several symposia, a tribute was paid to Walter J. Freeman (January 30, 1927–April 24, 2016) for his groundbreaking contributions to cognitive neurodynamics.

The topics of the conference covered almost all the branches of cognitive neurodynamics, from micro-, meso-, to macro-level dynamics, their applications, and some related topics, especially including neural coding, neural population dynamics, sensory and motor dynamics, EEG, fMRI and brain imaging, global cognitive functions, realistic neural networks, oscillation and synchronization, neural computing, brain computer interface, cognition disorder, multiscale neurodynamics, and also the coordination dynamics from neural-to-mental-to-social systems.

This volume fairly well reflects the large span of research presented at ICCN2017 conference. The papers in this volume (51 chapters by a total of 147 authors) were organized in the following five parts: (I) Neural Dynamics in Motor and Sensory Systems and in Cognitive Functions (10 chapters); (II) Cognitive Network and Multi-Scale Neural Network Dynamics (10 chapters); (III) Neuroengineering, Neuroinformation, and Brain Computer Interaction (10 chapters); (IV) Modelling Higher-Order Functions and Dysfunctions (10 chapters); and (V) Oscillation, Synchronization, Neural Plasticity, and Coordination Dynamics from Neural to Social Systems (11 chapters). All submitted papers were peer-reviewed by experts in the field based on originality, significance, quality, and clarity, under the coordination of the contact volume editor Dr. Raudel Sánchez-Campusano (Pablo de Olavide University). From the organizing committee, we thank all the authors for the outstanding quality of the contributions to ICCN2017 conference proceedings.

Finally, we wish to express our gratitude to all those who made ICCN2017 conference and this proceedings volume possible. In addition to all the contributing authors, we especially thank the plenary speakers, the symposium organizers, and the helpful students who assisted during the conference. We gratefully acknowledge sponsorship from CeslatiC Foundation, Carmona City Hall, Cibertec S.A., Univerlab S.L., BioAvan I+D+I, Olavide en Carmona Center, and Pablo de Olavide University, for the ICCN2017 conference. Also we thank the journal "Cognitive Neurodynamics" by Springer for the publication of this book series.

The 7th conference in the series – ICCN2019 – will be held in Alghero, Sardinia (Italy), September 29–October 2, 2019; organized by Prof. Alessandro E.P. Villa and colleagues (NeuroHeuristic Research Group and LABEX – HEC Lausanne, University of Lausanne, Switzerland). We have no doubt that ICCN2019 will be as successful as the previous ones.

Seville, Spain

José M. Delgado-García
Raudel Sánchez-Campusano

Organizers and Sponsors

This conference is organized by

Division of Neuroscience, Pablo de Olavide University, Seville, Spain

Sponsored by

CeslatiC Foundation (Center of Latin American Studies for the Science and Culture), (Centro de Estudios Latinoamericanos para la Ciencia y la Cultura)

Cosponsored by

Carmona City Hall
Cibertec S.A.
Univerlab S.L.
BioAvan I+D+i
Olavide en Carmona Center
Pablo de Olavide University

Committees

General Chair

José M. Delgado-García (Pablo de Olavide University, Spain)

International Committee

Jan Lauwereyns (Kyushu University, Japan)
Hans Liljenström (SLU and Agora for Biosystems, Sweden)
IchiroTsuda (Hokkaido University, Japan)
Minoru Tsukada (Tamagawa University, Japan)
José M. Delgado-García (Pablo de Olavide University, Spain)
Rubin Wang (East China University of Science and Technology, China)

Local Organizing Committee

José M. Delgado-García (Chairman)
Agnès Gruart
Miguel Merchán
Juan Carlos López-Ramos
Rocío Leal-Campanario
Raudel Sánchez-Campusano

Young Local Organizing Committee

Ana Rocío Conde Moro
Florbela Da Rocha Almeida
Mar Reus García
José Antonio García Moreno

Scientific Committee

José M. Delgado-García (Pablo de Olavide University, Spain)
Pierre-Paul Vidal (COGNAC-G Université Paris Descartes-CNRS, Paris, France)
Alberto Ferrús (Instituto Cajal, CSIC, Madrid, Spain)
Jan Lauwereyns (Kyushu University, Nishi-ku, Fukuoka, Japan)
Salvador Martínez (INA, CSIC and Miguel Hernández University, Alicante, Spain)
Laura Roa (University of Seville, Seville, Spain)
Agnès Gruart (Pablo de Olavide University, Seville, Spain)
Chris De Zeeuw (Netherlands Institute for Neuroscience, Amsterdam, Netherlands)
José L. Cantero (Pablo de Olavide University, Sevilla, Spain)
Yutaka Yamaguti (Fukuoka Institute for Technology, Japan)
Akihiro Yamaguchi (Fukuoka Institute for Technology, Japan)
Ichiro Tsuda (Hokkaido University, Japan)
Yoshikazu Isomura (Brain Science Institute, Tamagawa University, Tokyo, Japan)
Juan de los Reyes Aguilar (Hospital Nacional de Parapléjicos, Toledo, Spain)
Guo-Qiang Bi (University of Science and Technology of China, China)
Hans Liljenström (SLU and Agora for Biosystems, Sweden)
Toshishisa Tanaka (Tokyo University of Agriculture and Technology, Japan)
Jianting Cao (Saitama Institute of Technology, Japan)
Wu Li (Beijing Normal University in China, China)
Steven L. Bressler (Florida Atlantic University, Florida, USA)
Raudel Sánchez-Campusano (Pablo de Olavide University, Seville, Spain)
Xu Lei (Southwest University in China, Chongqing, China)
Xiaochuan Pan (East China University of Science and Technology, China)
Alessandro E. P. Villa (University of Lausanne, Switzerland)
Rubin Wang (East China University of Science and Technology, China)

Paper Acceptance and Contact Volume Editor

Raudel Sánchez-Campusano (Division of Neuroscience, Pablo de Olavide University)

Secretariat

Secretary General

Antonio Quetglas (Division of Neuroscience, Pablo de Olavide University)

Technical Secretary

Antonio Vázquez (Grupo Pacífico)
José María Ávila (Grupo Pacífico)

Home Page

http://iccn2017.pacifico-meetings.com/index.php

Contents

Part I
Neural Dynamics in Motor and Sensory Systems and in Cognitive Functions

Chapter 1
Decomposition of Superimposed Chaotic Spike Sequences by Using the Bifurcating Neuron

Akihiro Yamaguchi, Yutaka Yamaguti, and Masao Kubo

Abstract In this study, decomposition of superimposed chaotic spike sequence was investigated from the view point of neural information coding. We construct simple network of bifurcating neuron and introduce the coupling model to decompose superimposed chaotic spike sequences. The decomposing performance was demonstrated by the numerical simulation and evaluated by the ratio of synchronized spikes. As a result, for the superimposed two chaotic spike sequences, approximately 90% of spikes were correctly decomposed.

Keywords Chaotic synchronization · Bifurcating neuron · Neural coding

1.1 Introduction

The temporal structure of spike firing timing is considered to play an important role in information processing in the brain. In our previous studies, we have shown segmentation and feature linking of input images by using the chaotic cellular neural network to achieve chaotic synchronization of evoked spike sequences [1, 2]. The neuron model used to generate spike sequences with chaotic inter-spike intervals was based on the bifurcating neuron [3] and described by the spike response model [4]. The bifurcating neuron is a chaotic integrate-and-fire neuron that was introduced by Lee and Farhat [3].

Advantages of a chaotic spike sequence include its diversity and exponential decay of correlation function. By using these properties, we were able to distinguish different chaotic spike sequences and link identical chaotic spike sequences. In

A. Yamaguchi (✉) · Y. Yamaguti
Faculty of Information Engineering, Fukuoka Institute of Technology, Fukuoka, Japan
e-mail: aki@fit.ac.jp

M. Kubo
Department of Computer Science, National Defense Academy of Japan, Yokosuka, Kanagawa, Japan

J. M. Delgado-García et al. (eds.), *Advances in Cognitive Neurodynamics (VI)*,
Advances in Cognitive Neurodynamics,
https://doi.org/10.1007/978-981-10-8854-4_1

this study, decomposition of superimposed chaotic spike sequences was investigated from the viewpoint of neural information coding by employing a simple network model that we constructed using the bifurcating neuron. In the following sections, we describe our network model to decompose superimposed chaotic spike sequences and present the results of the numerical simulations.

1.2 Simple Network Model to Decompose Superimposed Chaotic Spike Sequences

In our model, the bifurcating neuron [3] is employed to generate and to decompose a chaotic spike sequence which inter-spike interval dynamics is chaotic. In this section, we explain the dynamics of the bifurcating neuron and our simple coupling model of bifurcating neuros to decompose superimposed chaotic spike sequences.

1.2.1 Bifurcating Neuron

In this study, we describe the bifurcating neuron as a form of spike response model (SRM) [4] to clarify the coupling term. Here, we denote the i-th neuron as $n^{(i)}$. Let $u^{(i)}(t)$ be an internal potential of $n^{(i)}$ at time t and its dynamics is defined as:

$$u^{(i)}(t) = u_{rest} + \eta^{(i)}(t) + \nu^{(i)}, \tag{1.1}$$

where u_{rest} is the resting potential, $\nu^{(i)} \in [-\nu_0, +\nu_1]$ is the uniform noise, and $\eta^{(i)}(t)$ is a kernel function of internal state dynamics. In the case of the bifurcating neuron, $\eta^{(i)}(t)$ is defined as:

$$\eta^{(i)}(t) = \eta_0\left(t_{last}^{(i)}, \phi^{(i)}\right) + \alpha\left(t - t_{last}^{(i)}\right); \tag{1.2}$$

$$\eta_0(t, \phi) = A_\eta \sin(2\pi\omega\, t + \phi), \tag{1.3}$$

where $t_{last}^{(i)}$ is the last firing time of $n^{(i)}$ and the constant α is the linearly increasing ratio of $\eta^{(i)}(t)$. The internal potential $u^{(i)}(t)$ is linearly increasing by the η kernel. When $u^{(i)}(t)$ exceeds the threshold value θ, $n^{(i)}$ is fired and $u^{(i)}(t)$ is reset to the initial potential given by the background oscillation $\eta_0(t, \phi)$. The constants A_η, ω, and ϕ are the amplitude, the frequency, and the phase of the background oscillation, respectively. The dynamics of the bifurcating neuron is shown in Fig. 1.1a.

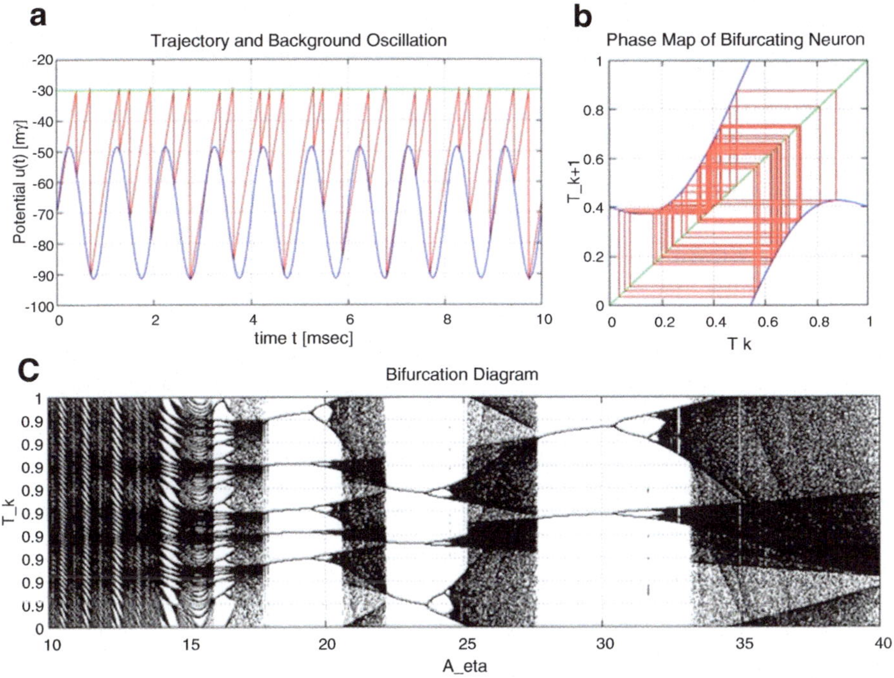

Fig. 1.1 The dynamics of the single bifurcating neuron. (**a**) Example of the time evolution where $\alpha = 100$, $\theta = -30$, $u_{rest} = -70$, $A_\eta = 21.5$, $\omega = 1$, and $\phi = 0$. The threshold value θ, the internal potential $u(t)$, and the background oscillation are represented by the green line, the red line, and the blue line, respectively. (**b**) The return map of the phase T_k of the firing time. (**c**) The bifurcation diagram of the single bifurcating neuron where the abscissa is the amplitude of background oscillation A_η and the ordinate is the phase T_k of the firing time

In the case without the noise term $\nu^{(i)}$, the $k+1$-th firing time $t_{k+1}^{(i)}$ of $n^{(i)}$ is simply determined by the map f and the previous firing time $t_k^{(i)}$ such as:

$$t_{k+1}^{(i)} = f\left(t_k^{(i)}; \phi^{(i)}\right) = t_k^{(i)} + \frac{\theta - u_{rest} - \eta_0\left(t_k^{(i)}, \phi^{(i)}\right)}{\alpha} \tag{1.4}$$

Furthermore, the phase $T_k^{(i)} = t_k^{(i)}$ *mod* 1 in the background oscillation is also determined by the one dimensional map:

$$T_{k+1}^{(i)} = F\left(T_k^{(i)}; \phi^{(i)}\right) = f\left(T_k^{(i)}; \phi^{(i)}\right) \ mod\ 1. \tag{1.5}$$

An example of map F is shown in Fig. 1.1b. As increasing A_η, dynamics of the phase $T_k^{(i)}$ shows various behavior including bifurcating one and chaotic one as shown in Fig. 1.1c.

1.2.2 Simple Network Model with Phase Response Coupling

Our network model consists of two types of neurons: a transmitter neuron and a receiver neuron. The transmitter neurons generate spike sequences with chaotic inter-spike intervals. The generated spike sequences are superimposed and inputted to the receiver neuron. The receiver neuron also generates spike sequences via its own dynamics and inputted sequences. These transmitter neurons with different inter-spike-interval dynamics are implemented by the bifurcating neuron (see Eq. (1.1)). In order to construct the network model, we introduce the coupling term to the bifurcating neuron. Let the set $\Gamma^{(i)}$ be a set of firing time of super imposed spike sequences inputted to the receiver neuron $n^{(i)}$ from transmitter neurons such that:

$$\Gamma^{(i)} = \left\{ s_0^{(i)}, s_1^{(i)}, s_2^{(i)}, \cdots \right\}, \tag{1.6}$$

where $s_j^{(i)}$ $(j = 0, 1, \cdots)$ is the firing time of the neurons coupled to $n^{(i)}$. The dynamics of the bifurcating neuron with phase response coupling is defined as:

$$u^{(i)}(t) = u_{rest} + \eta^{(i)}(t) + \xi_-^{(i)}(t) + \xi_+^{(i)}(t) + \nu^{(i)}, \tag{1.7}$$

where $\xi_-^{(i)}(t)$ and $\xi_+^{(i)}(t)$ are the negative coupling term and the positive one, respectively.

These coupling terms are designed to synchronize to the input spikes if its own dynamics is the same with the dynamics of input spike sequences. The definition of the negative and positive coupling terms are as follows:

$$\xi_-^{(i)}(t) = \sum_{s \in \Gamma^{(i)},\ t_{last}^{(i)} \le s < t} \varepsilon_- \left(s, t_{last}^{(i)} \right); \tag{1.8}$$

$$\varepsilon_-^{(i)} \left(s, t_{last}^{(i)} \right) = \begin{cases} 0 & s \le t_{last}^{(i)} \\ -\beta_- \frac{s - t_{last}^{(i)}}{\Delta_\varepsilon} & t_{last}^{(i)} < s \le t_{last}^{(i)} + \Delta_\varepsilon \\ 0 & t_{last}^{(i)} + \Delta_\varepsilon < s \end{cases}, \tag{1.9}$$

and

$$\xi_+^{(i)}(t) = \sum_{s \in \Gamma^{(i)},\ t_{last}^{(i)} \le s < t} \varepsilon_+ \left(s, \widehat{t}_{next}^{(i)}\right); \tag{1.10}$$

$$\varepsilon_+^{(i)} \left(s, \widehat{t}_{next}^{(i)}\right) = \begin{cases} 0 & s < \widehat{t}_{next}^{(i)} - \Delta_\varepsilon \\ +\beta_+ & \widehat{t}_{next}^{(i)} - \Delta_\varepsilon \le s < \widehat{t}_{next}^{(i)} \\ 0 & \widehat{t}_{next}^{(i)} \le s \end{cases}, \tag{1.11}$$

where β_- and β_+ are nonnegative coupling constant, Δ_ε is coupling time range where input spike is affective, $\varepsilon_-^{(i)}$ and $\varepsilon_+^{(i)}$ are phase response curves, and $\widehat{t}_{next}^{(i)}$ is a predicted next firing time such that:

$$\widehat{t}_{next}^{(i)} = t + \frac{\theta - u(t)}{\alpha}. \tag{1.12}$$

If the time s of the arrived spike is within the range Δ_ε from the last spike firing time $t_{last}^{(i)}$, then the phase response is negative to delay the next firing time. Otherwise, if the time s is within the range Δ_ε from the predicted next firing time $\widehat{t}_{next}^{(i)}$, then the phase response is positive to hasten the next firing time.

1.3 Numerical Experiments

In order to examine the decomposing performance of the proposed network, we numerically simulate the four neurons network where $n^{(0)}$ and $n^{(1)}$ are transmitter neurons Eq. (1.1) and $n^{(2)}$ and $n^{(3)}$ are receiver neurons (see Eq. (1.7)).

The generated spike sequences of $n^{(0)}$ and $n^{(1)}$ are superimposed and input to the receiver $n^{(2)}$ and $n^{(3)}$. The parameter values of these four neurons are identical without the phase shift value $\phi^{(i)}$. For the decomposition, the phase shift values are chosen as $\phi^{(0)} = \phi^{(2)}$ and $\phi^{(1)} = \phi^{(3)}$. Since the phase shift value characterizes the shape of the return map of firing phase (Fig. 1.1b), the internal dynamics of $n^{(2)}$ and $n^{(3)}$ are the same with $n^{(0)}$ and $n^{(1)}$, respectively.

Numerical simulations were performed for three cases such as (1) $\beta_- > 0$ and $\beta_+ = 0$, (2) $\beta_- = 0$ and $\beta_+ > 0$, and (3) $\beta_- > \beta_+ > 0$. Results of the numerical simulation for the case (3) are shown in Fig. 1.2. As shown in Fig. 1.2d–e, the receiver $n^{(2)}$ and $n^{(3)}$ synchronizes to the transmitter $n^{(0)}$ and $n^{(1)}$, respectively. The degree of synchronization is evaluated by the ratio of synchronized spikes between two neurons (Table 1.1). Here, the ratio of synchronized spikes is estimated by 10 trials of simulation and approximately 10,000 spikes are generated for each trial. For the case (3), approximately 90% of spikes are correctly decomposed by the synchronized response of receiver neurons.

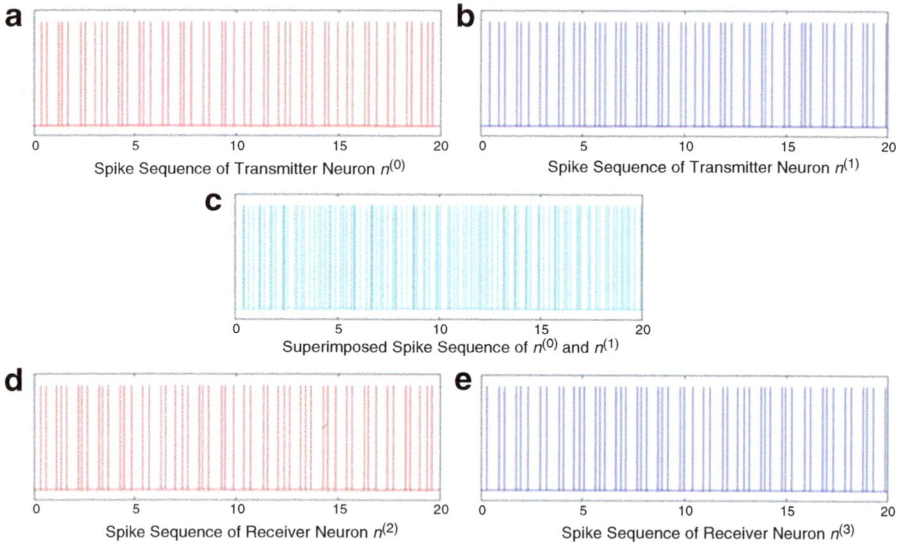

Fig. 1.2 Example of the numerical simulation of the proposed network model to decompose superimposed spike sequences, where $\beta_- = \beta_+ = 2.1$, $\Delta_\varepsilon = 0.05$, $\phi^{(0)} = \phi^{(2)} = 0$, $\phi^{(1)} = \phi^{(3)} = \pi$, and other parameters are same with Fig. 1.1a. (**a**) The spike sequence of the transmitter $n^{(0)}$. (**b**) The spike sequence of the transmitter $n^{(1)}$. (**c**) The superimposed spike sequence of $n^{(0)}$ and $n^{(1)}$. (**d**) The spike sequence of the receiver $n^{(2)}$. (**e**) The spike sequence of the receiver $n^{(3)}$

Table 1.1 Ratio of synchronized spikes between two neurons

	Ratio of synchronized spikes		
	$\beta_- = 2.1$	$\beta_- = 0$	$\beta_- = 2.1$
Target neurons	$\beta_+ = 0$	$\beta_+ = 2.1$	$\beta_+ = 2.1$
$n^{(0)}$ and $n^{(2)}$	$50.2 \pm 1.4\%$	$62.3 \pm 2.2\%$	$88.9 \pm 1.3\%$
$n^{(0)}$ and $n^{(3)}$	$19.8 \pm 0.7\%$	$20.9 \pm 0.5\%$	$20.1 \pm 0.6\%$
$n^{(1)}$ and $n^{(2)}$	$19.4 \pm 0.9\%$	$21.0 \pm 0.7\%$	$20.0 \pm 0.5\%$
$n^{(1)}$ and $n^{(3)}$	$49.2 \pm 1.6\%$	$63.6 \pm 2.1\%$	$89.4 \pm 2.1\%$
$n^{(0)}$ and $n^{(1)}$	$19.9 \pm 0.7\%$	$19.9 \pm 0.6\%$	$19.9 \pm 0.4\%$
$n^{(2)}$ and $n^{(3)}$	$19.3 \pm 0.5\%$	$23.5 \pm 0.5\%$	$22.5 \pm 0.6\%$

1.4 Summary and Discussion

In this study, we proposed the coupling model to decompose superimposed chaotic spike sequences generated by the bifurcating neuron. As a result, we demonstrated that two chaotic spike sequences with the different phase shift values are able to decompose by the proposed coupling model of the bifurcating neuron.

This result indicates two possibilities. One is that multiple information are simultaneously representable by the superimposed chaotic spike sequences. The other is that neural activity of different neurons is linkable by their selective synchronization if they obey the same chaotic dynamics. Although the proposed coupling model might be too artificial in order to apply the neural information coding in the real brain, we could demonstrate the possibility of chaotic spike sequence as a carrier of information. Further analyses of decomposing mechanism and performance are our future work.

References

1. Yamaguchi, A., Arakane, S., Kubo, M.: Feature linking using synchronized responses in chaotic cellar neural networks for visual stimulus of moving objects. J. Robot. Netw. Artif. Life. **2**, 230–233 (2016)
2. Fujiwara, M., Yamaguchi, A., Kubo, M.: Synchronized response to grayscale image inputs in chaotic cellular neural network. J. Robot. Netw. Artif. Life. **2**, 26–29 (2016)
3. Lee, G., Farhat, N.H.: The bifurcating neuron network 1. Neural Netw. **14**, 115–131 (2001)
4. Gerstner, W., Kistler, W.: Spiking neuron models: single neurons populations plasticity. Cambridge University Press, Cambridge (2002)

Chapter 2
Neural Energy Properties and Mental Exploration Based on Neural Energy Field Gradient

Yihong Wang, Xuying Xu, and Rubin Wang

Abstract Neural coding problem is one of the most important basic problems of cognitive neuroscience. The classic coding theories based on firing rate now encounter their own bottlenecks. Energy coding method studies the coding problem by the energy characteristics of neural systems which possesses the advantages of globality and economy. This research analyzed the energy coding theory in computational level and applied it to mental exploration and path optimization. First, we defined and calculated the neural energy supply and consumption based on the Hodgkin-Huxley model during two activity states using ion-counting and power integral method. Then the energy properties of each ion channel are analyzed. The energy efficiency of a neuron is 76% and above 100% under these two circumstances. Finally, we study the mental exploration by energy method and constructed an effective model to find and optimize the path to the target.

Keywords Energy coding · Mental exploration · Neural energy field · Place cells

2.1 Introduction

It is one of the most important questions in cognitive neural science that how the neural systems code and decode neural information [1]. Scientists have established phase coding, frequency coding, and group coding to encounter this problem. Unfortunately, the scope of these techniques is limited, and the definitions are vague [2]. Currently, no complete theory for neural coding and decoding has been accomplished to direct the research of global brain activities. One reason is that

Y. Wang · R. Wang (✉)
East China University of Science and Technology, Science School, Shanghai, China

X. Xu
Institute of Cognitive Neurodynamics, East China University of Science and Technology, Shanghai, China

J. M. Delgado-García et al. (eds.), *Advances in Cognitive Neurodynamics (VI)*, Advances in Cognitive Neurodynamics,
https://doi.org/10.1007/978-981-10-8854-4_2

these coding theories are focusing on local neural activities and do not include the cross influence of large-scale neural activities. Furthermore, due to the nonlinear property of the neurodynamics, it is very hard to perfectly analyze the neural coding and decoding problem by classical coding methods. Neural activities and neural information processes should follow the principles of energy minimization and information transmission efficiency maximization [3], and neural system should be restricted by energy minimization regardless of suprathreshold or subthreshold activity. This is the economical essence of neural system because of evolution. Information transmission efficiency must maximize the energy utilization in a neural system; this property reflects the high efficiency of neural system for information processing. However, it is difficult to define and describe neural metabolic energy, neural electric energy, and the relationship between them. Some researches helped to understand the neural energy consumption and transformation [4], but they are not related to information coding by neuron group activity.

Some researchers have proposed a new method to study neural coding by energy [3]. In order to describe the relationship between bioenergy of the brain and the neural information processes of the prefrontal cortex, a biophysical model concerning neural circuit has been constructed. Furthermore, quantitative relationship between firing patterns and neural energy evolutionary process has been discovered. Based on these unique relationship, researchers developed the concept of energy coding and further calculated the energy of a single neuron [3]. Some interesting findings have been discovered during the study of the energy distribution properties of structural neural networks. These ideas have laid the foundation for energy coding research of the functional neural network.

Although many scientists achieved remarkable works studying neural energy, a basic question has been ignored, which is how to define and distinguish neural energy supply and consumption. The neural energy concept is quite vague in many research; as a result, we need to clarify the different type of neural energy. In this research, we will analyze this problem by energy coding method.

Energy coding method can be used to study variety of cognitive activity, such as spatial representation and learning. The concept of the cognitive map can be used to solve the navigation problems in environment such as self-locating, target-searching, and pathfinding. Place cells in hippocampus are the biological foundation of cognitive map, which are firstly found by the Nobel Prize winner O'Keefe in the hippocampus with an electrophysiological method [5]. Redish and Touretzky found that the hippocampus possesses ability of spatial memory and spatial navigation in rodent animal [6]. However, the deficiency of cognitive map model is that it took tremendous of physical explorations to form path vector. The agent needs to explore the actual spatial environment continually through the physical movements, which waste much time and energy. Our study can make up the defects, and physical exploration can be improved to mental exploration. Mental exploration was firstly introduced by Hopfield [7]. He adapted plane attractor and substituted the mental exploration in the virtual space for the heavy process of physical exploration. Mental exploration has some obvious advantages compared to physical exploration [7]. However, it was first based on the artificial neural network, without direct physiological significance. Furthermore, during the process of pathfinding, there is no demand for learning speed and path efficiency. In our work, based on Hopfield's theory, neural energy coding method with clearer biological meanings is adopted, and the firing power of place cell is the key to guide mental exploration. An efficient mental exploration path can be achieved by this method, which also possesses the function of path optimization. It is an effective application of neural energy coding method.

2.2 Neural Energy Properties

In order to study the neural energy and its reflection of neural information, we first should solve basic question that has been ignored for a long time, which is how to define and distinguish neural energy supply and consumption. Let us consider the energy transformation in the neuron. First, ATP hydrolyzes to provide chemical energy to ion pump, especially the Na^+/K^+ pump. Then the ion pump works to transport ions against the concentration gradient to preserve electrical potential. It ejects Na^+ and injects K^+ across the cell membrane. This process is equivalent to charging a battery, during which chemical energy is transformed to electric potential energy. When the stimulus occurs, ions flow through ion channels pushed by the electric field force, the potential energy preserved in the membrane capacitor is released and turned into joule heat due to the resistance effect of ion channels. During this process, an action potential fired or subthreshold activity occurs. Finally, ion pump must transport the ions again to recover the membrane potential, and the chemical energy of the ATP will be consumed again. This is an energy cycle of a neuron. To conclude, the chemical energy of ATP is the energy supply for the neuron, and the electric energy carried by ion currents to transmit neural signal is the energy consumption by the neuron. Apparently, energy should be conserved during lager scale of time, but in small time interval energy supply and consumption are not really matched in every moment. This property makes it possible to study brain activity status based on energy supply and consumption properties.

From the former discussion, it can be deducted that energy supplied to a neuron is the energy released by ATP which consumed by the ion pump. The energy consumed by a neuron is the joule heat transformed from electric potential energy. It is also known that every 3 Na^+ ions pumped out of a cell membrane, one ATP molecule is consumed [8]; each mole of ATP molecules can release between 46 and 62 kJ free energy. After Na^+ flow into neuron during neural activity, the Na^+/K^+ pump will expel the same amount of Na^+ to reset the resting membrane potential. Thus, if the amount of Na^+ flow into neuron can be counted, the ATP consumption could be calculated [9]. And based on a proper neuron ion channel model, joule heat can be obtained [4]. Fortunately, all these characters can be deduced by the classical Hodgkin-Huxley model (H-H model) as shown below (Fig. 2.1).

The differential equation is

$$C_m \frac{dV_m}{dt} = g_l\left(E_l - V_m\right) + g_{Na} m^3 h\left(E_{Na} - V_m\right) + g_K n^4\left(E_K - V_m\right) + I \quad (2.1)$$

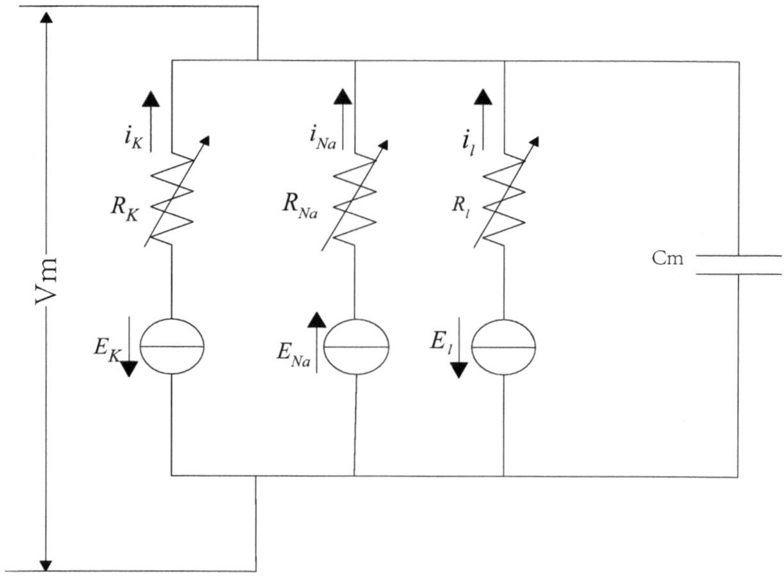

Fig. 2.1 Circuit of Hodgkin-Huxley model (H-H model)

where C_m is membrane capacitance of a neuron, V_m is membrane potential, E_{Na} and E_K are Nernst potentials of Na^+ and K^+, and El is the potential, while leakage current is zero. g_l, g_{Na}, and g_K are, respectively, leakage conductance, Na^+ channel conductance, and K^+ channel conductance.

Energy supplied by ATP can be calculated based on the H-H model:

$$E_s = \frac{\lambda}{3\ eN_A} \int_t g_{Na} m^3 h \left(E_{Na} - V_m\right) dt \tag{2.2}$$

where λ is amount of energy released by one mole ATP; e is the elementary charge, which is 1.6×10^{-19} coulomb; and NA is Avogadro constant, and the integrand is the current of the Na^+ channel [4]. By integrating the H-H equation at a particular time interval, we are able to calculate the energy consumed by a neuron during this time period [4]:

$$E_c = \int_t \left[V_m I + i_{Na}\left(E_{Na} - V_m\right) + i_K\left(E_K - V_m\right) + i_l\left(E_l - V_m\right)\right] dt \tag{2.3}$$

As soon as energy supply and consumption are calculated, energy efficiency can be defined by percentage of energy consumption over supply:

$$\eta = \frac{E_c}{E_s} \times 100\% \tag{2.4}$$

We can also calculate the synchronicity of energy consumption and currents of different ion channels. As shown in Fig. 2.2, Na^+ (red) and K^+ channels consumed most of the total electrical power (green), and the energy consumption of leakage (yellow) and stimulus currents (fuchsia) are relatively small. Figure 2.2a is the energy during action potential, and Fig. 2.2b is subthreshold activity.

Integrating the power shown in Fig. 2.2, we can get the energy consumed by a neuron during these periods. Meanwhile, energy supplied to a neuron can also be calculated by integrating Na^+ current and counting the ions. Results are shown in Table 2.1.

In conclusion, the energy properties of a neuron are significant under two states; these differences may provide an insight to further understanding neural information coding and processing problem.

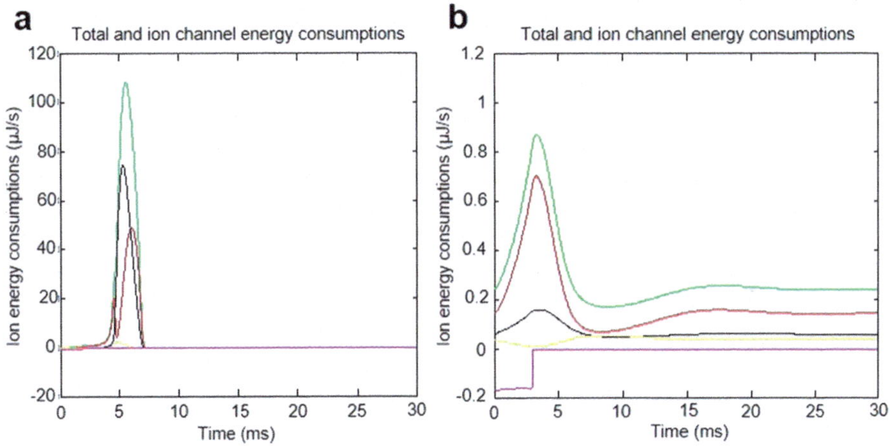

Fig. 2.2 Energy consumption of ion currents [10]

Table 2.1 Energy properties of a neuron [10]

	Super-threshold activity	Subthreshold activity
Energy supplied	2.468×10^{-7} J/cm^2	8.75×10^{-9} J/cm^2
Energy consumed	1.879×10^{-7} J/cm^2	8.31×10^{-9} J/cm^2
Synchronicity	0.782	0.96
Phase difference	38.5°	16.26°
Energy efficiency	76%	105.3%

2.3 Mental Exploration Based on Neural Energy Field

Since neural energy can express the neural information, we further apply the energy method to study mental exploration and solve the pathfinding problem. Place cells represent space, and every place cell has a unique firing power at a certain moment. As a result, we can construct a mapping from a subset of space to the firing power of place cell. This mapping associate with the space is called a neural energy field. After normalized with the maximum power, the field has the following form (Eq. 2.5):

$$\begin{aligned} P_k(x,y) &= P_{k0}(x,y) + P_{kt}(x,y); \\ P_{k0}(x,y) &= \left(1+\sum_{j\neq i}^{n}\omega_{ji}\right)\frac{1}{2\pi\sigma_1\sigma_2}\exp\left\{-\frac{1}{2}\left[\frac{(x-x_p)^2}{\sigma_1{}^2}+\frac{(y-y_p)^2}{\sigma_2{}^2}\right]\right\}; \\ P_{kt}(x,y) &= \frac{1}{2\pi\sigma_1\sigma_2}\exp\left\{-\frac{1}{2}\left[\frac{(x-x_t)^2}{\sigma_1{}^2}+\frac{(y-y_t)^2}{\sigma_2{}^2}\right]\right\} \\ &(\mathrm{k}=1,2,3\ldots) \end{aligned} \tag{2.5}$$

While applying the gradient vector to direct the mental exploration, the system can find the target by random search and optimize the path through change of synapsis. The results are shown in Fig. 2.3.

After ten times exploration, the system obtains an optimal path to the target (Fig. 2.3a), and the gradient vector serves as a navigation vector (Fig. 2.3b). This result implies that this model is effective and neural energy is a promising method to study neural system.

Our study suggests that the energy properties of a neuron vary greatly between different neural activity states, and it can serve as an effective tool to study the coding and information processing of the neural systems. Based on these results,

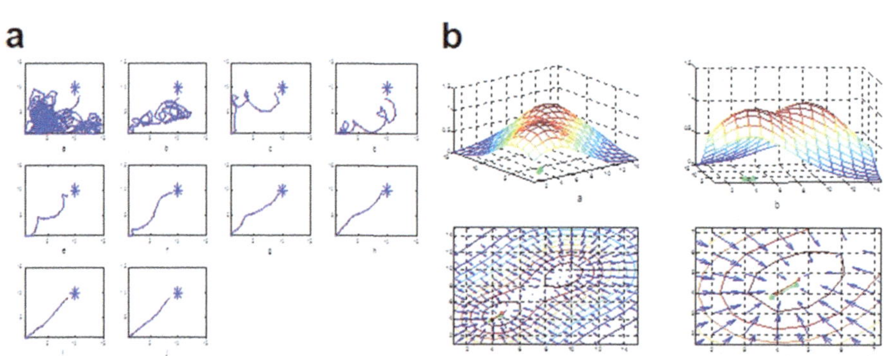

Fig. 2.3 Mental exploration process and neural energy field [11]

we further construct a mental exploration model to study spatial environment so as to solve pathfinding problems; energy field and its gradient are calculated which can be used to optimize the path to the target in the virtual plane. The study shows that the neural energy method is a possible solution to the obstacle in front of coding theory, and it will give us a new comprehensive insight to the brain.

References

1. Borst, A., Theunissen, F.E.: Information theory and neural coding. Nat. Neurosci. **2**, 947–957 (1999)
2. Jacobs, A., et al.: Ruling out and ruling in neural codes. Proc. Natl. Acad. Sci. U. S. A. **106**, 5936–5941 (2009)
3. Wang, R., Tsuda, I., Zhang, Z.: A new work mechanism on neuronal activity. Int. J. Neural Sys. **25**, 1450037 (2015)
4. Moujahid, A., d'Anjou, A., Torrealdea, F.J.: Energy and information in Hodgkin-Huxley neurons. Phys. Rev. E. **83**, 031912 (2011)
5. O'Keefe, J., Nadel, L.: The Hippocampus as a Cognitive Map. Oxford University Press, Oxford (1978)
6. Redish, A.D.: Beyond the Cognitive Map. The MIT Press, Cambridge (1999)
7. Hopfield, J.J.: Neurodynamics of mental exploration. Proc. Natl. Acad. Sci. U. S. A. **107**, 1648–1653 (2010)
8. Attwell, D., Laughlin, S.B.: An energy budget for signaling in the grey matter of the brain. J. Cereb. Blood Flow Metab. **21**, 1133–1145 (2001)
9. Moujahid, A., D'Anjou, A., Graña, M.: Energy demands of diverse spiking cells from the neocortex, hippocampus, and thalamus. Front. Comput. Neurosci. **8**, 1–12 (2014)
10. Wang, Y., Wang, R., Xu, X.: Neural energy supply-consumption properties based on Hodgkin-Huxley model. Neural Plast. **2017**, 6207141 (2017)
11. Wang, Y., Wang, R., Zhu, Y.: Optimal path-finding through mental exploration based on neural energy field gradients. Cogn. Neurodyn. **11**, 99–111 (2017)

Chapter 3
Information Coded in the Striatum During Decision-Making

Makoto Ito and Kenji Doya

Abstract The basal ganglia are known to play an essential role in decision-making. The striatum, the major input site of the basal ganglia, has a dorsal-ventral gradient in the input modality: the more dorsolateral part receives sensorimotor-related information and the more ventral part receives associative and motivational information. Previous lesion studies have suggested that subareas of the striatum have distinct roles: the dorsolateral striatum (DLS) functions in habitual action, the dorsomedial striatum (DMS) in goal-directed actions, and the ventral striatum (VS) in motivation. However, it has not been investigated what kind of roles are taken by these subareas concurrently during the same process of decision-making. In this study, we systematically investigated information represented by phasically active neurons in DLS, DMS, and VS during two types of choice tasks: fixed- and free-choice tasks. In both tasks, rats were required to perform nose poking to either the left or right hole after cue-tone presentation. A food pellet was delivered probabilistically depending on the presented cue and the selected action. The reward probability was fixed in fixed-choice task and varied in a block-wise manner in free-choice task. We found the following: (1) before rats started a trial, a majority of VS neurons increased their firing rates and information regarding task type and state value was most strongly represented in VS; (2) during action selection, information of action and action values was most strongly represented in DMS; and (3) activity peaks of DLS neurons were sharper and more uniformly distributed than those of DMS and VS. To explain our results, we proposed a hierarchical reinforcement learning hypothesis that VS, DMS, and DLS are hierarchical learning modules in charge of actions at different physical and temporal scales. VS is the coarsest module

M. Ito (✉)
Neural Computation Unit, Okinawa Institute of Science and Technology, Okinawa, Japan

PROGRESS TECHNOLOGIES, Inc., Tokyo, Japan
e-mail: ito@oist.jp

K. Doya
Neural Computation Unit, Okinawa Institute of Science and Technology Graduate University, Okinawa, Japan

J. M. Delgado-García et al. (eds.), *Advances in Cognitive Neurodynamics (VI)*,
Advances in Cognitive Neurodynamics,
https://doi.org/10.1007/978-981-10-8854-4_3

governing objects of actions, such as aiming for a goal, avoiding a danger, or just taking a rest. DMS is the middle module in charge of abstract actions, such as turn left, turn right, or go straight, by taking into account contextual information. DLS is the module in charge of the finest control of physical actions, such as the control of each limb.

Keywords Reinforcement learning · Decision-making · Basal ganglia · Striatum · Action value · State value · Q-learning model

3.1 Introduction

The basal ganglia are known to play an essential role in decision-making. The striatum, the major input site of the basal ganglia, has a dorsal-ventral gradient in the input modality: the more dorsolateral part receives sensorimotor-related information and the more ventral part receives associative and motivational information. Previous lesion studies have suggested that the dorsomedial striatum (DMS) and the dorsolateral striatum (DLS) contribute differently to goal-directed actions (DMS) and habitual actions (DLS), respectively. Lesion and recording studies of the ventral striatum (VS) suggested its role in motivation in response to reward-predicting cues. However, it was still unclear what kind of roles were assigned to these subareas in the same process of decision-making. So far, we have systematically investigated information represented by phasically active neurons in DLS, DMS, and VS of rats during choice tasks. In this paper, we summarize some of our previous findings [1, 2] and introduced our hypothesis that we previously proposed to explain our result [3].

3.2 Activity Patterns of Striatal Neurons During Choice Tasks

3.2.1 Fixed-Choice Task and Free-Choice Task

We recorded the neuronal activity in the striatum when rats were performing fixed-choice and free-choice tasks (Figs. 3.1 and 3.2). In fixed-choice task, left tone (900 Hz) or right tone (6500 Hz) was presented, and the rats were required to perform a nose poke in the left hole or the right hole, respectively. In free-choice task, after choice tone (white noise), rats were required to select left or the right nose poke, and then a reward pellet was delivered stochastically depending on the selected action.

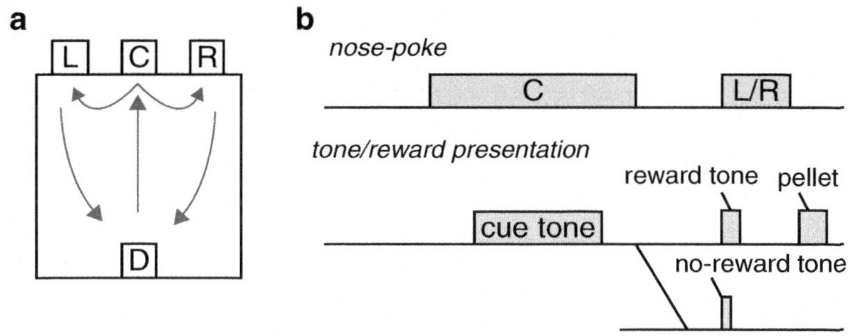

Fig. 3.1 Task design. (**a**) Schematic illustration of the experimental chamber. The chamber was equipped with three holes for nose poking (*L*, left hole; *C*, center hole; *R*, right hole) and a pellet dish (*D*). (**b**) Time sequence of choice tasks. After a rat poked its nose into the center hole, one of three cue tones was presented (left tone, right tone, or choice tone). The rat had to maintain the nose poke in the center hole during presentation of the cue tone. After offset of the cue tone, the rat was required to perform a nose poke in either the left or right hole, and either a reward tone or a no-reward tone was presented. The reward tone was followed by delivery of a sucrose pellet in the food dish. The reward probability was determined by the given cue tone and the chosen action. For the left tone, the reward probabilities were (left, right) = (50%, 0%). For the right tone, the probabilities were (left, right) = (0%, 50%). These probabilities were fixed throughout the experiments. Left tone or right tone were presented in fixed-choice task. For the choice tone, reward probabilities were varied: one of four pairs of reward probabilities [(left, right) = (90%, 50%), (50%, 90%), (50%, 10%), and (10%, 50%)] was used for each block. Choice tone was presented in free-choice task [1]

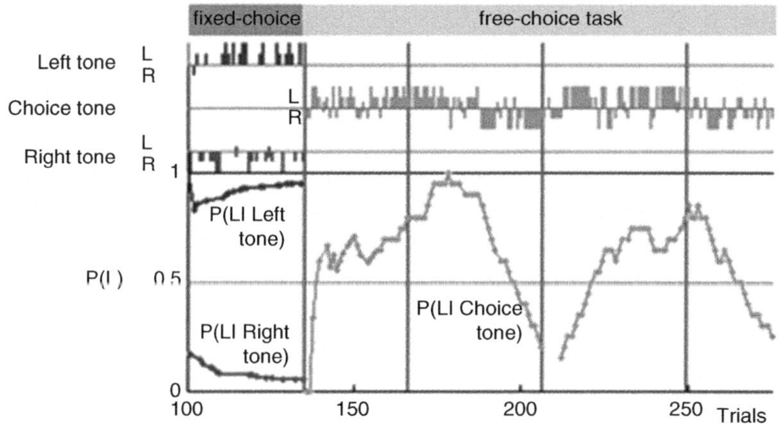

Fig. 3.2 Representative example of a rat's performance of fixed-choice task (the first block) and free-choice task (2nd to 5th blocks). The vertical bars indicate rat's choices, and the lines represent the frequencies that the rat selected left for the last 20 trials. Reward probabilities were fixed in fixed-choice task and varied in free-choice task [1]

3.2.2 Activity Patterns and State, Action, and Reward Coding

We isolated 190, 105, and 119 neurons from DLS, DMS, and VS, respectively, as phasically active neurons (PANs, putative medial spiny projection neurons) based on statistics of interspike intervals and waveforms. These striatal neurons were activated at different task events and task phases between events, and most neurons changed their activity patterns depending on upcoming actions, selected actions, reward outcomes, and types of tasks. For instance, a DLS neuron shown in Fig. 3.3a had three peaks before a trial (phase 1), during movement to left or right hole (phase 5) and after a trial (phase 7), and the peaks in phase 5 and 7 were changed by action, reward, and tasks. A DMS neuron shown in Fig. 3.3b increased the activity when the rat selected right but not left. A VS neuron shown in Fig. 3.3c increased the activity when the rat started a trial (phase 1) and finished a trial (phase 7). These both peaks were higher in fixed-choice task than in free-choice task. The second peak also modulated by action and reward.

First, we investigated the activity patterns of all recorded neurons. We found that the proportion of neurons that increased their activity as a rat approached the center hole (phase 1) was more than 60% in VS, significantly larger than in DLS and DMS.

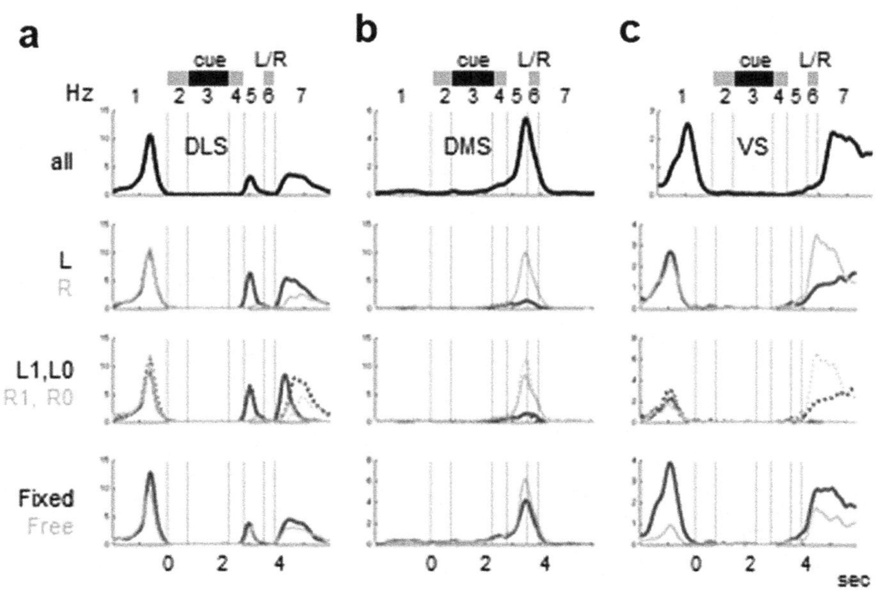

Fig. 3.3 Representative activity patterns of phasic active neurons in the DLS (**a**), DMS (**b**), and VS (**c**). Event-aligned spike histograms (EASHs) with 10 ms bins were calculated for different conditions: "all" represents the averages of EASHs for all trials. L and R represent EASHs for left or right selected trials, respectively. L1, L0, R1, and R0 represent EASHs for four different action-reward combinations. Fixed and Free represent EASHs for fixed- and free-choice trials. The numbers from 1 to 7 represent task phases in one trial [1]

After a rat's exit from the center hole until its entrance into the L/R hole (phase 5), more than 60% of DMS neurons were activated, which was significantly larger than the proportions of DLS and VS. Activity peaks of DLS neurons were not only sharper than those of DMS and VS, but also uniformly distributed compared with DMS and VS, without specific preferred task events.

Second, we quantified the proportion of event-coding neurons, namely, how many neurons changed their firing rates for selected actions, reward outcomes, and the different choice blocks (states). The proportion of state-coding (fixed- or free-choice block) neurons during cue presentation (phase 3) was the largest in VS, and the proportion of action-coding neurons was the largest in DMS during action execution (phase 5). The proportion of reward-coding neurons was the largest in VS during the L/R poking. These proportions were similar in fixed-choice and free-choice blocks.

3.2.3 Action-Value and State-Value Coding

Based on reinforcement learning theory, it has been hypothesized that the patch compartment in the striatum learns reward prediction for each stimulus (state) in the form of a "state value", and the matrix compartment in the striatum learns expected reward after selecting each action candidate, in the form of an "action value" [4].

We estimated action values and state values for each trial under the assumption that all actions were selected by a generalized Q-learning model with a forgetting effect [2]. We then identified the neurons coding action values and state values, as well as action and reward, by a liner regression analysis (Fig. 3.4). The proportion of action-value coding neurons was highest in DMS during action execution (phase 5), while the proportion of state-value coding neurons was the highest in VS in the most of the all task phases (phase 1–6).

3.3 Hierarchical Reinforcement Learning Hypothesis

To explain our findings, we proposed a working hypothesis that VS, DMS, and DLS are hierarchical learning modules in charge of actions at different physical and temporal scales [3]. VS is the coarsest module governing actions of the whole animal, such as aiming for a goal, avoiding a danger, or just taking a rest. DMS is the middle module in charge of abstract actions, such as turn left, turn right, or go straight. DLS is the module in charge of the finest control of physical actions, such as the control of each limb.

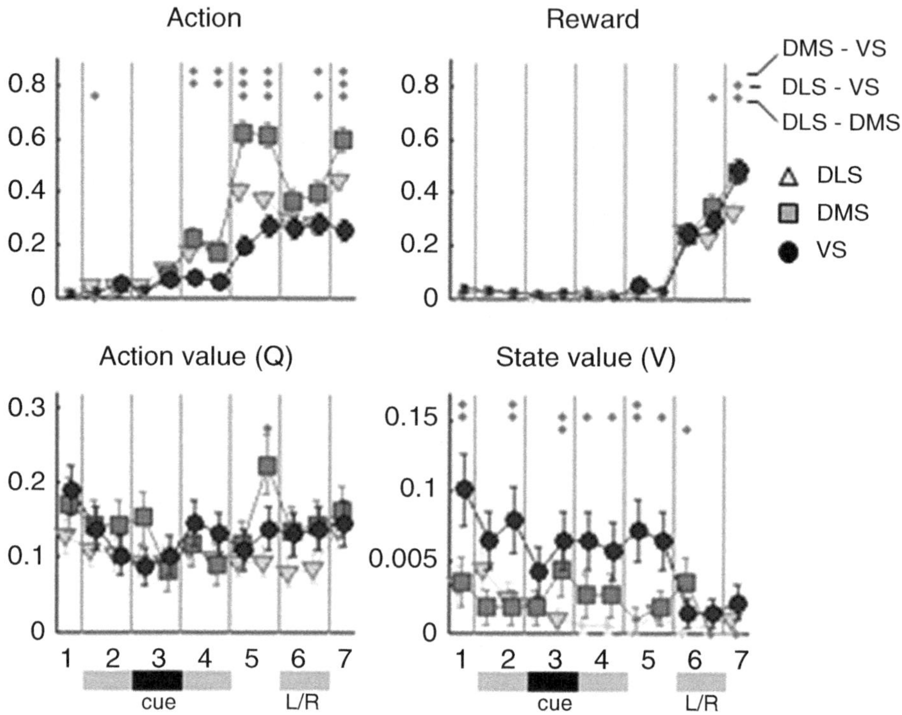

Fig. 3.4 Proportions of neurons coding action, reward, action value, and state value ($p < 0.01$, t test). Dots in the upper area indicate significant differences in the proportions between subareas ($p < 0.05$, Mann-Whitney U-test) [1]

Consistent with this hypothesis, the average firing duration (exceeding the half of the peak firing rate) was longest in VS neurons among three subareas, that of DMS neurons was the second, and DLS neurons showed the shortest firing duration. A large majority of VS neurons was activated while rats headed to the center hole to start a trial. This might be interpreted to mean that VS is involved in higher-order decisions to initiate a trial. State values coded in VS might be an action value for higher-ordered action, "do a trial". In DMS, most neurons were activated during execution of the action selection, and at that time, action and action-value information was strongly represented. These findings suggest that DMS is the site most likely to be involved in decisions regarding abstract actions, such as "select the left hole" or "select the right hole". Activity peaks of DLS neurons were not only sharper than those of DMS and VS but also uniformly distributed over different task phases. Each activity peak might help to control the body and limbs during a brief time window (Fig. 3.5).

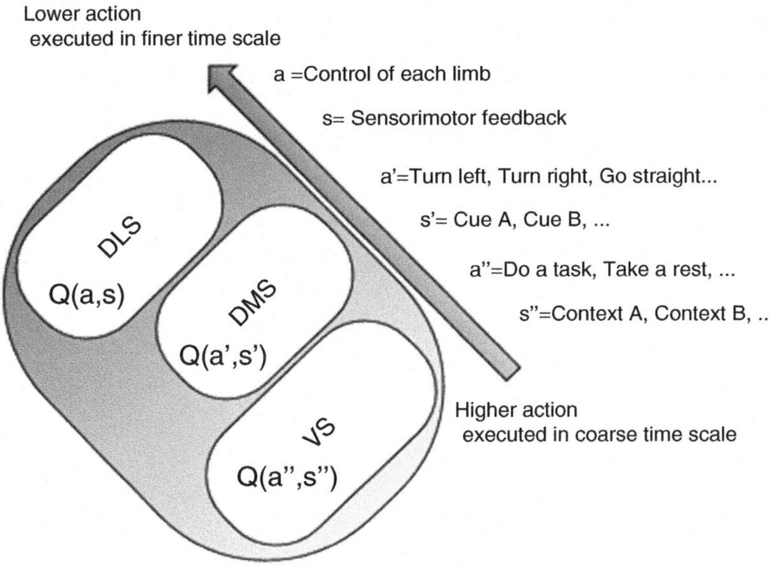

Fig. 3.5 A hypothesis that the dorsolateral (DLS), the dorsomedial (DMS), and the ventral striatum are parallel and hierarchical Q-learning modules that are in charge of actions at different physical and temporal scales [3]

Acknowledgments This work was supported by Ministry of Education, Culture, Sports, Science and Technology KAKENHI Grants 23120007 and 26120729, Japan Society for the Promotion of Science KAKENHI Grant 25430017, and Okinawa Institute of Science and Technology Graduate University research support to K.D.

References

1. Ito, M., Doya, K.: Distinct neural representation in the dorsolateral, dorsomedial, and ventral parts of the striatum during fixed- and free-choice tasks. J. Neurosci. **35**, 3499–3514 (2015)
2. Ito, M., Doya, K.: Validation of decision-making models and analysis of decision variables in the rat basal ganglia. J. Neurosci. **29**, 9861–9874 (2009)
3. Ito, M., Doya, K.: Multiple representations and algorithms for reinforcement learning in the cortico-basal ganglia circuit. Curr. Opin. Neurobiol. **21**, 368–373 (2011)
4. Doya, K.: What are the computations of the cerebellum, the basal ganglia and the cerebral cortex? Neural Netw. **12**, 961–974 (1999)

Chapter 4
A Comparison of Reward Values Encoding Function Between the Prefrontal Cortex and Striatum in Monkey

Zaizhi Wen, Jianhua Zhang, and Xiaochuan Pan

Abstract Reward prediction is essential for learning behavior and decision-making process in the brain. It is well known that neurons in both prefrontal cortex (PFC) and striatum are involved in encoding reward values. The difference in reward coding function between these two brain regions remains unclear. In this work, local field potentials (LFPs) were recorded in the lateral PFC and striatum of a male monkey while performing a reward prediction task. A pattern classification method was used to characterize the function of PFC and striatum for encoding reward values. We used two different feature extraction methods to extract input features to two different classifiers, including random forest (RF) and support vector machine (SVM). We optimized the SVM using the particle swarm optimization (PSO) algorithm. The results suggested that even in a model-based process, the neurons in striatum are capable of encoding more reward information than those in PFC.

Keywords Prefrontal cortex · Striatum · Reward prediction · Support vector machine · Random forest · ApEn

---

Z. Wen
School of Information Science and Engineering, East China University of Science and Technology, Shanghai, People's Republic of China

J. Zhang (✉)
School of Information Science and Engineering, East China University of Science and Technology, Shanghai, People's Republic of China

Institute of Cognitive Neurodynamics, East China University of Science and Technology, Shanghai, People's Republic of China
e-mail: zhangjh@ecust.edu.cn

X. Pan
Institute of Cognitive Neurodynamics, East China University of Science and Technology, Shanghai, China

J. M. Delgado-García et al. (eds.), *Advances in Cognitive Neurodynamics (VI)*,
Advances in Cognitive Neurodynamics,
https://doi.org/10.1007/978-981-10-8854-4_4

4.1 Introduction

Reward prediction is paramount for learning behavior and decision-making processing in the brain. Many researches have shown that many brain regions are involved in reward prediction [1]. Several fMRI studies have demonstrated the importance of both the PFC and the striatum in the basal ganglia for reward prediction and have compared the functional difference in reward prediction between them [2]. The hypothesis that relates the difference between "model-based vs. model-free" processes to the difference in functions of the PFC and striatum was supported by the results of studies on humans and primates [3]. However, several later studies had shown different results which made the hypothesis above seems dubious [4].

Pan et al. [4, 5] found that even striatal neurons could correctly predict rewards right from the first sequential paired-association task (SPAT) immediately after reward instruction trials (RITs), which indicates that striatal neurons also possess some information about state transition of stimuli in the SPAT task [4]. Therefore, it is reasonable to suggest that besides PFC neurons, the striatal neurons also have the ability to perform reward prediction in a model-based manner.

In this study, by analyzing LFPs recorded from PFC and striatal neurons in the reward prediction task, we found that striatal neurons may encode even more reward information than PFC neurons. Our results also suggested that besides commonly used energy-based feature, the ApEn, which is based on complexity, could also reflect the difference of neurons activity under different reward conditions.

4.2 Materials and Methods

4.2.1 Behavioral Task

One male Japanese monkey served as a subject in this study (Tom, 8.5 kg). A detailed description of the experimental procedure and behavioral task can be found in [5]. Briefly, the monkey was first trained to learn two associative sequences (Fig. 4.1a) in a SPAT (Fig. 4.1b). After that, an asymmetric reward schedule was introduced using RITs (Fig. 4.1c). RITs and SPATs were arranged in one block, first reward instruction trials (three trials) then followed by SPATs. In a given block, a correct choice of A1 $\rightarrow$ B1 $\rightarrow$ C1 would get the subject monkey a large reward while A2 $\rightarrow$ B2 $\rightarrow$ C2 would be associated with the small one. The stimulus-reward contingency was pseudo-randomized between blocks. A trial in which the subject monkey made correct choice, whether in large or small reward condition, was considered a correct trial. We recorded the choices of the subject monkey in each trial using recorded LFPs in the correct trials for further analysis.

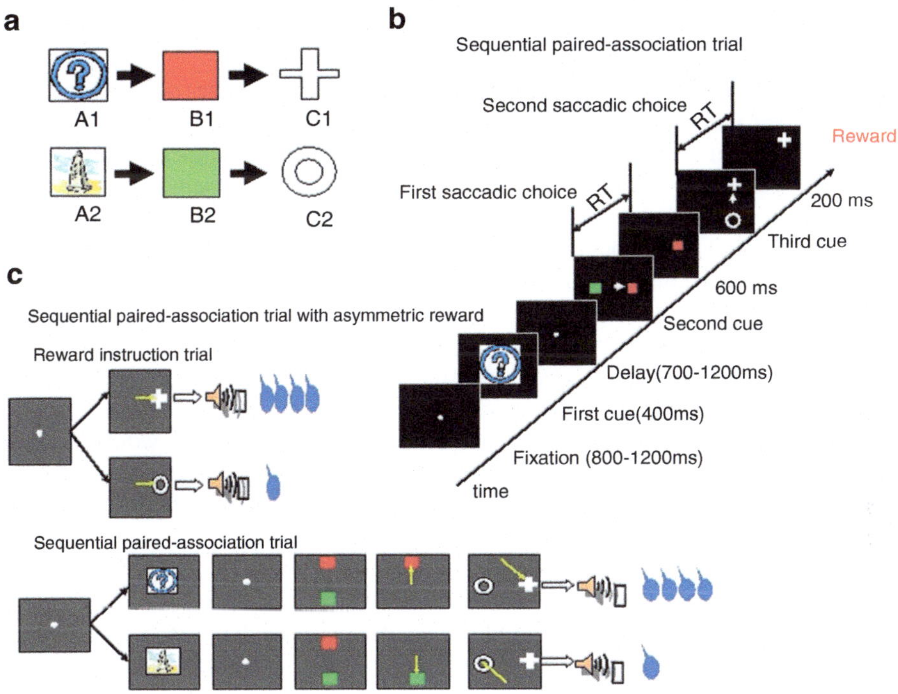

Fig. 4.1 The sequential paired-association task with an asymmetric reward scheme

4.2.2 *Data Acquisition*

Extracellular recordings were conducted using linear array multi-contact electrodes (U-probe, Plexon, USA) to obtain LFPs. In our experiment, each electrode contained eight recording contacts (impedance, 0.3–0.5 MΩ at 1 kHz) with an inter-contact spacing of 150 or 300 μm. Neuronal activity was measured against a local reference that was close to the electrode contacts (a stainless guide tube or the tube of U-probe). We performed data amplification, filtering, and acquisition with a Multichannel Acquisition Processor (Plexon, USA). The acquired signal from each contact (channel) was passed through a head stage and then split to extract the spike and the LFP components separately. The extracted LFP signals were then filtered with a passband of 0.7–170 Hz, further amplified, digitized at 1 kHz, and saved in Plexon files. In each recording session, these two U-probe electrodes were inserted simultaneously into the PFC and striatum. Once the two electrodes reached the target positions, we did not move them any more throughout the whole session. However, for different sessions, the positions of these two electrodes were different.

4.2.3 Data Analysis Methods

Off-line analysis was performed using custom-made MATLAB programs on a PC. As described before, we recorded the choices of the monkey in SPATs in large and small reward trials and used only the LFPs recorded in correct trials for later analysis. In total, 50 sessions of LFP data were recorded from the monkey. For each session, the number of correct trials varies between 94 and 119, while large and small reward trials have roughly equivalent number. For the recorded LFPs of each session, we adopted the same analysis process described below.

Unlike commonly used statistical method, to analyze the difference between PFC and striatal neurons in reward prediction process, a pattern classification method was introduced. Here is the basic idea: given the hypothesis that both PFC and striatal neurons encode the reward information in the behavioral task, the recorded LFPs may have some underlying pattern when the monkey predicted it would get big reward in some trials and some different pattern with small reward trials. If this is the case, we could design a classifier which uses the recorded LFPs as input and corresponding reward condition as category label.

Given the LFPs from different trials, the trained classifier should be able to tell the corresponding reward conditions, and the accuracy rate should remain above 50% (random guess possibility). Based on this idea, we used the LFPs recorded from PFC and striatal neurons, respectively, to train and test our classifier and compared the difference of accuracy rates between the PFC and striatum. At first, only the LFPs of the cue period (400 ms after the first cue onset) were analyzed.

Considering it may take a while before the first visual cue in the behavioral task trial was transferred and made the PFC and striatal neurons of the monkey truly respond to it, we applied a time window of 400 ms (the length of cue period), starting from the first cue onset, then moved it with 100 ms each time until the 1000 ms period after the first cue onset. The full cue period and partial delay period were covered, such that for each session, we got seven different data sets and then applied the same schedule below.

For each data set of every session, we used two different feature extraction methods to extract input features from raw LFPs and constructed feature vectors which served as actual input for the two different kinds of classifiers, namely, Random Forest and SVM. The original data of each session was randomly partitioned into five equal-sized subsamples, and a single subsample was retained as the validation data for testing the model, and the remaining four subsamples were used as training data. The above process was then repeated five times, with each of the five subsamples used exactly once as the validation data. This so-called five-folder cross-validation method was used to get a valid classification accuracy rate.

We repeated the classification process discussed above with two kinds of feature extraction methods combined with two classifiers and compared the difference between them. As discussed earlier, we have seven data sets for each session, and the one with the highest accuracy was finally selected as the actual data we used for analysis.

A. Approximate entropy (ApEn)

Approximate entropy is a measure that quantifies the regularity or complexity of a time series. It considers the temporal order of points in a time sequence and is therefore a preferred measure of randomness or regularity [6]. In this work, the ApEn was calculated through the LFPs recorded from each electrode on each trial and served as the extracted feature vector. To calculate the ApEn value, the embedding dimension, vector comparison distance, and time delay were set to 2, 0.2 times the standard deviation of the data and 1, respectively, as suggested by Pincus [7].

B. Discrete wavelet transform (DWT)

Discrete wavelet transforms are widely applied in many engineering fields for solving various real-life problems. DWT provides a more flexible way of time-frequency representation of a signal by allowing the use of variable-sized windows. In our work, sixth level wavelet decomposition was applied to LFPs recorded from each electrode of each trial. Then we got the coefficients of nodes A4, D4, A5, D5, A6, and D6 and calculated the energy of each node, respectively. After that, we combined them together and constructed the feature vectors we need.

C. Random Forest (RF)

Random forest is one of the most successful classifiers based on assemble learning algorithm. RF is very user-friendly in the sense that it has only two parameters – the number of variables in the random subset at each node (mtry) and the number of trees in the forest (ntree) – and is usually not very sensitive to their values [8, 9].

D. Support Vector Machine (SVM)

SVM is a powerful and well-known method for binary classification tasks in machine learning [10]. In our work, we adopted the Gaussian kernel function, and the PSO method was used to choose the suitable regularization parameter and kernel parameter for each session's data [11].

4.3 LFP Data Analysis Results

We recorded LFPs simultaneously in the PFC and striatum using two U-probe electrodes while the monkey was performing the sequential paired-association task with the asymmetric reward schedule. A classification method was used to compare the difference of reward information encoding ability between the PFC and striatum. We used two kinds of feature extraction methods combined with two different classifiers and compared their ability for distinguishing LFPs of PFC and striatum under different reward conditions. Figs. 4.2 and 4.3 show the classification accuracy rates for different feature extraction methods of PFC and striatum with RF and

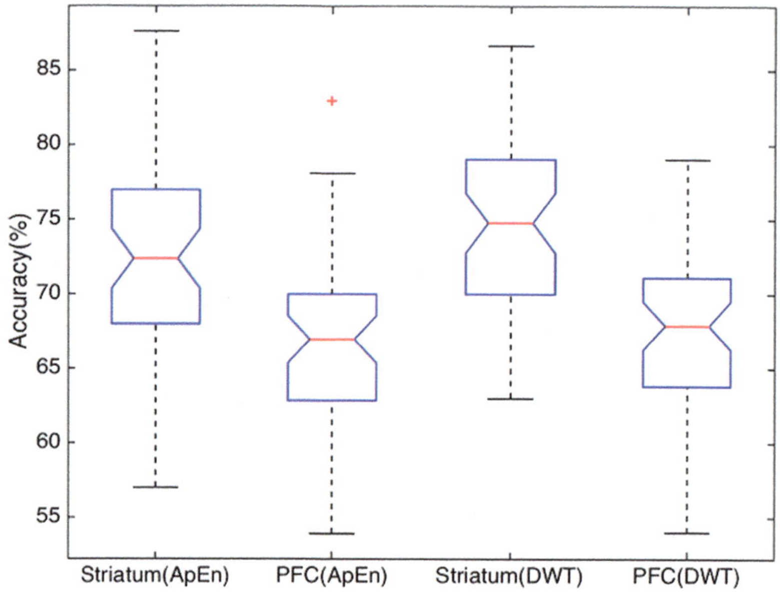

Fig. 4.2 Comparison of the classification accuracy for striatum and PFC regions with RF classifier

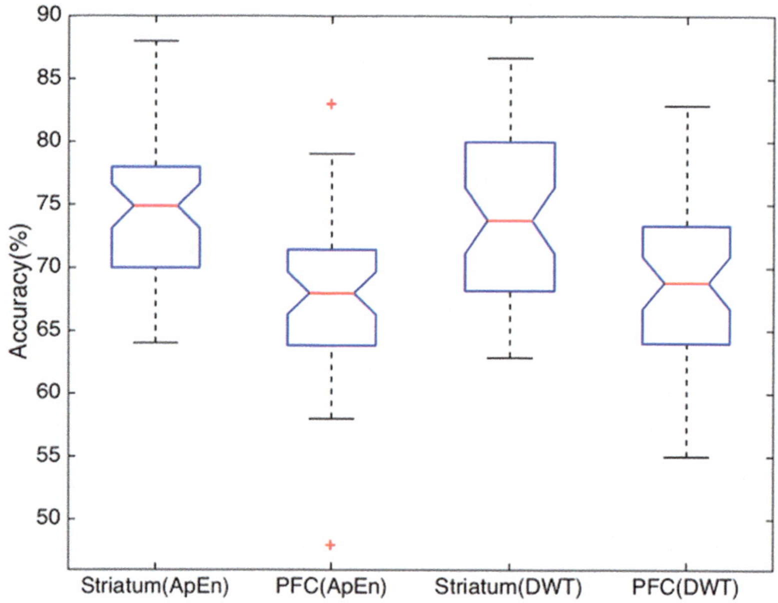

Fig. 4.3 Comparison of the classification accuracy for striatum and PFC regions with SVM classifier

SVM classifiers, respectively. The box plots in these figures show the distributions of classification accuracy in different conditions. As shown in these figures, the classification accuracy rates calculated from striatum were significantly higher compared to that calculated from PFC regardless which feature extraction method or classifier was used. When using RF classifier, as shown in Fig. 4.2, with the ApEn feature, the average classification accuracy for striatum among 50 sessions is 72.2% while that for PFC is just 66.8% (ANOVA F-test: $F = 19.35.79, p = 2.76 \times 10^{-5}$); as for the energy-based feature, that average accuracy rates for striatum and PFC are 74.7% and 67.7%, respectively (ANOVA F-test: $F = 35.62, p = 3.83 \times 10^{-8}$). Fig. 4.3 shows similar results for the SVM classifier. Specially, with the ApEn feature, the average classification accuracy rates for striatum and PFC are 74.4% and 67.1% (ANOVA F-test: $F = 36.74$, $p = 2.42 \times 10^{-8}$); these two figures for the energy-based feature are 74.1% and 68.7% (ANOVA F-test: $F = 16.81, P = 8.52 \times 10^{-5}$). As a higher classification accuracy implies more encoded reward information in the LFPs, these results suggest that striatum may encode more reward information than PFC. From Figs. 4.2 and 4.3, we can also find that the accuracy rates vary a lot on different sessions, for both PFC and striatum, which suggests that different areas of PFC or striatum have varying reward information encoding ability given the fact that electrodes were not fixed exactly in the same area of PFC or striatum for each session.

4.4 Conclusion

In this study, we introduced a pattern classification method for the analysis of LFPs recorded in a reward prediction task and found significant classification accuracy difference between the PFC and striatal neurons. Recent researches have suggested that both PFC and striatal neurons perform reward prediction in a model-based manner and striatum may not simply use model-free learning rule to predict reward [3]. Supporting this, we found that we can use LFPs recorded from striatal neurons as well as PFC neurons to distinguish different reward conditions. Furthermore, we found significant classification accuracy difference between the PFC and striatal neurons, indicating that striatum neurons may encode more reward information than PFC neurons, even in a model-based reward prediction process. These results suggest that striatum may play a more important role for predicting reward than we thought before. Our results also suggested that, besides commonly used energy-based feature, the ApEn, which is based on complexity, could also reflect the difference of neurons activity under different reward conditions.

References

1. Garrison, J., Erdeniz, B., Done, J.: Prediction error in reinforcement learning: a meta-analysis of neuroimaging studies. Neurosci. Biobehav. Rev. **37**, 1297–1310 (2013)
2. Kahnt, T., Heinzle, J., Park, S.Q., Haynes, J.D.: Decoding the formation of reward predictions across learning. J. Neurosci. **31**, 14624–14630 (2011)
3. Shingo, T., Pan, X., Mineki, O., Jessica, E.T., Masamichi, S.: Dissociable functions of reward inference in the lateral prefrontal cortex and the striatum. Front. Psychol. **6**, 995 (2015)
4. Pan, X., Fan, H., Sawa, K., Tsuda, I., Tsukada, M., Sakagami, M.: Reward inference by primate prefrontal and striatal neurons. J. Neurosci. **34**, 1380–1396 (2014)
5. Pan, X., Sawa, K., Tsuda, I., Tsukada, M., Sakagami, M.: Reward prediction based on stimulus categorization in primate lateral prefrontal cortex. Nat. Neurosci. **11**, 703–712 (2008)
6. Kumar, Y., Dewal, M.L., Anand, R.S.: Epileptic seizure detection using DWT based fuzzy approximate entropy and support vector machine. Neurocomputing. **133**, 271–279 (2014)
7. Pincus, S.M.: Approximate entropy as a measure of system complexity. Proc. Natl. Acad. Sci. U.S.A. **88**, 2297–2301 (1991)
8. Ocak, H.: Automatic detection of epileptic seizures in EEG using discrete wavelet transform and approximate entropy. Expert Syst. Appl. **36**, 2027–2036 (2009)
9. Early Seizure Detection Algorithm Based on Intracranial EEG and Random Forest Classification: Int. J. Neural Syst. **25**, 1550023 (2015)
10. Abe, S.: Fuzzy support vector machines for multilabel classification. Pattern Recognit. **48**, 2110–2117 (2015)
11. Chang, B.-M., Tsai, H.-H., Yen, C.-Y.: SVM-PSO based rotation-invariant image texture classification in SVD and DWT domains. Eng. Appl. Artif. Intell. **52**, 96–107 (2016)

Chapter 5
Injection of Muscimol into Prefrontal Cortex Impairs Monkey's Reward Transitive Inference

Xiaochuan Pan, Rubin Wang, and Masamichi Sakagami

Abstract It is known that both the prefrontal cortex and striatum are involved in reward processing, but neurons in the two areas may utilize distinct strategies to predict reward. It was reported in a reward inference task that neurons in the lateral prefrontal cortex (LPFC) predict reward value for a stimulus using transitive inference. Striatal neurons predict reward on the basis of directly experienced stimulus-reward associations. We hypothesized that inactivation of the LPFC could impair reward predictive ability based on transitive inference but did not based on associations. To test this hypothesis, muscimol was injected in LPFCs bilaterally while a monkey was performing the reward inference task. The reward prediction behavior based on stimulus-reward associations was not affected by muscimol, but the behavior based on inference to predict reward was altered. The behavioral results with muscimol indicated that local inactivation in the LPFC impaired only the reward predictive ability that requires transitive inference, suggesting that the LPFC is an important region to make inference.

Keywords Prefrontal cortex · Muscimol · Reward · Model-based · Model-free

5.1 Introduction

The prefrontal cortex (PFC) and the striatum have mutual connections. PFC neurons directly project to the striatum [1], as the input of basal ganglia. The output of the striatum finally reaches the thalamus through direct and indirect pathways and is feedback to the LPFC, forming a closed loop [2]. Their tightly anatomical

X. Pan (✉) · R. Wang
Institute of Cognitive Neurodynamics, East China University of Science and Technology, Shanghai, China
e-mail: pxc@ecust.edu.cn

M. Sakagami
Brain Science Institute, Tamagawa University, Tokyo, Japan

J. M. Delgado-García et al. (eds.), *Advances in Cognitive Neurodynamics (VI)*,
Advances in Cognitive Neurodynamics,
https://doi.org/10.1007/978-981-10-8854-4_5

connections imply that the two areas have close relation in cognitive functions [3], involving learning, reward processing, behavior control, and so no. A computational model suggested that the PFC used model-based learning rule to control goal-directed behavior, while the striatum applied model-free rule to guide habitual behavior [4]. Several lines of evidences in neurophysiological and fMRI studies support this hypothesis, finding model-free signals in the striatum and model-based signals in the PFC [5, 6]. But some fMRI experiments reported both model-free and model-based signals in the striatum [7]. Due to anatomical connections between the PFC and striatum, it is not clear the source of model-based signals found in the striatum. Is the signal transferred from the PFC or generated in striatal circuits?

It is known that the PFC and the striatum are involved in reward prediction processes [3]. In a reward inference task [8], it was found that PFC neurons could utilize transitive inference to predict reward value of a stimulus without the requirement to experience the stimulus-reward association directly. While striatal neurons didn't have such ability, instead, they used exclusive inference or directly experienced stimulus-reward associations to predict reward. On the basis of these results, we hypothesized that inactivation of the LPFC could impair reward predictive ability based on transitive inference but did not impair the ability based on exclusive inference or associations. To test this hypothesis, muscimol was injected in LPFCs bilaterally while a monkey was performing the reward inference task. Muscimol acts as a potent, selective agonist for the $GABA_A$ receptors and alters neuronal activity in its injected location. The behavior results confirmed the hypothesis, suggesting that the LPFC and striatum predict reward based on distinct mechanisms.

5.2 Materials and Methods

One male Japanese monkey (*Macaca fuscata*) served as a subject in this study (Tap, 6.5 kg). A head-holder and two recording chambers were implanted under aseptic techniques with ketamine (4.6–6.0 mg kg^{-1} i.m.) and sodium pentobarbital (Nembutal, 4.5–6.0 mg kg^{-1} i.v.) anesthesia. All surgical and experimental protocols were approved by the Animal Care and Use Committees in Tamagawa University and conducted in accordance with the National Institutes of Health's Guide for Care and Use of Laboratory Animals.

During experimental sessions, the monkey was seated in a primate chair with its head fixed inside a completely enclosed sound-attenuated and electrically shielded room. All visual stimuli were presented on a 21-inch CRT display (FE220, NEC, Japan) with 60 Hz refresh rate, at a distance of 60.0 cm in front of the monkey. Eye movements were monitored by the Eyelink2 system (SR Research Ltd., Mississauga, Canada) with 500 Hz sample rate.

5.2.1 Behavioral Task

The detail description of the task can be found in Pan et al. [8]. Briefly, the monkey was trained to learn two stimulus-stimulus associative sequences (Fig. 5.1a) in a sequential paired-association task (SPAT, Fig. 5.1b). The two correct sequential associations were A1➔B1➔C1 and A2➔B2➔C2. After completing the stimulus-stimulus associations, the monkey performed the task with an asymmetric reward schedule in a blocked paradigm (Fig. 5.1c). In a given block, the monkey first learned one stimulus (e.g., C1) was paired with a large amount of water (0.4 ml) and the other stimulus with a small amount of water (0.2 ml) in reward instruction trials. The SPAT trials followed reward instruction trials. The stimulus-reward contingency was the same between SPAT trials and reward instruction trials.

One sequence (e.g., A1➔B1➔C1) is associated with the large reward and the other sequence (A2➔B2➔C2) with the small reward. In another block, the stimulus-reward contingency might be reversed. The stimulus-reward contingency between blocks was pseudo-randomized.

The stimuli A1, A2, B1, B2, C1, and C2 were well experienced in the asymmetric reward task by the monkey. These stimuli were referred to as "old stimuli." We

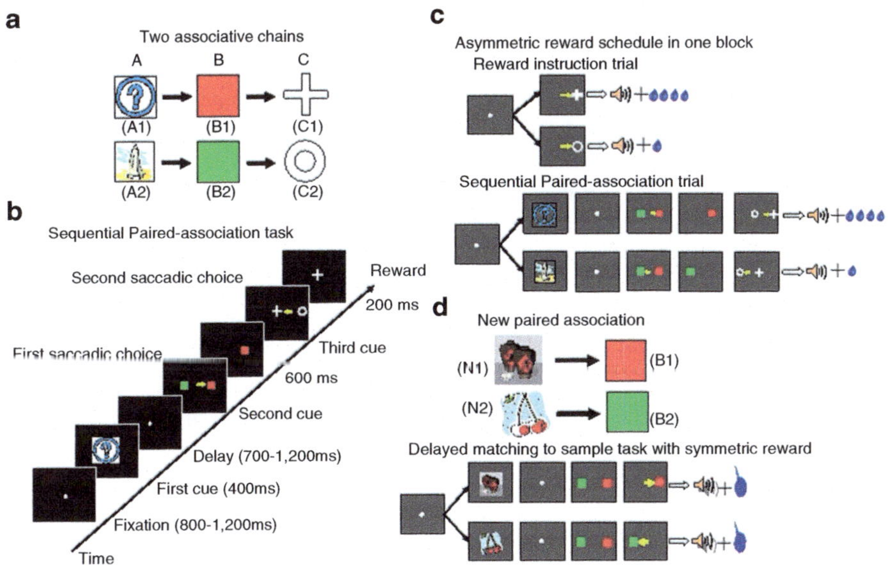

Fig. 5.1 The reward prediction task. (**a**) Two associative stimulus-stimulus sequences (A1➔B1➔C1 and A2➔B2➔C2) learned by the monkey. (**b**) Schematic illustration of time events in the sequential paired-association trial (SPAT). The monkey made a choice by a saccadic eye movement, as indicated by small yellow arrows. (**c**) An asymmetric reward schedule with the old stimuli used in one block. (**d**) A pair of new stimuli learned in the delay matching-to-sample task with symmetric reward

downloaded 200 icons from the Internet and classified them into 100 pairs of stimuli. Each pair of stimuli was associated with the two color patches, B1 and B2, in a delayed matching-to-sample task with a symmetric reward schedule (Fig. 5.1d). These newly learned stimuli will hereafter be referred to as "new stimuli." There is no any asymmetric reward information associated with the new stimuli till now.

After having fully acquired the new associations, the monkey performed the reward-instructed sequential paired-association task with new stimuli. This was identical to the reward-instructed SPATs with old stimuli (Fig. 5.1c) except that, in these SPATs, a newly learned stimulus was presented as the first cue instead of the old stimulus (A1 or A2).

5.2.2 Injection and Task Procedures

We made muscimol solution of 5 μg/μl. Two syringes with needles were attached to the recording chambers on the left and right hemispheres for bilateral injection. In each session, 4 μl muscimol (or saline) was injected into each hemisphere. In order to put the needle into the neural layer correctly, we first used two FHC single electrodes to identify positions of neural layer (first neuron) at each penetration and then set the needle; its tip was 2 mm under the first neuron. After the needle reached the target position, the syringe was pushed slowly by hand to inject the liquid into the LPFCs. The procedure cost about 10 min.

The task procedure in one session was shown as the following. Before each task session, we assigned a daily session to train the monkey to learn five pairs of associations between the new stimuli and B1 and B2 with symmetric reward (Fig. 5.1d). So at the beginning of each task session, the monkey rehearsed five pairs of new associations and then performed four blocks of the asymmetric reward SPAT with old stimuli. After that, muscimol (or saline) was injected into each LPFC. After waiting for 40 min, the monkey started performing the asymmetric reward SPAT, four blocks of old stimuli and four blocks of new stimuli (pair 1) and then four blocks of old stimuli and new stimuli (pair 2), till the completion of five pairs of new stimuli. The new stimuli were used only one time in a task session. Before the next task session, the monkey had to learn another five pairs of new stimuli associated with B1 and B2.

5.3 Results

The monkey performed the asymmetric reward SPAT in eight saline and seven muscimol sessions. The injection map was presented in Fig. 5.2.

We first examined the correct choice rate (the choice of B1 and B2 based on the first stimulus) in response to the initial presentation of the new stimuli in SPAT trials after reward instruction of C1 and C2 (Fig. 5.3a).

In these specific trials, the new stimuli were presented for the first time, so the monkeys had therefore had no previous opportunity to pair the new stimuli directly

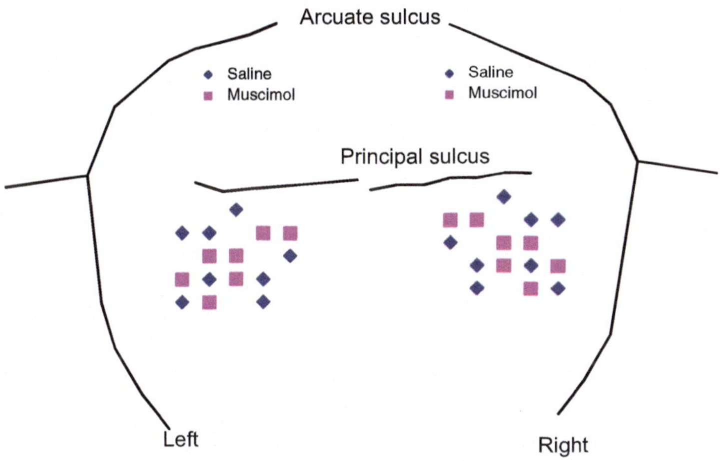

Fig. 5.2 Injection locations of saline (blue diamond) and muscimol (purple squire) in the two hemispheres

with a particular (large or small) amount of reward. In saline sessions, the correct rate in large reward trials was significantly higher than that in small reward trials (Fisher exact test, $p < 0.001$), consistent with the behavior results reported in Pan et al. [8], suggesting the monkey could infer reward values for the new stimuli at their first presentation in the SPAT.

In muscimol sessions, the correct rate for the new stimuli did not significantly differ between the two reward conditions (Fisher exact test, $p = 0.711$), indicating local inactivation in the LPFCs impaired reward prediction ability for the new stimuli. We examined response times in reward instruction trials. In both saline and muscimol sessions, the response times of last reward instruction trial in each block were significantly smaller in large reward trails than in small reward trials (Mann-Whitney U-test, $p < 0.01$), indicating the monkey correctly learned the stimulus-reward contingency in both saline and muscimol session.

The overall performances of old and new stimuli (including all trials) in saline and muscimol sessions were presented in Fig. 5.3b, c. A two-way ANOVA (two factors: session (saline vs. muscimol) and reward (large vs. small)) revealed that the correct rates of old stimuli had a main effect of session ($p = 0.014$) and the correct rate in saline was higher. There was a main effect of reward ($p < 0.001$). The correct rates in large reward trials were higher in both saline and muscimol sessions. There was no significant interaction between the two factors.

The correct rate of new stimuli had no main effect of session, significant main effect of reward, and no significant interaction. In saline and muscimol sessions, the correct rates of new stimuli were higher in large than in small reward trials. Local inactivation in the LPFCs did not impair the overall reward prediction performance for new stimuli.

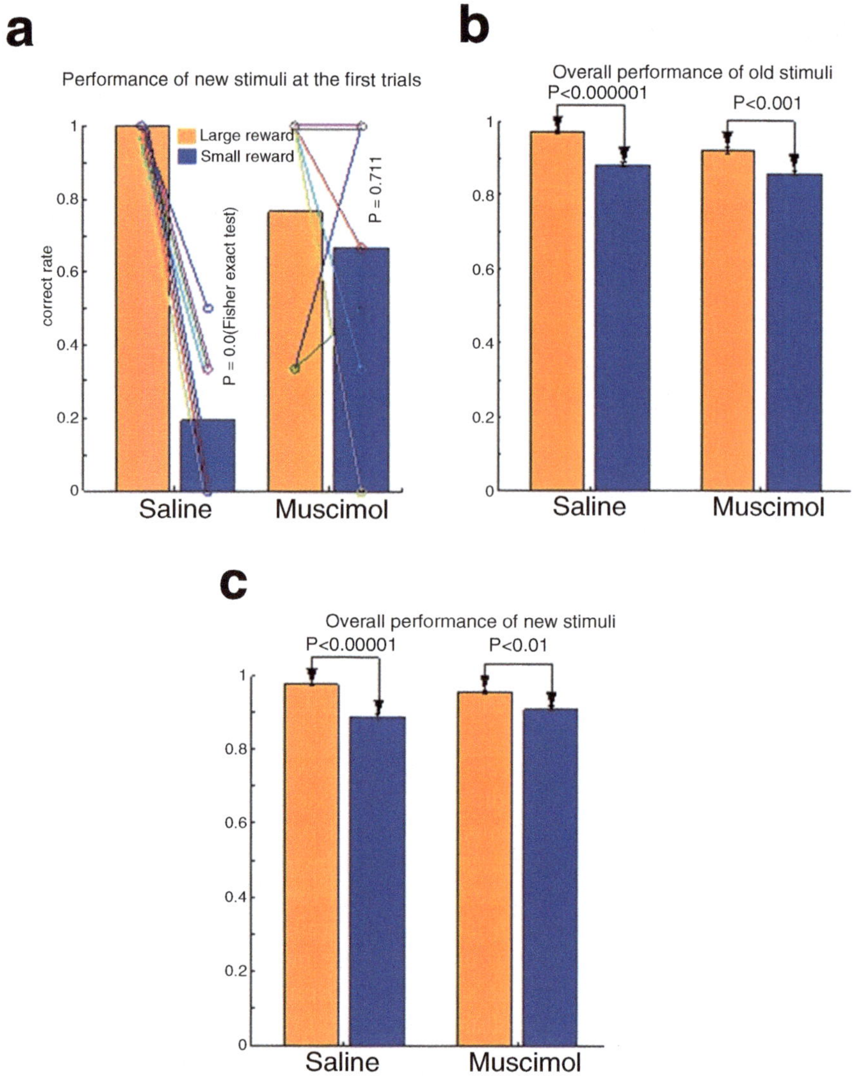

Fig. 5.3 The correct rates in saline and muscimol sessions. (**a**) The correct rates of the new stimuli presented at the first time in SPATs. Each line indicates the performance in each session. Statistical significance was tested by Fisher exact test. (**b**) The correct rates of the old stimuli in all trials. (**c**) The correct rates of the new stimuli in all trials. Statistical significance was examined by Mann-Whitney U-test. Brown bars indicate large reward trials and blue bars indicate small reward trials

5.4 Discussion

Muscimol that inactivated the LPFC locally could impair reward predictive ability for the new stimuli at their first presentation but did not alter the monkey's ability to learn stimulus-reward associations, the overall performances for old and new stimuli. The behavioral results are consistent with the prediction from the model that the PFC and striatum represent different aspects of the task structure. The PFC forms categorical representation for the stimuli that are associated in a sequence. The striatum stores each stimulus-stimulus and stimulus-reward associations to form a look-up table.

The PFC predicts reward values for old and new stimuli based on category inference (a model-based method), while the striatum looks for stimulus-reward associations in the look-up table to predict reward (a model-free method). Inactivation in the LPFC impairs the inference ability to infer reward values of the new stimuli at their first presentation and at the same time the striatum could not find the new stimulus-reward associations in the look-up table, which may cause the behavior impairment of the new stimuli in muscimol sessions.

After the monkey experiences the stimulus-reward association in the first trials, the striatum stores this information and uses it to predict reward for the new stimuli in next trials. The intact striatum might compensate impaired functions in the LPFC to recover the overall performance for old and new stimuli. The muscimol behavioral results together with single-unit results in Pan et al. [8] suggest that the LPFC and striatum predict reward independently with distinct mechanisms.

It has been suggested that cooperation and competition between the model-based system in the PFC and the model-free system in the striatum are important to maintain normal cognitive functions [9]. The imbalance between the two systems was associated with mental disorders. Our results indicate that the striatum might not directly receive model-based signals from the PFC. "Model-based signals" found in the striatum are likely generated in local striatal circuits after the subject experiences and learns the task. The information sent from the PFC to the striatum might be used for controlling the balance of model-based and model-free systems [10].

Acknowledgments This work was supported by Grant-in-Aid for Scientific Research on Innovative Areas (4303) from MEXT (Ministry of Education, Culture, Sports, Science and Technology) of Japan (http://decisions.naist.jp/index.html). It is supported by National Natural Science Foundation of China (No. 11232005, No. 11472104) and sponsored by Shanghai Pujiang Program (No. 13PJ1402000).

References

1. Haber, S.N., Kim, K.S., Mailly, P., Calzavara, R.: Reward-related cortical inputs define a large striatal region in primates that interface with associative cortical connections, providing a substrate for incentive-based learning. J. Neurosci. **26**, 8368–8376 (2006)

2. Alexander, G.E., Delong, M.R., Strick, P.L.: Parallel organization of functionally segregated circuits linking basal ganglia and cortex. Annu. Rev. Neurosci. **9**, 357–381 (1986)
3. Assad, W.F., Eskandar, E.N.: Encoding of both positive and negative reward prediction errors by neurons of the primate lateral prefrontal cortex and caudate nucleus. J. Neurosci. **31**, 17772–17787 (2011)
4. Daw, N.D., Niv, Y., Dayan, P.: Uncertainty-based competition between prefrontal and dorsolateral striatal systems for behavioural control. Nat. Neurosci. **8**, 1704–1711 (2005)
5. Samejima, K., Ueda, Y., Doya, K., Kimura, M.: Representation of action-specific reward values in the striatum. Science. **310**, 1337–1340 (2005)
6. Glascher, J., Daw, N., Dayan, P., O'Doherty, J.P.: States versus rewards: dissociable neural prediction error signals underlying model-based and model-free reinforcement learning. Neuron. **66**, 585–595 (2010)
7. Daw, N.D., Gershman, S.J., Seymour, B., Dayan, P., Dolan, R.J.: Model-based influences on humans' choices and striatal prediction errors. Neuron. **69**, 1204–1215 (2011)
8. Pan, X., Fan, H., Sawa, K., Tsuda, I., Tsukada, M., Sakagami, M.: Reward inference by primate prefrontal and striatal neurons. J. Neurosci. **34**, 1380–1396 (2014)
9. Tanaka, S., Pan, X., Oguchi, M., Taylor, J.E., Sakagami, M.: Dissociable functions of reward inference in the lateral prefrontal cortex and the striatum. Front. Psychol. **6**, 995 (2015)
10. Lee, S.W., Shimojo, S., O'Doherty, J.P.: Neural computations underlying arbitration between model-based and model-free learning. Neuron. **81**, 687–699 (2014)

Chapter 6
Behavioral and Cognitive Impairments Induced by Low Doses of MK-801 and Ketamine

Marta Lovera-Ulecía, Lucía Moreno-Lama, María Ángeles Gómez-Climent, José M. Delgado-García, and Agnès Gruart

Abstract N-methyl-D-aspartate receptors (NMDARs) are ionotropic glutamate receptors with a key role in behavioral and cognitive processes. Disruption of NMDARs has traditionally been linked to several neurological disorders, including schizophrenia. NMDAR antagonists can be used as experimental models of symptoms associated with the neural disorders caused by NMDAR dysfunctions, as well as in preclinical studies, to evaluate the effectiveness of potential antipsychotic drugs or cognitive enhancers. The effects of low doses (0.05, 0.1, and 0.2 mg/kg) of MK-801 (a noncompetitive NMDAR antagonist) on motor and cognitive functions were assessed in adult mice. The three doses increased motor activities and evoked inverted-U prepulse inhibition changes, but only the two higher doses impaired associative learning, therefore allowing its application in preclinical studies of cognitive-related deficits. In addition, this study was aimed at determining the motor and behavioral effects produced by subanesthetic doses of ketamine (a non-specific NMDA antagonist) in adult mice and the possibility of generating a mild cognitive impairment model for pharmacological purposes. We evaluated how low doses (10, 15, and 20 mg/kg) of ketamine affected the acquisition of an instrumental conditioning task, as well as their effects on motor and prepulse inhibition capabilities. Results of ketamine administration indicate a clear dose-dependent decrease of learning abilities and motor and prepulse inhibitory effects at the highest dose. Thus, ketamine administration at these three doses can be used as a model of cognitive impairment and for the induction of schizophrenic symptoms.

Marta Lovera-Ulecía and Lucía Moreno-Lama have contributed equally with all other contributors.

M. Lovera-Ulecía · L. Moreno-Lama · M. Á. Gómez-Climent ·
J. M. Delgado-García · A. Gruart (✉)
Division of Neuroscience, Pablo de Olavide University, Seville, Spain
e-mail: agrumas@upo.es

J. M. Delgado-García et al. (eds.), *Advances in Cognitive Neurodynamics (VI)*,
Advances in Cognitive Neurodynamics,
https://doi.org/10.1007/978-981-10-8854-4_6

The use of these two drugs at low doses in experimental models of selected cognitive disorders is discussed.

Keywords Cognitive functions · Ketamine · MK-801 · Motor activities · NMDA receptors · Operant conditioning · Prepulse inhibition · Schizophrenia

6.1 I ntroduction

NMDARs are a specific type of ionotropic glutamate receptor with key roles in the development of the central nervous system as well as in many different higher functions such as locomotion, learning, and memory [1]. Disruptions in the level or function of NMDARs have traditionally been linked to several neurological and cognitive-related disorders, including schizophrenia and other psychoses [2]. Research on drugs that specifically act on these receptors has grown considerably over the last few decades, mostly aimed at putative treatments of related disorders.

In contrast, the attractiveness of NMDAR antagonists is based firstly on their application as models of symptoms more or less related to neurological disorders caused by NMDAR dysfunctions. Indeed, there is a possibility of using them in preclinical studies, thus allowing the evaluation of the effectiveness of potential antipsychotics and/or cognitive enhancers which would reverse the transient effects induced by the antagonist.

MK-801 is a noncompetitive antagonist of NMDARs with a high affinity and selectivity for a site located in the NMDA channel [3]. Our aim was using this drug at doses lower than those evoking ataxic disorders, noticeable derangements of learning and memory capabilities, and even neural lesions [4]. Because of its interactions with NMDA and dopamine receptor functions at prefrontal and striatal levels, MK-801 could represent an interesting experimental model for the study of higher cognitive functions [5]. It has already been reported that MK-801 administered at doses >0.05 mg/kg in mice evokes overt behavioral and cognitive deficits [6]. In this regard, we have checked here the effects of very low doses of MK-801 (<0.2 mg/kg) on selective motor (open-field), behavioral (prepulse inhibition), and cognitive (operant conditioning) functions.

Ketamine is a hypnotic, analgesic, and amnesic substance usually considered as a dissociative anesthetic [7]. Used at anesthetic doses, ketamine induces a dissociative or cataleptic state characterized by analgesia, amnesia, and changes in the attentional state, but not necessarily loss of consciousness [8]. Ketamine is an antagonist of NMDA receptors but also of non-glutamatergic (muscarinic, opioid) receptors [7]. Administered at subanesthetic doses, ketamine can evoke psycho-dyslectic and psychotic symptoms. We consider that it would be interesting to study in detail the motor, behavioral, and cognitive effects of the administration of subanesthetic doses ($\leq$20 mg/kg) of ketamine to determine whether this procedure can be used as an experimental model of cognitive disorders, including schizophrenia and other related psychiatric disorders.

6.2 Methods

6.2.1 *Experimental Animals*

Experiments were carried out in a total of 80 C57Bl/6 male adult mice (3–5 months old; 25–30 g) obtained from an official supplier (University of Granada Animal House, Granada, Spain). Upon arrival, animals were housed in separate cages ($n = 5$ per cage), but they were switched to individual cages 1 week before the beginning of the experimental study. Mice were kept on a 12-h light/dark cycle with constant ambient temperature (21.5 ± 1 °C) and humidity (55 ± 8%). Food and water were available ad libitum, except for the operant conditioning procedures. Experiments were carried out in accordance with the guidelines of the European Union (2010/63/EU) and Spanish regulations (BOE 34/11370-421, 2013) for the use of laboratory animals in chronic studies. All experimental protocols were also approved by the local Ethics Committee.

Animals were divided in eight experimental groups ($n = 10$ animals/group). In the case of MK-801, we prepared a control (saline) group and three experimental groups with increasing (0.05, 0.1, and 0.2 mg/kg) doses of the drug dissolved in saline. For ketamine, we prepared a control (saline) group and three experimental groups with increasing (10, 15, and 20 mg/kg) doses of the drug.

6.2.2 *Open-Field Test*

As illustrated in Fig. 6.1a, mice were placed in the center of the open-field apparatus (a box of 28 × 28 × 21 cm from Cibertec S.A., Madrid, Spain) and observed for 15 min. The apparatus was provided with infrared lights, located every 2 cm, in the three (X, Y, Z) spatial axes. Animals' displacements in the open field were quantified automatically with the help of a computer program (MUX_XYZ16L), also from Cibertec S.A. The apparatus was located in a soundproof room, and the experimental area was dimly and homogeneously illuminated. In order to avoid any interference with the following experimental animal, the whole apparatus was cleaned with alcohol (70° proof) after each use.

6.2.3 *Prepulse Inhibition Task*

Animals were placed individually inside a startle chamber (Cibertec S.A.; Fig. 6.2a–c). The startle response was measured using a piezoelectric accelerometer controlled by a computer, using the protocol described elsewhere [9, 10]. The digitized signal was averaged from 25 to 30 recordings. For training, the mouse was placed in the startle chamber for an acclimation period of 3 min.

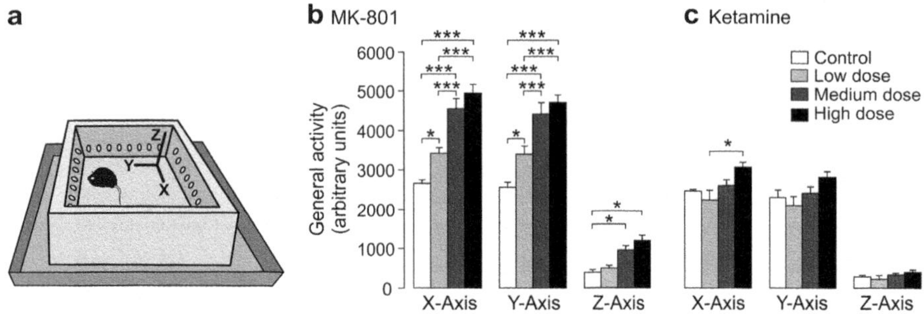

Fig. 6.1 Effects of MK-801 and ketamine administration on motor activities performed by adult mice in an open field. (**a**) A diagram of the open field. The recording area was provided by infrared lights (2 cm apart) in the three axes (X, Y, and Z). (**b**) Results collected from the four groups of MK-801 animals for the three spatial axes. Note the significant increase in motor activity induced by the three selected doses of MK-801. (**c**) The same for the four groups of ketamine mice. Values are Mean ± SEM of the collected data. $*p < 0.05$; $***p < 0.001$

Baseline responses were averaged after the presentation of 20 sounds (125 dB, 100 ms long). During prepulse inhibition trials, the same 125-dB 100-ms burst was preceded (250 ms) by a prepulse stimulus of 85 dB, lasting for 50 ms. Trials including prepulse stimuli were randomly presented with normal startle stimuli, the final total being 25 of each. The ambient background noise was 70 dB. The total startle response area (mV × ms) was recorded and quantified. Following [11], data were computed in accordance with the Eq. (6.1):

$$\left[(\text{startle/prepulse ratio}) \times 100\right]/\text{baseline value.} \tag{6.1}$$

6.2.4 Operant Conditioning Procedures

Following previous descriptions by [12, 13], operant conditioning took place in five Skinner box modules measuring 12.5 × 13.5 × 18.5 cm (MED Associates, St. Albans, VT, USA; Fig. 6.3a). Each Skinner box was housed within a sound-attenuating chamber (90 × 55 × 60 cm), which was constantly illuminated (19 W lamp) and exposed to a 45 dB white noise (Cibertec S.A.). Each Skinner box was equipped with a food dispenser from which pellets (MLabRodent Tablet, 20 mg; Test Diet, Richmond, IN, USA) could be delivered by pressing a lever. Before training, mice were handled daily for 7 days and food-deprived to 85–90% of their free-feeding weight.

For operant conditioning, animals were trained to press the lever to receive pellets from the feeder using a fixed-ratio (1:1) schedule (Fig. 6.3b). Sessions lasted for 20 min. Animals were maintained on this 1:1 schedule until they reached

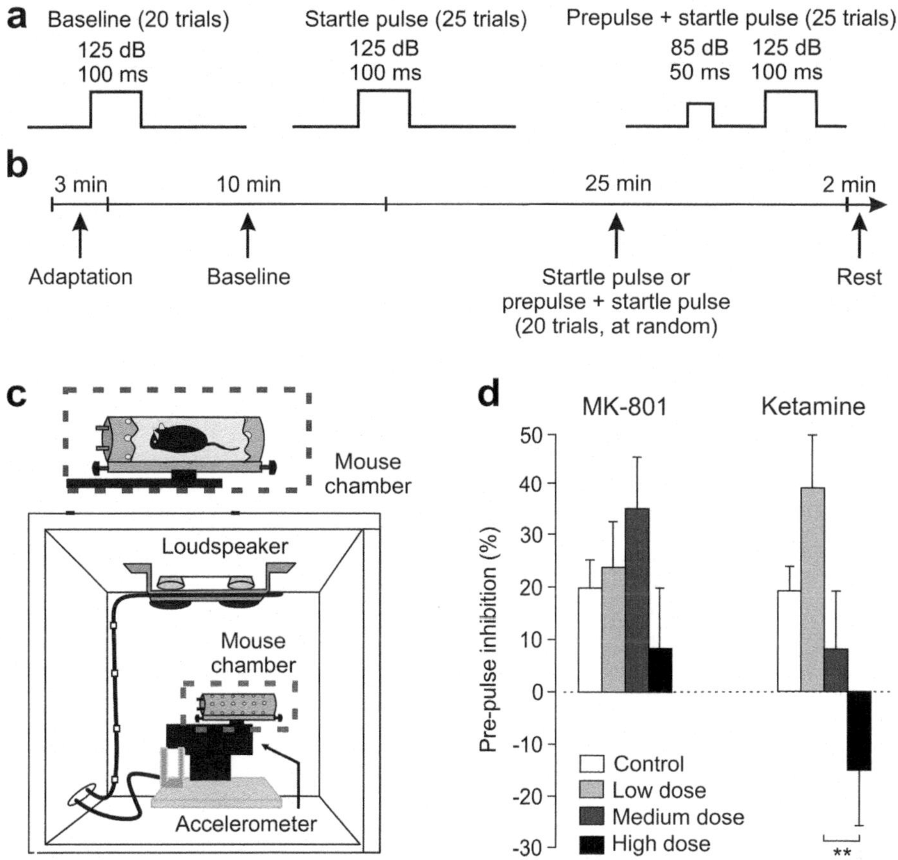

Fig. 6.2 Effects of MK-801 and ketamine administration on the execution of the acoustic startle response and prepulse inhibition test by adult mice. (**a**) The three types of tone used in this study. (**b**) Experimental design of the startle response and prepulse inhibition test. (**c**) Startle chamber. (**d**) Graphical representation of the prepulse inhibition (% of the startle response) obtained from the four MK-801 and ketamine groups. Values are Mean ± SEM of the collected data. $^{**}p < 0.01$

the selected criterion—namely, until they were able to obtain ≥20 pellets/session for two successive sessions. With these experimental procedures, wild-type mice reached criterion after 4–7 days of training [14].

Once criterion was reached, animals were further conditioned using a light/dark protocol for 10 additional days (Fig. 6.3c). In this protocol, only lever presses during the light period (20 s) were reinforced with a pellet. Lever presses performed during the dark period (20 ± 10 s) were not reinforced but restarted the dark protocol for an additional random (1–10 s) time.

The number of lever presses during light and dark periods was quantified for each training session. Conditioning programs, lever presses, and delivered reinforcements were monitored and recorded by a computer, using a MED-PC program (MED Associates, St. Albans, VT, USA).

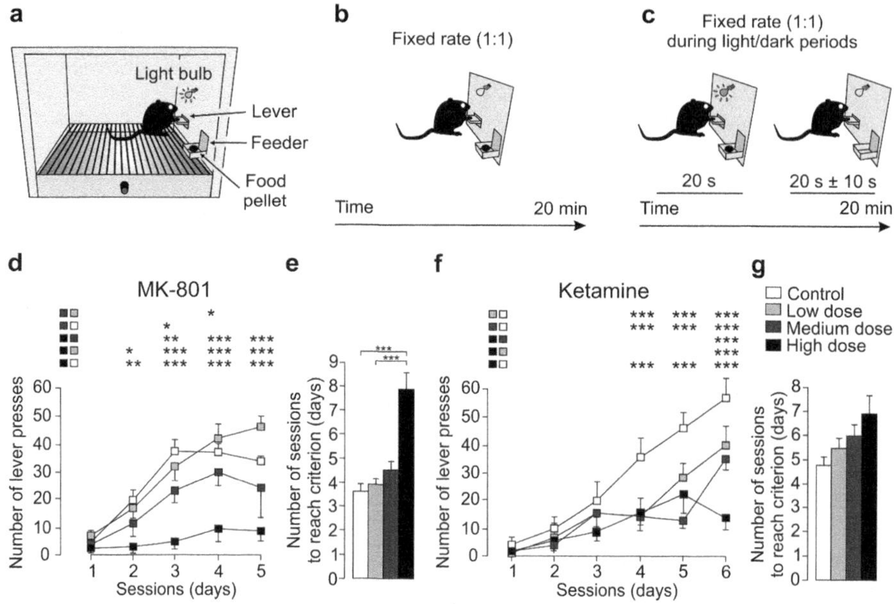

Fig. 6.3 Effects of MK-801 and ketamine administration on the execution of an operant conditioning task using a fixed-ratio (1:1) schedule. (**a**–**c**) Mice were trained in a Skinner box to press a lever to obtain a food pellet (**a**). For operant conditioning, we used two paradigms of increasing difficulty. In the first paradigm (a fixed-ratio of 1:1), the selected criterion was that the mouse had to press the lever 20 times per 20 min session for 2 successive sessions to successfully complete the task (**b**). In the second paradigm (a fixed-ratio of 1:1 during light/dark periods), lever presses were rewarded only when a light bulb was switched on. In this case, lever presses while the bulb was off were punished with a time penalty of up to 10 s during which the bulb would not turn on (**c**). (**d** and **e**) Performance of mice during the first 5 days of training with the fixed-ratio (1:1) schedule and following the administration of the three selected doses of MK-801. Significant differences between groups are indicated (**d**). Days to reach the selected criterion by each experimental group are illustrated in (**e**). (**f** and **g**) Same as in (**d** and **e**) but for data collected from mice injected with the three selected doses of ketamine. The code for the four experimental groups in (**g**) is also for (**d**–**f**). *$p < 0.05$; **$p < 0.01$; ***$p < 0.001$

6.2.5 Drug Administration

(+)-MK-801 (Dizocilpine, Sigma-Aldrich, Steinheim, Germany) was dissolved in a 0.9% NaCl solution and administered i.p. in a total volume of 0.3 mL/mouse. The selected doses were 0.05 mg/kg, 0.1 mg/kg, and 0.2 mg/kg [6]. Ketamine [(2R)-2-(2-chlorophenyl)-2-(methylamino)cyclohexanone; Sigma-Aldrich, Steinheim, Germany] was dissolved in a 0.9% NaCl solution and administered i.p. in a total volume of 0.3 mL/mouse. The respective control groups were injected with the saline solution at the indicated volume. The selected doses were 10 mg/kg, 15 mg/kg, and 20 mg/kg. Drugs were administered 30 min before the performance of the selected test.

6.2.6 *Data Analysis*

Collected data were translated into Excel spreadsheets for further analysis and representation. Unless otherwise indicated, data are represented as mean ± SEM. Statistical analyses were carried out with the help of the SigmaPlot 11.0 program (Systat Software Inc., San Jose, CA, USA). Acquired data from the open field and the startle chamber were analyzed using one-way repeated measures ANOVA with all pairwise multiple comparison procedures (Holm-Sidak method) when statistical differences were found. Kruskal-Wallis one-way ANOVA on ranks was used when the normality test or the equal variance test failed. Data from the operant conditioning test were analyzed using two-way repeated measures ANOVA, with the Kruskal-Wallis-Tukey test and the Holm-Sidak method as needed. Data collected from the light/dark test were analyzed with linear regression lines. The linear equation and coefficient of correlation corresponding to each set of data were calculated.

6.3 Results

6.3.1 *Performance of the Open-Field Test by Mice Treated with MK-801 and Ketamine*

As described in Methods, animals of each group were placed individually in an open-field apparatus (Fig. 6.1a) to determine their motor activities for a single 15-min period. The number of light-beam crossings in the three spatial axes (X, Y, Z) was quantified and totalized for each experimental group. The administration of increasing doses of MK-801 significantly increased the spontaneous motor activities of the experimental animals in the three axes (X, $F_{[3, 36]} = 25.55$, $p < 0.001$; Y, $F_{[3, 36]} = 23.44$, $p < 0.001$; and Z, $H = 16.48$ with three degrees of freedom, $p < 0.05$). In contrast, the administration of the selected doses of ketamine evoked no significantly different motor activities in the open field, apart from the highest dose (20 mg/kg) and only for the X-axis ($F_{[3, 36]} = 3.33$, $p < 0.05$).

On the whole, the selected doses of MK-801, but not those of ketamine, produced an evident increase in the motor activity of the animals in the open field, suggesting an increase in their exploratory behaviors.

6.3.2 *Prepulse Inhibition of the Startle Response by Mice Treated with MK-801 and Ketamine*

In Fig. 6.2a–c is illustrated the apparatus used to evoke a startle response in individual mice and the protocol followed to evoke its inhibition. The startle response was evoked by the presentation of a 125-dB 100-ms tone, while its inhibition was achieved by the presentation of an 85-dB 50-ms tone 250 ms in advance of the stronger tone. The administration of increasing doses of MK-801 evoked an inverted-U modification of the prepulse inhibition obtained in the control group (Fig. 6.2d). However, the collected results did not reach significant differences. In contrast, the highest (20 mg/kg) dose of ketamine used in these experiments produced a significant decrease in the evoked prepulse inhibition. In conclusion, only the highest dose of ketamine evoked a significant decrease in the amount of prepulse inhibition of the startle response reached by the control groups.

6.3.3 Operant Conditioning of Animals Treated with MK-801 and Ketamine

Mice were trained in Skinner boxes to obtain a food pellet every time they pressed a lever in daily sessions of 20 min, using a fixed-ratio (1:1) schedule (Fig. 6.3a, b). Mice were considered to complete the task when pressing the lever ≥20 times in 2 successive sessions (i.e., the criterion; Fig. 6.3d–g). As shown in Fig. 6.3d, the administration of increasing doses of MK-801 significantly ($F_{[12,144]} = 6.06$, $p < 0.001$) affected the animals' proper performance of the operant conditioning task, particularly for mice administered with the highest dose (0.2 mg/kg).

In addition, control MK-801 mice and those receiving the two lower doses (0.05 and 0.1 mg/kg) reached the selected criterion in ≈4 days, but those injected with the highest dose (0.2 mg/kg) reached criterion significantly later ($p < 0.001$; $H = 21.39$ with three degrees of freedom; Fig. 6.3e). The administration of increasing doses of ketamine also significantly ($F_{[15,180]} = 5.08$, $p < 0.001$) decreased performance of the mice in the Skinner box during the first six training sessions (Fig. 6.3f).

Although the administration of increasing doses of ketamine increased the number of sessions necessary to reach the selected criterion, no significant differences between groups were observed ($p = 0.9092$; $H = 6.45$ with three degrees of freedom; Fig. 6.3g).

Mice that successfully reached the above criterion were subjected to a more complex operant conditioning task. In this case, pressing the lever was rewarded with a food pellet only during periods of 20 s in which a light bulb above the lever was switched on (light/dark, Fig. 6.3c). As illustrated in Fig. 6.4a, control MK-801 decreased the number of lever presses during the dark periods across the ten training sessions. In contrast, the administration of increasing doses of MK-801 prevented the proper acquisition of this operant conditioning task (Fig. 6.4b, c) and even reversed it (Fig. 6.4d). In the same way, the administration of the highest dose of ketamine disturbed the proper acquisition of the light/dark task (Fig. 6.4e–h).

In summary, the administration of low doses of MK-801 and ketamine affected the proper acquisition of an operant conditioning task, mainly for the MK-801 and ketamine groups administered the highest doses (0.2 mg/kg and 20 mg/kg,

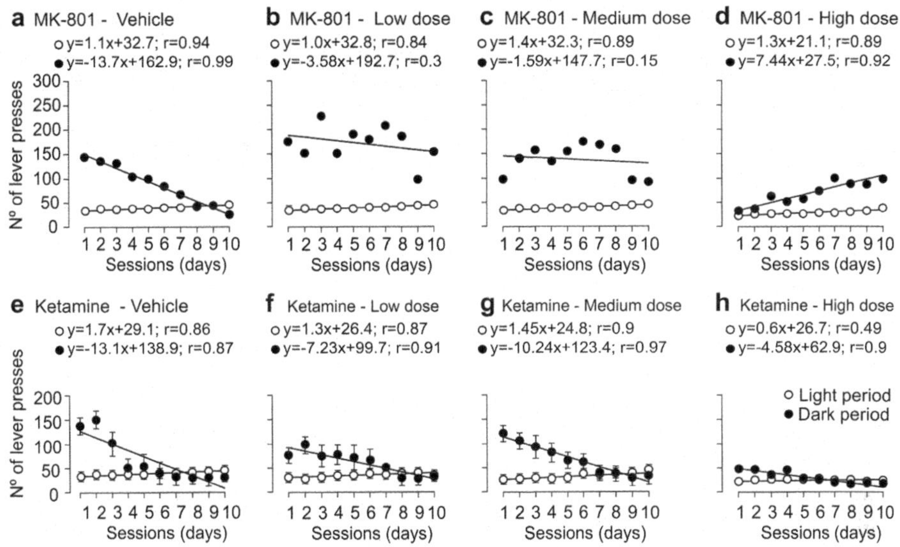

Fig. 6.4 Effects of MK-801 and ketamine administration on the execution of an operant conditioning task using a fixed-ratio (1:1) schedule in a go/non-go situation. In this situation, only lever presses carried out during the light period were rewarded. Lever presses carried out while the bulb was off were punished with up to 10 additional seconds during which the bulb would not turn on (light/dark paradigm; see Fig. 6.3c). (**a–d**) Lever presses performed by mice of the four (control and three doses) MK-801 groups across ten successive sessions using the light/dark paradigm. (**e–h**) Same for experiments carried out with the four ketamine groups. As indicated in (**h**): white circles, lever presses when the light bulb was on (light); black circles, lever presses when the bulb was off (dark). Regression lines and their coefficient of correlation (*r*) are indicated above each representation

respectively). The three groups administered with MK-801 and the ketamine group administered with the highest dose were unable to acquire an operant light/dark task entailing the specific inhibition of natural appetitive behaviors.

6.4 Discussion

We have studied here the effects of low doses of MK-801 and of ketamine on the spontaneous motor activities, prepulse inhibition of a tone-evoked startle response, and associative learning capabilities of young adult mice. The aim was to determine whether those drugs, administered at low doses, could evoke symptoms related with cognition-related disorders such as schizophrenia and related psychoses.

The three doses of MK-801 used here evoked significant increases in the motor activities of the experimental animals in the open-field apparatus. It can be proposed that this increase in exploratory activities also disturbed the proper acquisition of the light/dark operant conditioning task presented by the three groups of mice injected with MK-801, particularly those administered with the highest dose.

As already described [15], mice injected with the highest dose used here (0.2 mg/kg) presented evident motor deficits (such as ataxic movements, vestibular misbalances, etc.); these motor effects could rule out using this dose for operant conditioning tasks. Finally, we have been unable to reproduce here the effects of MK-801 on prepulse inhibition described elsewhere [16].

According to the present results, ketamine administration at subanesthetic doses did not evoke the noticeable hyperactive motor responses described in mice following the acute administration of a much higher dose (100 mg/kg, i.p.) [17]. However, the highest dose of ketamine used here (20 mg/kg) significantly increased motor activities in the open field and decreased the inhibitory effects of a prepulse on the tone-evoked startle response but did not prevent the proper acquisition of an operant conditioning task. In this sense, this low dose of ketamine can be used to evoke mild cognitive impairments potentially related to some psychotic states. For example, it is well known that the absence of a proper prepulse inhibition mechanism is considered a symptom related to a schizophrenic condition [9, 18, 19].

The administration of low doses ($\leq$12 mg/kg) of ketamine also disturbed the acquisition of an instrumental conditioning in rats, using fixed-interval (1:50 s and 1:200 s) schedules [20]. In this regard, it is possible that a fixed interval could represent a task more difficult to acquire than the fixed-ratio (1:1) schedule used here.

In conclusion, the highest dose of MK-801 used here could be a useful experimental tool for evoking cognitive-related impairments involving operant conditioning tasks, probably related to the increase in interspecific exploratory activities. Similarly, the highest dose of ketamine used here could be useful for creating deficits in the prepulse inhibition test.

Acknowledgments We thank Ms. Olga Palomo-Casillas, Ms. María Sánchez-Enciso, and Mr. José M. González-Martin for their collaboration in animal experiments and data analysis. We also thank Mr. Roger Churchill for his editorial help.

References

1. Zhou, H.X., Wollmuth, L.P.: Advancing NMDA receptor physiology by integrating multiple approaches. Trends Neurosci. **40**, 129–137 (2017)
2. Wiescholleck, V., Manahan-Vaughan, D.: Long-lasting changes in hippocampal synaptic plasticity and cognition in an animal model of NMDA receptor dysfunction in psychosis. Neuropharmacology. **74**, 48–58 (2013)
3. Vanderschuren, L.J., Schoffelmeer, A.N., Mulder, A.H., De Vries, T.J.: Dizocilpine (MK801): use or abuse? Trends Pharmacol. Sci. **19**, 79–81 (1998)
4. Fix, A.S., Horn, J.W., Wightman, K.A., Johnson, C.A., Long, G.G., Storts, R.W., Olney, J.W.: Neuronal vacuolization and necrosis induced by the noncompetitive N-methyl-D-aspartate (NMDA) antagonist MK (+) 801 (dizocilpine maleate): a light and electron microscopic evaluation of the rat retrosplenial cortex. Exp. Neurol. **123**, 204–215 (1993)

5. Collingridge, G.L., Volianskis, A., Bannister, N., France, G., Hanna, L., Mercier, M., Tidball, P., Fang, G., Irvine, M.W., Costa, B.M., Monaghan, D.T., Bortolotto, Z.A., Molnár, E., Lodge, D., Jane, D.E.: The NMDA receptor as a target for cognitive enhancement. Neuropharmacology. **64**, 13–26 (2013)
6. Miyamoto, S., Leipzig, J.N., Lieberman, J.A., Duncan, G.E.: Effects of Ketamine, MK-801, and amphetamine on regional brain 2-Deoxyglucose uptake in freely moving mice. Neuropsycopharmacology. **22**, 400–412 (2000)
7. Mion, G., Villevieille, T.: Ketamine pharmacology: an update (pharmacodynamics and molecular aspects, recent findings). CNS Neurosci. Ther. **19**, 370–380 (2013)
8. Domino, E., Chodoff, P., Corssen, G.: Pharmacologic effects of CI-581, a new dissociative anaesthetic in human. Clin. Pharmacol. Ther. **6**, 279–291 (1965)
9. Borrell, J., Vela, J.M., Arévalo-Martin, A., Molina-Holgado, E., Guaza, C.: Prenatal immune challenge disrupts sensorimotor gating in adult rats: implications for the etiopathogenesis of schizophrenia. Neuropsychopharmacology. **26**, 204–215 (2002)
10. Porras-García, E., Cendelin, J., Domínguez-del-Toro, E., Vožeh, F., Delgado-García, J.M.: Purkinje cell loss affects differentially the execution, acquisition and prepulse inhibition of skeletal and facial motor responses in Lurcher mice. Eur. J. Neurosci. **21**, 979–988 (2005)
11. Eleore, L., López-Ramos, J.C., Yi, P.J., Delgado-García, J.M.: The cognitive enhancer T-588 partially compensates the motor associative learning impairments induced by scopolamine injection in mice. Behav. Neurosci. **121**, 1203–1214 (2007)
12. Jurado-Parras, M.T., Gruart, A., Delgado-García, J.M.: Observational learning in mice can be prevented by medial prefrontal cortex stimulation and enhanced by nucleus accumbens stimulation. Learn. Mem. **19**, 99–106 (2012)
13. Jurado-Parras, M.T., Delgado-García, J.M., Sánchez-Campusano, R., Gassmann, M., Bettler, B., Gruart, A.: Presynaptic GABAB receptors regulate hippocampal synapses during associative learning in behaving mice. PLoS One. **11**, e0148800 (2016)
14. Madroñal, N., López-Aracil, C., Rangel, A., del Río, J.A., Delgado-García, J.M., Gruart, A.: Effect of enriched physical and social environments on motor performance, associative learning, and hippocampal neurogenesis in mice. PLoS One. **5**, e11130 (2010)
15. Liljequist, S., Ossowska, K., Grabowska-Anden, M., Anden, N.E.: Effect of the NMDA receptor antagonist, MK-801, on locomotor activity and on the metabolism of dopamine in various brain areas of mice. Eur. J. Pharmacol. **195**, 55–61 (1991)
16. Yee, B.K., Chang, D.T., Feldon, J.: The effects of dizocilpine and phencyclidine on prepulse inhibition of the acoustic startle reflex and on prepulse-elicited reactivity in C57BL6 mice. Neuropsychopharmacology. **29**, 1865–1877 (2004)
17. Chatterjee, M., Ganguly, S., Srivastava, M., Palit, G.: Effect of 'chronic' versus 'acute' ketamine administration and its 'withdrawal' effect on behavioural alterations in mice: implications for experimental psychosis. Behav. Brain Res. **216**, 247–254 (2011)
18. Cilia, J., Hatcher, P., Reavill, C., Jones, D.N.: Ketamine-induced prepulse inhibition deficits of an acoustic startle response in rats are not reversed by antipsychotics. J. Psychopharmacol. **2**, 302–311 (2007)
19. Hsu, H.-R., Mei, Y.-Y., Wu, C.-Y., Chiu, P.-H., Chen, H.-H.: Behavioural and toxic interaction profile of ketamine in combination with caffeine. J. Compil. Basic Clin. Pharmacol. Toxicol. **104**, 379–383 (2009)
20. Pallarés, M.A., Nadal, R.A., Silvestre, J.S., Ferré, N.S.: Effects of ketamine, a noncompetitive NMDA antagonist, on the acquisition of the lever-press response in rats. Physiol. Behav. **57**, 389–392 (1995)

Chapter 7
Changes in Brain Activity During Instrumental Behavior After Additional Learning in Rats

Vladimir Gavrilov

Abstract To study how additional learning may change the brain activity during realization of initially learnt behavior, we compared EEG potentials in rats performing an instrumental lever-pressing task before and after additional training in a slightly different environment. EEG was recorded over motor, retrosplenial posterior, and visual areas of the cortex. A similar configuration of behavior-related EEG potentials was observed during performing the instrumental task before and after additional learning, which suggests that, in general, common brain processes underlie the behavior in the compared conditions. However, differences in the amplitudes and latencies of components of behavior-related potentials shown in this work support the hypothesis that the composition of the elements of individual experience underlying this behavior changes after additional learning.

Keywords EEG · Instrumental behavior · Learning · Individual experience

7.1 Introduction

Single-unit recording studies in behaving animals revealed a stable relation of impulse activityof neurons in different brain areas with acts of behavior, i.e., behavioral specialization of neurons [1–4] (Fig. 7.1). It has been shown that the behavioral specialization is permanent and does not change after additional learning [3, 5]. Types of neuronal specializations depend on the history of formation of behavior [3]. Molecular bases of neuronal specialization have also been established [6]. These previous findings suggest that any behavior is based on cooperative activity of neurons in different brain areas which form functional systems of behavioral acts. The formation of these systems – systemogenesis – occurs during learning in problem situations when previous experience does not provide a way

V. Gavrilov (✉)
Institute of Psychology of Russian Academy of Sciences, Moscow, Russia

J. M. Delgado-García et al. (eds.), *Advances in Cognitive Neurodynamics (VI)*,
Advances in Cognitive Neurodynamics,
https://doi.org/10.1007/978-981-10-8854-4_7

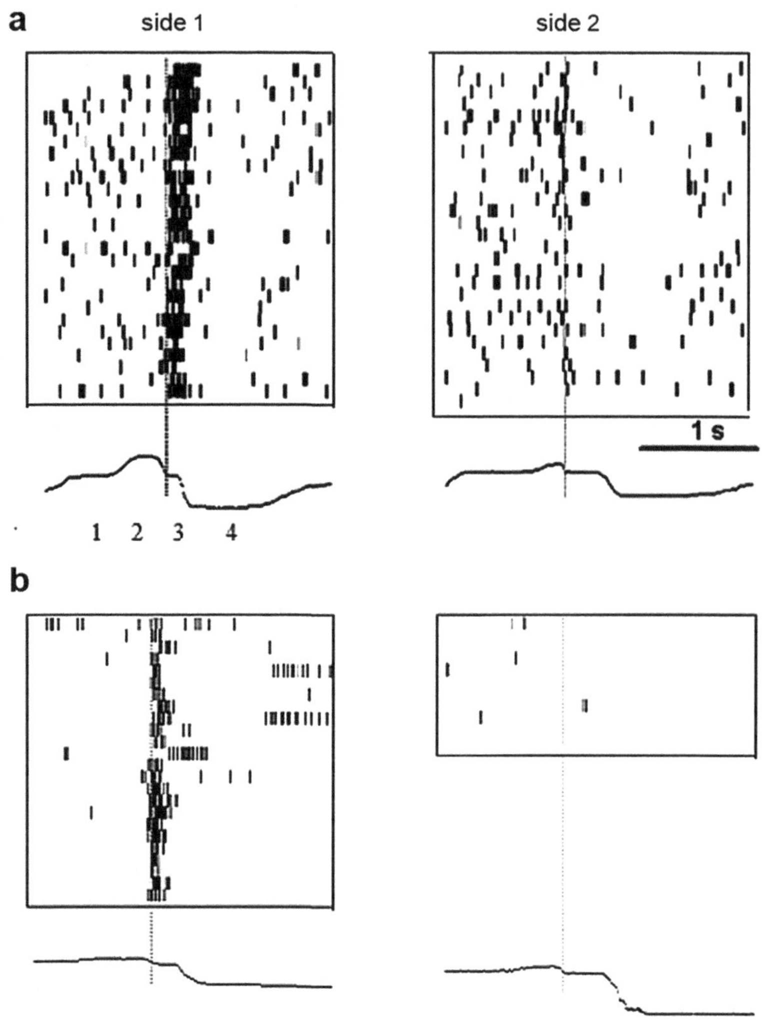

Fig. 7.1 The same behavioral specialization of neurons in different brain areas. Examples of single units from motor cortex (**a**) and retrosplenial cortex (**b**) increasing their firing rates when a rat moved from a lever to a feeder on one side of the experimental cage but not on the other. The rasters are plotted against the end of lever pressing (dashed lines). Averaged actograms of behavior are shown under the rasters. On the basis of sets of behaviorally specialized neurons, the following behavioral acts were defined: (1) approaching the lever, (2) lever pressing, (3) approaching the feeder, and (4) lowering the head into the feeder to capture a portion of food

of achieving successful outcomes of behavior. Activity of a system of neurons underlies implementation of the corresponding behavioral act which is followed by the next behavioral act supported by activity of another system of neurons [1, 3, 4] (Fig. 7.2). Thus, any complex behavior can be viewed as a continuum of behavioral acts subserved by systems of neurons distributed across the brain and specialized in relation to the results of these behavioral acts.

To study the system organization of behavior, we use a model of instrumental behavior where an animal is trained to press a lever which activates a feeder delivering a portion of food into the experimental cage. Using this model, we showed that EEG potentials averaged from behavioral events (lever pressings and moments of lowering the head into the feeder) had a similar configuration over different areas of the brain [9]. This configuration reflected the system organization of brain processes: negative and positive EEG waves corresponded to realization and changes of behavioral acts. Similar results were shown in other studies in animals and human [7, 8].

Our previous work has demonstrated that additional training to perform already learned instrumental lever-pressing behavior in a slightly different environment (another compartment of an experimental cage with different placement of a lever) leads to specialization of additional neurons and their embedding into the existing system structure of experience [1–4]. In this work, we aimed to study changes in the organization of brain activity during instrumental lever-pressing behavior after additional training to perform this behavior in another similar environment.

7.2 Methods

Long Evans adult male rats ($n = 8$) were trained in an experimental cage supplied with two levers and two feeders in the corners. The experimental cage was divided into two equal compartments with a transparent plastic partition. First, animals were trained in one compartment and then additionally trained in the other EEG, which was recorded during instrumental lever-pressing behavior before and after the additional training. Rats were initially trained in four stages: (1) capturing food in the feeder, (2) turning head and walking away from the feeder, (3) approaching to the lever, and (4) pressing the lever. After this initial training, EEG was recorded during performing the instrumental behavior in the first compartment of the experimental cage. Then rats were placed into the other compartment where they learned to perform the same task without the help of the experimenter. After such additional learning, rats were retested in the first compartment, and their EEG was recorded. We analyzed components of event-related potentials before and after additional learning. EEG was recorded during three 30-min sessions before and after additional learning, to evaluate variability of slow brain potentials in these conditions. Monopolar EEG was recorded with silver electrodes implanted epidurally over motor (a. 3.0, l. 3.0), retrosplenial posterior (p. 4.5, l. 1.0), and visual

Fig. 7.2 An example of three motor cortex neurons (**a**–**c**) increasing their firing rates in consecutive behavioral acts: during lever pressing (**a**) (26 trials), approaching the feeder (**b**) (35 trials), and capturing food in the feeder (**c**) (29 trials). The rasters are plotted against lever pressings (in **a** and **b**) and against the moments of lowering the head into the feeder (**c**) (dashed lines)

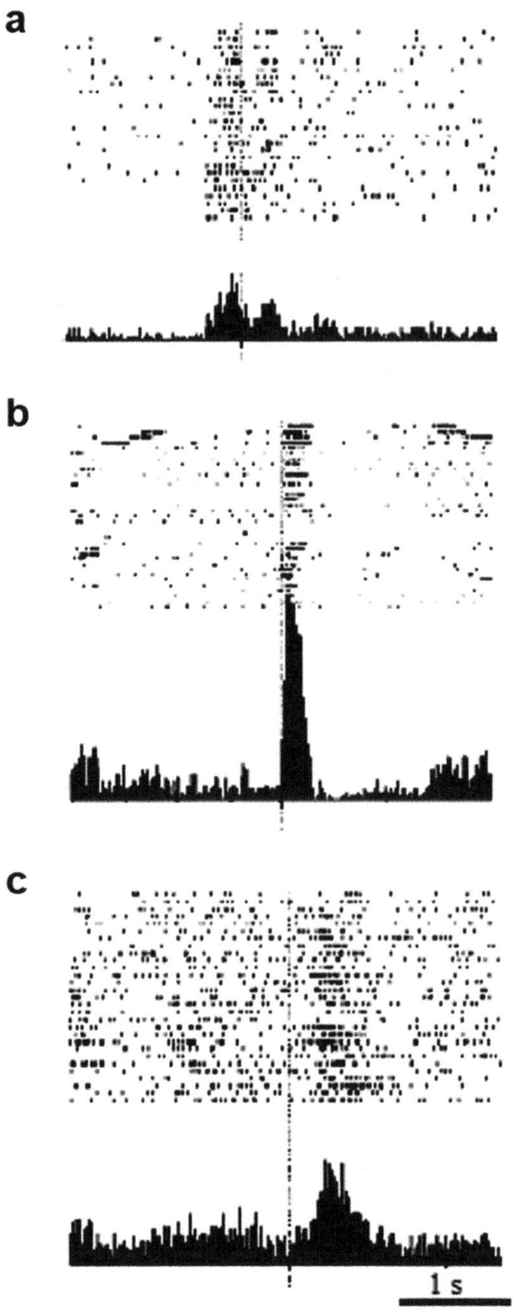

(p. 7.0, l. 4.5) areas of the cerebral cortex. Brain potentials were averaged from the moments of lever pressings and lowering the head into the feeder.

7.3 Results

A similar configuration of behavior-related potentials was observed in all studied cortical areas: negative and positive EEG waves corresponded to realization and changes of behavioral acts, respectively (Fig. 7.3). Taken together with the previously shown results [7–9], our data suggest that behavior-related EEG poten-

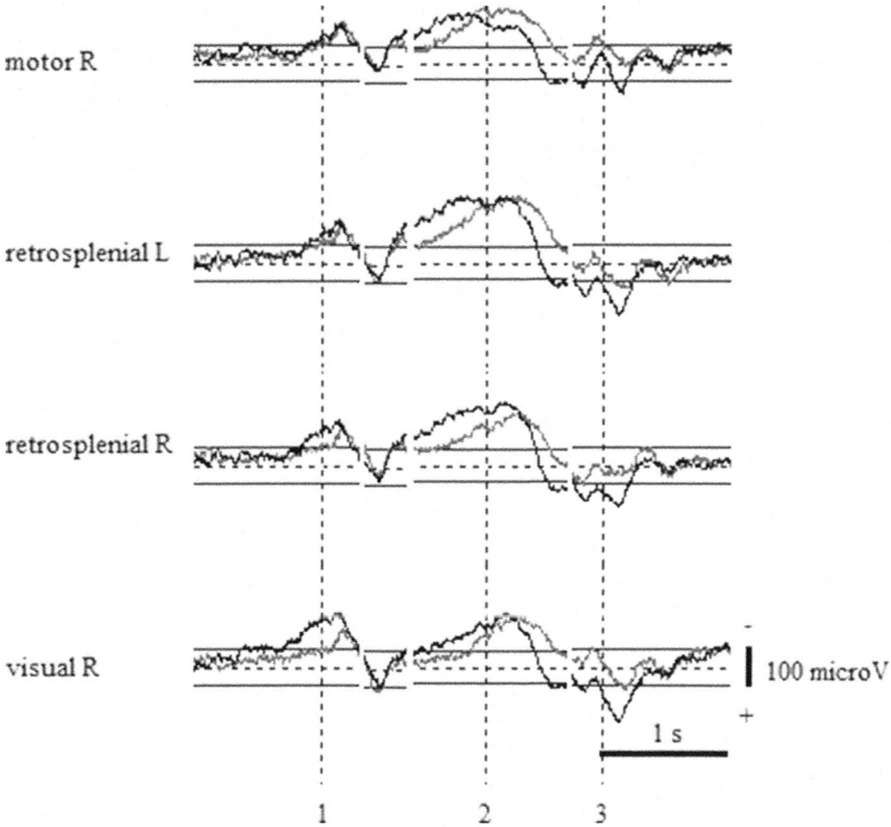

Fig. 7.3 An example of behavior-related potentials during the same instrumental lever-pressing behavior in experimental sessions before (black) ($n = 251$ trials) and after (gray) ($n = 171$ trials) additional learning in another compartment of the experimental cage. Behavior-related potentials were averaged from lever pressing (1), lowering head into the feeder (2), and taking the head out of the feeder (3). R (right) and L (left) indicate the position of the electrodes over the right or left hemispheres. Visible differences are statistically significant (t-test, $p < 0.05$)

tials reflect the systemic brain organization of behavior. Similar configuration of behavior-related potentials was also shown during performing the initially learned behavior in the first compartment of the experimental cage before and after additional learning. These results indicate that the organization of the brain processes in the compared conditions was generally analogous (Fig. 7.3). At the same time, we found differences in amplitudes and latencies of components of behavior-related potentials. This is in line with the hypothesis about the changes in the composition of the elements of experience after additional learning (Fig. 7.3). Variability of slow brain potentials in test sessions before additional learning and re-test sessions after was not significant (*t*-test, $p < 0.05$).

7.4 Conclusions

Additional learning changes the brain activity during realization of initially learnt behavior. Although the general brain processes reflected in a similar configuration of behavior-related EEG potentials underlie the realization of the instrumental task before and after additional learning, the observed differences in the amplitudes and latencies of their components support the hypothesis that the composition of the elements of individual experience underlying this behavior changes after additional learning.

Acknowledgment This work was supported by Russian Science Foundation (Grant №. 14-28-00229).

References

1. Shvyrkov, V.B.: Behavioral specialization of neurons and the system selection hypothesis of learning. In: Klix, F., Hagendorf, H. (eds.) Human memory and cognitive capabilities, pp. 599–611. Elsevier, Amsterdam (1986)
2. Alexandrov, Y.I.: Psychophysiological significance of activity of central and peripheral neurons in behavior. Nauka, Moscow (1989). (in Russian)
3. Alexandrov, Y.I., Grechenko, T.N., Gavrilov, V.V., Gorkin, A.G., Shevchenko, D.G., Grinchenko, Y.V., Aleksandrov, I.O., Maksimova, N.E., Bezdenezhnych, B.N., Bodunov, M.V.: Formation and realization of individual experience: a psychophysiological approach. In: Miller, R., Ivanitsky, A.M., Balaban, P.M. (eds.) Complex brain functions: conceptual advances in russian neuroscience. Series: Conceptual advances in brain research, vol. 2, pp. 181–200. Harwood Academic Publishers, Amsterdam (2000)
4. Gavrilov, V.V., Grinchenko, Y.V., Alexandrov, Y.I.: Comparisons of the sets of behaviorally specialized limbic cortex neurons in rats and rabbits. Eur. J. Neurosci. **10**, 154 (1998)
5. Gorkin, A.G., Shevchenko, D.G.: The stability of units behavioral specialization. Neurosci. Behav. Physiol. **21**, 222–229 (1991)
6. Svarnik, O.E., Alexandrov, Y.I., Gavrilov, V.V., Grinchenko, Y.V., Anokhin, K.V.: Fos- expression and task-related neuronal activity in rat cerebral cortex after instrumental learning. Neuroscience. **136**, 33–42 (2005)

7. Maksimova, N.E., Aleksandrov, I.O.: Slow brain potentials and their relation to the structure of behavior. In: Johnson, R., Rohrbaugh, J.W., Parasuraman, R. (eds.) Current trends in event-related potential research (EEG suppl. 40), pp. 4–7. Elsevier, Amsterdam (1987)
8. Aleksandrov, I.O., Maksimova, N.E.: Typology of brain slow potentials, neuronal activity and dynamics of system organization of behavior. In: Shvyrkov, V.B., Rusalov, V.M., Shevchenko, D.G. (eds.) EEG and neuronal activity in psychophysiological experiments, pp. 44–72. Moscow, Nauka (1987). (in Russian)
9. Gavrilov, V.V.: Relation of EEG and unit activity in rabbit's behavior. In: Shvyrkov, V.B., Rusalov, V.M., Shevchenko, D.G. (eds.) EEG and neuronal activity in psychophysiological experiments, pp. 33–44. Nauka, Moscow (1987). (in Russian)

Chapter 8
Coincidence Detection and Absolute Threshold in the Auditory Brainstem

Ray Meddis

Abstract In psychophysics, absolute threshold is explained using the concept of a "leaky integrator." However, this is difficult to reconcile with our knowledge of the physiology of the auditory periphery. A computer model is used to explore the potential of coincidence detection neurons to emulate absolute threshold phenomena when two layers of coincidence detection neurons are used. This arrangement is able to distinguish acoustically driven auditory nerve action potentials from spontaneous activity and identify absolute threshold with a dependence of threshold on signal duration as found in psychophysics.

Keywords Absolute threshold · Coincidence detection · Auditory brainstem

8.1 Introduction

"Absolute auditory threshold" is one of the simplest concepts in the psychophysics of hearing, i.e., the quietest sound that can be heard. It is also the most common measurement in auditory science and audiological practice. In principle, it should be easy to relate this to the occurrence of action potentials in the auditory nerve (AN), but this is not the case because most AN fibers are spontaneously active and there is no simple way of knowing whether a spike indicates an acoustic event or whether it merely occurred spontaneously.

Of course, some fibers, such as the so-called low spontaneous rate (LSR) auditory nerve fibers, are less spontaneously active than others. In this case, a spike in a LSR fiber is more likely to indicate a real sound. However, the rate thresholds of these fibers (defined as the level of the least intense pure tone capable of eliciting a firing rate greater than the spontaneous rate) are higher than the rate thresholds of high spontaneous rate (HSR) fibers. For this reason LSR fibers are unlikely to form the

---

R. Meddis (✉)
Department of Psychology, Essex University, Colchester, UK
e-mail: rmeddis@essex.ac.uk

J. M. Delgado-García et al. (eds.), *Advances in Cognitive Neurodynamics (VI)*,
Advances in Cognitive Neurodynamics,
https://doi.org/10.1007/978-981-10-8854-4_8

basis of an explanation of absolute threshold. In any case they are less common than HSR fibers, and they rarely have firing rates low enough to allow an unambiguous detection of an acoustic event based purely on a single spike. Even if we did have zero spontaneous rate fibers with thresholds as low as HSR fibers, we would still be left with the need to explain how the brainstem "knows" which fibers can be trusted to deliver information about near-threshold acoustic events.

From a physiological point of view, there is no difficulty in defining the auditory nerve *rate* threshold. Below rate threshold, the rate of firing is completely unrelated to signal level and that is why it is called "spontaneous." However, the transition from spontaneous rate to a *driven* rate is sudden, and, as a result, the rate/level function appears to indicate a real "threshold," a clear discontinuity. We might therefore feel that "detection" is based on some method for monitoring the *rate* of firing. However, a detailed examination of the timing of spikes in a HSR fiber shows that spikes *below* the rate threshold can be in phase with the acoustic stimulation for low-frequency tones. If this is the case, the spikes must be responding to the presence of the physical stimulus even below the rate threshold. This means that we have at least two possible explanations of the detection of near-threshold stimulation, firing rate increases or increased temporal organization of spiking activity.

In both cases we are left with the problem of discovering how the nervous system "knows" that the rate of firing or the timing of action potentials has changed. Among psychophysicists, it is generally believed that thresholds involve a "leaky integrator." This works by integrating some quantity over time until the integrand exceeds some criterion before announcing that an acoustic event has occurred. In psychophysics the quantity being accumulated is normally signal energy, but in a neuronal context, we might be tempted to substitute AN action potentials. What could be simpler?

In its simplest form, a spike integrator will not be a satisfactory detector because, even in silence, spontaneous spiking activity will quickly fill the integrator and give rise to a false detection. For this reason, the integrator must be "leaky" and *leak at exactly the same rate* as the spontaneous spiking activity so that detection is not triggered during silence. While the scheme works well in psychophysics, it does not work in physiology because it is not at all obvious where in the nervous system the integrator might be located because, in psychophysics at least, the time constant of integration exceeds hundreds of milliseconds. There are no obvious candidates for this in the auditory brainstem nor is it clear how such a system would know how to set an appropriate leakage rate in order to make error-free judgments.

The long time constant of integration is necessary to explain the psychophysical observation that the threshold for a short sound, say 20 ms, is considerably higher than that for a long sound, say 200 ms. This relationship between threshold and duration extends asymptotically well beyond 100 ms. The leaky integrator theory works well in psychophysics where there are no conceptual constraints on the value of the time constant of the integrator; it is simply chosen to fit the data. However, it can only work in physiology if we can find a real neuronal integrator somewhere in the auditory brainstem capable of integrating over hundreds of milliseconds. No such integrator exists there, and, consequently, we must find a different principle for detecting very quiet sounds.

8.1.1 Coincidence Detection

The leaky integrator model is very seductive, and it is difficult to change one's focus to look elsewhere for a different kind of detector. Nevertheless, a mechanism might be possible if we can find a method to distinguish between spontaneous spiking activity and acoustically driven spikes. In other words, if we could remove spontaneous spikes from the stream of AN action potentials, the remaining spikes could be depended on to indicate that some acoustic event has occurred.

"Coincidence detection" is one way of eliminating spontaneous or random events. A neuronal coincidence detector is a cell that receives inputs from multiple sources but generates an action potential only when two or more input spikes arrive almost simultaneously. If we assume that the input action potentials are statistically independent, then the likelihood of coincidence will depend on their rate of occurrence. If we assume that the window for detecting coincidence is very small (of the order of 1 ms, say), then coincidences will be rare for spontaneous activity but more frequent for higher *driven* input rates. This makes coincidence detection a possible candidate for detecting increases in firing rate. This is not a perfect arrangement because the coincidence detector will still fire occasionally in silence. However, making the coincidence detection window smaller can reduce the likelihood of a false positive, and it can be further reduced by defining coincidence as the simultaneous arrival of inputs from *three or more* inputs.

If the input consists of a mixture of spontaneous and driven spikes, the cell will respond preferentially to driven spikes for another reason. Coincidence detection takes advantage of the randomness of spontaneous activity by ignoring input spikes that are uncorrelated across time. If the inputs are not random but correlated because their time of occurrence is determined by a real acoustic event, coincidental firing will be more likely. This temporal principle will apply even if the input rate is no higher than the spontaneous rate.

We now have two good reasons for exploring the coincidence principle based on AN spike activity as an explanation of absolute threshold. The simplest version of this principle is to propose that a single spike emitted from the coincidence detector can be taken to mean that an acoustic event has occurred. False positives can occur, but their rate can be controlled and traded against the overall sensitivity of the system by changing the criterion for coincidence detection or by *adding extra layers of coincidence detectors*. The first layer of coincidence detection can be located in the cochlear nucleus where there are a number of different types of cell that receive multiple inputs from the auditory nerve.

The idea of using a single spike in the coincidence detector as the basis of distinguishing a stimulus-driven spike from spontaneous activity at its input may seem unrealistic, but it does allow us to explore the general principle and the other important aspect of "absolute" threshold, namely, its dependence on duration. If the rate of occurrence of coincidences is very low because the stimulus has low intensity, the likelihood of "at least one spike" will depend on the duration of the stimulus. In this way we can expect a dependence of threshold on duration that does not depend on a physical integrator.

8.2 Neural Implementation and Computer Model

Fortunately, there are many examples of neuronal circuits in the auditory brainstem that act as coincidence detectors. However, to act as a detector, it is important that they feature little spontaneous activity. Candidate cells can already be found in the cochlear nucleus, the first processing station after the auditory nerve. One example candidate is the sustained chopper cell in the ventral cochlear nucleus, VCN(chop-S). It is an excitatory neuron with little spontaneous activity that projects to the central nucleus of the inferior colliculus (IC) which, in turn, projects to the thalamus and more locally to cells that control action or send efferent signals back toward the cochlea.

A computer model of the auditory brainstem was used to simulate this arrangement where AN(HSR) spiking activity feeds multiple models of the VCN(chop-S) neurons whose output feeds a model of IC units. The VCN units received inputs from 30 AN(HSR) fibers, and the IC units received inputs from 30 VCN(chop-S) units.

All units had best frequencies (BF) of 2 kHz and were simulated using Hodgkin/Huxley equations. A model of the auditory periphery was used by Meddis and O'Mard (2005) [1] to drive the AN. All brainstem units were configured as type I cells as described by Manis and Marx (1991) [2].

8.3 Results

Figure 8.1 shows the spiking activity of the three units in conditions of silence and when a 2-kHz tone was presented just above threshold at 10 dB SPL, respectively. In silence (Fig. 8.1a), the model AN fiber is spontaneously very active (50 spikes/s), while the model VCN unit is much less active (5 spikes/s), and the model IC unit is silent. When the signal is presented at 0 dB SPL (Fig. 8.1b), a level just above threshold, the AN fiber is more active, the VCN unit is also more active, and the IC unit shows a single spike. In terms of the simple criterion adopted above, this constitutes "detection."

Figure 8.1c shows the rate/level function for each unit measured over 100 trials. Both VCN(chop-S) and IC units show little or no spontaneous activity at the lowest tone levels. As a result any activity is an indication that a stimulus has been

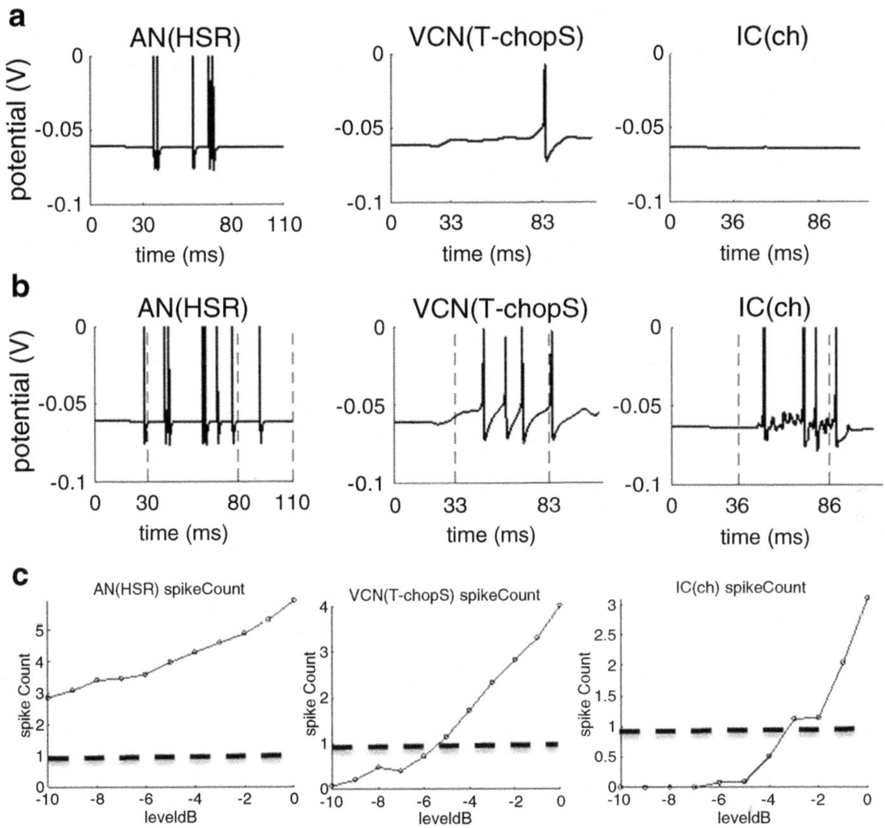

Fig. 8.1 Spiking activity at three stages of a computational model featuring AN(HSR) fibers, VCN(chop-S) units, and an IC unit. (**a**) Spontaneous activity (no tone): typical example of the spiking response of the three unit types in the absence of stimulation. (**b**) -6 dB-SPL, 2-kHz tone: spiking response to a 2 kHz, 50 ms/2 kHz tone presented at −6 dB SPL (just above threshold). Vertical dotted lines indicate tone onset and offset corrected for conduction delays. (**c**) Rate/level function: mean spike counts for all three units at a range of signal levels between −10 and 0 dB SPL, i.e., just below and just above threshold. A horizontal dotted line indicates a mean spike count of 1 spike and can be used to indicate detection

presented. The CN unit has a small amount of spontaneous activity, and this could give rise to some false-positive identifications of a sound. However, the IC unit has virtually no spontaneous activity and is a more reliable guide to the presence or absence of acoustic stimulation. The second layer of coincidence reduction reduces the likelihood of a false detection. The IC rate/level function shows no spiking activity below 4-dB SPL.

Figure 8.2 explores the consequences of increasing the duration of the tone. It shows the rate/level functions for four different durations (25, 50, 100, and 200 ms) at a range of signal levels. The mean spike count per fiber rises more quickly for

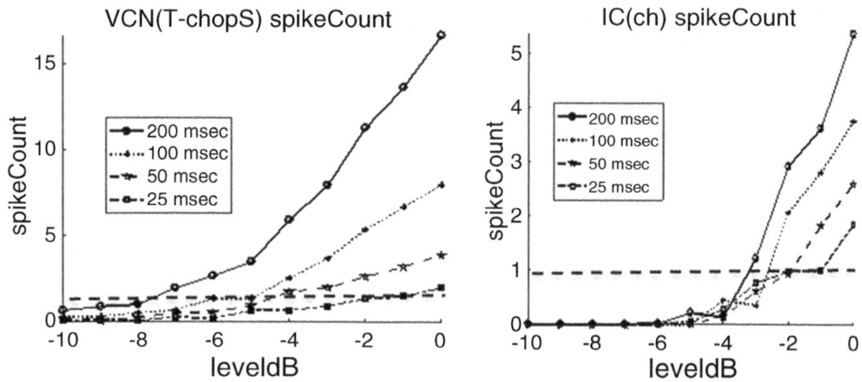

Fig. 8.2 Mean spike count across 200 observations at 11 tone levels (x-axis) and four different tone durations (25, 50, 100 and 200 ms). Note that the spike count is greater for longer durations

the longer-duration stimuli. This is not surprising because longer tones offer more opportunity for spikes to occur. A horizontal dotted line is used to indicate a mean spike count of 1 spike during the presentation of the tone.

We can use this as a proxy for psychophysical threshold. It shows clearly that longer tones will meet the criterion of a single spike at a lower stimulus intensity. In this respect, the model predictions are similar to the psychophysical results.

8.4 Discussion

The simple model described above shows how the physiology of the auditory brainstem can be reconciled with the psychophysics of absolute threshold. It does so using only coincidence detection without the need for a "leaky integrator" with a long time constant. This provides an important bridge between physiology and psychophysics that allows for building physiological models of psychophysical observations. Our own research has modeled hearing impairment by introducing peripheral dysfunction into the models such as loss of cochlear gain, inner hair cell loss, and reductions in endocochlear potential [3] to show how impairment is reflected in psychophysical measurements with patients.

The patient testing procedures were designed so that their yes/no decisions were always for unmasked tones, i.e., in a background of silence. In each case, the estimated threshold was based on the single-spike criterion suggested above. In this way, we could extend the technique to predict tuning curves and residual compression by employing forward masking methods. Because we used target tones in a silent background, we avoided the more difficult question of how the nervous system can detect tones against a simultaneous noisy background. How it does this remains mysterious and is a question for future research.

References

1. Meddis, R., O'Mard, L.P.: A computer model of the auditory-nerve response to forward-masking stimuli. J. Acoust. Soc. Am. **117**, 3787–3798 (2005)
2. Manis, P.B., Marx, S.O.: Outward currents in isolated ventral cochlear nucleus neurons. J. Neurosci. **11**, 2865–2880 (1991)
3. Panda, M.R., Lecluyse, W., Tan, C.M., Jurgens, T., Meddis, R.: Hearing dummies: individualized computer models of hearing impairment. Int. J. Audiol. **53**, 699–709 (2014)

Chapter 9
Simultaneous Observation and Imagery of Hand Movement Enhance Event-Related Desynchronization of Stroke Patients

Atsuhiro Ichidi, Yuka Hanafusa, Tatsunori Itakura, and Toshihisa Tanaka

Abstract During voluntary movement, motor imagery (MI), or action observation (AO), the short-lasting attenuation or blocking of rhythms within alpha or beta band of electroencephalogram (EEG) called event-related desynchronization (ERD) is observed over the central area. Some studies showed that with the increase of the ERD during MI, impaired motor function after stroke was improved. Recently, it has been reported that the ERD of healthy subjects during combined AO and MI (AO+MI) was stronger than that during either MI or AO individually. However, it is unclear how AO+MI affects stroke patients in terms of the ERD. To investigate this, in this paper, EEG signals during the three tasks, gazing at a still picture (termed Gaze), MI, and AO+MI, of stroke patients and healthy subjects were analyzed. Statistical analyses showed that the ERD of AO+MI was stronger than that of Gaze or MI. This implies that AO+MI may be more effective approach to recover the motor function in terms of neurorehabilitation.

Keywords Neurorehabilitation · electroencephalogram · event-related desynchronization · motor imagery · brain-computer interfacing

9.1 Introduction

Stroke occurs when blood flow to the brain is critically reduced. After the stroke occurred, patients may suffer from paralysis. The main mechanism underlying motor recovery involves enhanced activity of the primary motor cortex induced by motor training [1]. However, the motor training has some limitations for patients

A. Ichidi · T. Itakura · T. Tanaka (✉)
Tokyo University of Agriculture and Technology, Tokyo, Japan
e-mail: tanakat@cc.tuat.ac.jp

Y. Hanafusa
Division of Rehabilitation, Okayama East Neurosurgery Hospital, Okayama, Japan

J. M. Delgado-García et al. (eds.), *Advances in Cognitive Neurodynamics (VI)*,
Advances in Cognitive Neurodynamics,
https://doi.org/10.1007/978-981-10-8854-4_9

with severe hemiparesis, for whom executing physical practice with their impaired limb is very difficult and sometimes impossible [2].

For patients with severe paresis after stroke, motor imagery (MI)-based brain–computer interface (BCI) technology is a novel and effective method of neurorehabilitation [3]. This type of BCI systems usually estimates the patient's motor intention from the variation in brain activity over the primary sensorimotor cortex and gives the patients the estimated result through feedback [3, 4]. The variation is typically observed as event-related desynchronization (ERD), which is the short-lasting attenuation or blocking of rhythms within the alpha (8–12 Hz) and beta (13–28 Hz) bands of electroencephalogram (EEG) during motor execution or MI [5]. The study in [4] has investigated the effectiveness of neurorehabilitative training using the MI-based BCI for hand paralysis following stroke. The study reported that motor function was improved together with the increase of the ERD during MI. As another approach of the neurorehabilitation, digital mirror therapy (DMT), which is similar to action observation (AO), has been proposed [6]. In the DMT, a monitor is used to form an illusory hand, which is visually superimposed on the impaired hand. The patients are asked to turn their head to observe the illusory hand in the monitor and to persuade themselves that the impaired hand can move. The study reported that the ERD of healthy subjects was stronger when the DMT was provided compared to when the DMT was not provided. Eaves et al. investigated an effect of combined AO and MI (AO+MI) on brain activity [7] and reported that the ERD of healthy subjects during AO+MI was stronger than that during either of individual MI and AO. However, it is unclear how AO+MI affects the brain activity of stroke patients.

This paper analyzes EEG signals of stroke patients and healthy subjects during three tasks based on the DMT concepts. The three tasks include the following: the subjects were asked to gaze a static picture of an open hand on a monitor (Gaze), to imagine opening and closing their hands according to speech cues (MI), or to watch a video clip depicting the dorsal view of a hand opening and closing and imagine their hands' movement in synchrony with the video clip displayed on the monitor and the speech cues (AO+MI). In the analysis, the ERD was compared between stroke patients and healthy subjects and also compared among the three tasks. As a result, the ERD of stroke patients was stronger than that of healthy subjects. Moreover, the ERD of AO+MI was stronger than that of Gaze and MI. This implies that AO+MI may be more effective approach to improve the motor function in terms of neurorehabilitation.

9.2 Method

9.2.1 Subjects

In the experiment, eight stroke patients (six males and two females; mean age 66 years, range 48–81 years; four right paralyzed and four left paralyzed) and eight healthy subjects (eight males; mean age 23 years, range 21–25 years; seven right-handed and one both-handed) participated. They all gave their written informed consent. This study was approved by Okayama East Neurosurgery Hospital and the Research Ethics Committee of Tokyo University of Agriculture and Technology.

9.2.2 Task

During the experiment, the subjects were seated in a chair, and a 19 inch LCD monitor with a resolution of 1280 × 1024 was arranged in front of them over their paralyzed forearm for patients and their right forearm for healthy subjects. Figure 9.1a, b illustrates the experimental condition.

The subjects took part in three tasks termed Gaze, MI, and AO+MI. In Gaze, the subjects were asked to gaze a static picture of an open hand displayed on the monitor. In MI, the subjects were asked to imagine opening and closing their paralyzed hands (their right hands for healthy subjects) according to cues in Japanese speech, "pah" (open) and "goo" (close), given every 1 s. In AO+MI, the subjects were asked to watch a video clip depicting the dorsal view of a hand opening and closing every 1 s and imagine their paralyzed hands movement in synchrony with the video clip displayed on the monitor and the speech cues given every 1 s.

Figure 9.1c illustrates a sequence in the experiment. Each task consisted of rest, preparation, and task periods, respectively. Immediately after the rest period, an instruction of task was displayed on the monitor to notify the subjects of the

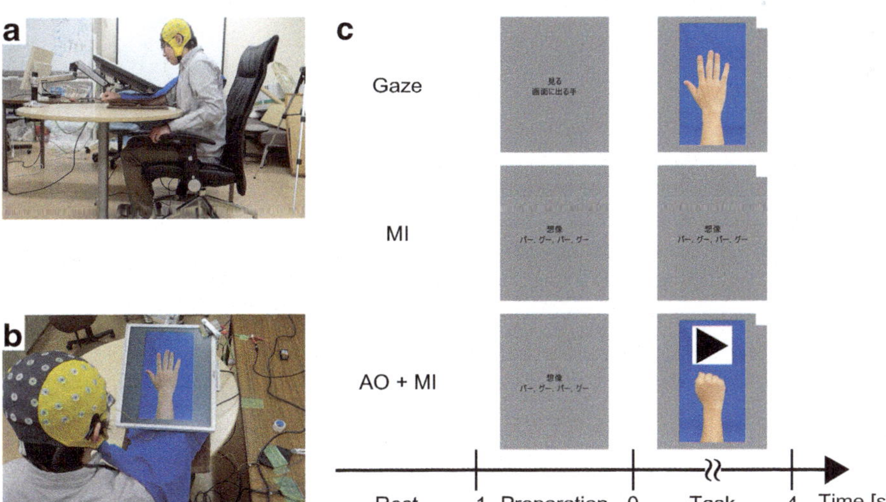

Fig. 9.1 The experimental condition (left panels; views from the left (**a**) and the top (**b**)) and schematic sequence of the experiment (right panel) (**c**)

preparation period (1 s). After that, the subjects were asked to perform the instructed task during the task period (4 s).

The experiment consisted of four sets for healthy subjects and two or three sets for patients. Each set included 15 trials (5 trials for each task). The first trial was always Gaze. After the first trial, the order of tasks was randomly provided over each set. There were breaks between sets and trials. The subjects were video recorded during experiment in order to eliminate those who moved their hands during the rest period.

9.2.3 EEG Recording

During the experiment, EEG signals were recorded using Ag/AgCl passive electrodes embedded in a cap named waveguard (ANT Neuro). For recording, six electrodes located at Fp1, Fp2, C3, C4, O1, and O2 following the international 10–20 system were used. To check eye movements, electrooculogram (EOG) signals were recorded with an Ag/AgCl flat passive electrode (Nihon Kohden) which was placed at the right side of right eye. To check hand movements, electromyograph (EMG) signals were recorded with a pair of the passive electrodes over the paralyzed extensor digitorum communis (EDC) muscle (their right EDC muscle for healthy subjects). The electrode for GND was located at AFz. The EEG, EOG, and EMG signals were referred for average signal recorded between the left and right mastoid leads. The signals were amplified by Polymate V (TEAC) that has a bandpass filter of 0.3–333 Hz and an A/D converter with a sampling rate of 1,000 Hz. A software named AP Monitor (TEAC) was used to record the signals.

9.2.4 Data Processing

The recorded EEG signals were bandpass filtered from 5 to 30 Hz using a zero-phase FIR filter and cut into epochs from −3.5 to 4.5 s with respect to the onset of the tasks. Each epoch was down sampled to 200 Hz. To remove the EOG artifact from the EEG, independent component analysis (ICA) was applied to the concatenated epochs, and independent components related to EOG artifact [8] were removed based on visual inspection. The epochs were re-referenced to the average signal across all EEG electrodes. If EEG epochs satisfied any of the following conditions, they were rejected: (1) by video inspection, any movement of a hand or an arm was recognized; (2) an amplitude of the EEG exceeded $\pm 50\,\mu V$; and (3) by visual inspection, an EMG artifact was observed.

After the preprocessing, the short-time Fourier transform with Hamming window of 1 s shifted by a step of 0.125 s was applied to the epochs to obtain the power spectrogram, which was averaged across the epochs for each task. Further, the power spectrogram was averaged in the alpha (8–12 Hz) or beta (13–28 Hz) bands and averaged in the time windows of the task (from 0 to 4 s) and rest period (from −3 to −1 s), respectively. For each frequency band, the ERD amplitude was defined as the $10 \log_{10}$ transform of the ratio of the power during the task period relative to the power during the rest period.

9.2.5 Statistical Analysis

To investigate the difference of the ERD amplitude between the groups and tasks, the ERD over the central area (the electrodes C3 and C4) was statistically analyzed using the three-way analysis of variance (ANOVA) for repeated measures at each of alpha and beta bands. The between-subject factor was group (healthy and patient), and the within-subject factors were task (Gaze, MI, and AO+MI) and hemisphere (contralateral and ipsilateral). Significant interaction effects were followed by tests of the simple interaction effects and the simple main effects. Post hoc comparisons were based on the Shaffer's modified sequentially rejective Bonferroni's procedure [9]. The Greenhouse–Geisser correction was used to reduce the degrees of freedom [10]. The significance level was set at $p < 0.1$.

9.3 Results

For the alpha band, a main effect of task was significant: $F(1.29, 15.48) = 3.996$, $p = 0.055$. However, the post hoc tests showed that there was no significant difference in the ERD amplitude between Gaze, MI, and AO+MI. No other effects were significant (all $p > 0.1$).

For the beta band, two main effects of group and task were significant. First, the main effect of group was significant: $F(1, 12) = 5.354$, $p = 0.039$. This result suggested that the ERD amplitude of patients was greater than that of healthy subjects (Fig. 9.2a). Second, the main effect of task was significant: $F(2, 24) = 15.697$, $p = 0.000044$. The post hoc tests showed that the ERD of AO+MI was significantly stronger than that of Gaze ($p = 0.0002$, Fig. 9.2b) and MI ($p = 0.0091$, Fig. 9.2b). Moreover, the ERD of MI was significantly stronger than that of Gaze ($p = 0.023$, Fig. 9.2b). No other effects were significant (all $p > 0.1$).

9.4 Discussion

During the tasks, the ERD of patients was stronger than that of healthy subjects. A possible factor causing this result may be stroke patients' motor impairment. It has been reported that during MI of the affected hand, stroke patients with higher impairment showed stronger ERD than stroke patients with weaker

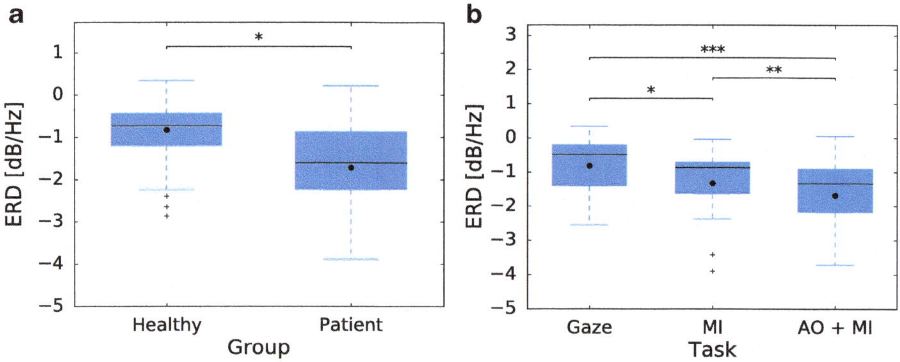

Fig. 9.2 The result of the three-way ANOVA for the beta band. The box plot of ERD, collapsed across the tasks and hemispheres, in healthy subjects and stroke patients (**a**). The box plot of ERD, collapsed across the groups and hemispheres, in Gaze, MI, and AO+MI (**b**). Statistical significance is indicated by asterisks ($* : p < 0.05$, $** : p < 0.01$, $*** : p < 0.001$)

impairment [11]. Another possible factor causing this result may be aging. Stroke patients were older than healthy subjects in the experiment. It has been reported that the ERD of elderly subjects was stronger than that of young subjects [12].

During AO+MI, a significantly stronger ERD was observed compared to the other tasks. The study in [4] showed that motor function was improved with the increase of the ERD during MI [4]. This finding together with our results suggest that AO+MI may be a more effective approach to improve the motor function than MI.

Acknowledgements The authors thank Hirotaka Nagai for technical assistance. This work was supported in part by the Project for Next-Generation Research, Tokyo University of Agriculture and Technology.

References

1. Calautti, C., Baron, J.C.: Functional neuroimaging studies of motor recovery after stroke in adults: a review. Stroke **34**(6), 1553–1566 (2003)
2. Lindberg, P., Schmitz, C., Forssberg, H., Engardt, M., Borg, J.: Effects of passive-active movement training on upper limb motor function and cortical activation in chronic patients with stroke: a pilot study. J. Rehabil. Med. **36**(3), 117–123 (2004)
3. Ang, K.K., Guan, C.: Brain–computer interface in stroke rehabilitation. J. Comput. Sci. Eng. **7**(2), 139–146 (2013)
4. Shindo, K., Kawashima, K., Ushiba, J., Ota, N., Ito, M., Ota, T., Kimura, A., Liu, M.: Effects of neurofeedback training with an electroencephalogram-based brain–computer interface for hand paralysis in patients with chronic stroke: a preliminary case series study. J. Rehabil. Med. **43**(10), 951–957 (2011)
5. Yuan, H., Perdoni, C., He, B.: Relationship between speed and EEG activity during imagined and executed hand movements. J. Neural Eng. **7**(2), 026001 (2010)

6. Lee, H.M., Li, P.C., Fan, S.C.: Delayed mirror visual feedback presented using a novel mirror therapy system enhances cortical activation in healthy adults. J. Neuroeng. Rehabil. **12**(1), 56 (2015)
7. Eaves, D.L., Behmer, L., Vogt, S.: EEG and behavioural correlates of different forms of motor imagery during action observation in rhythmical actions. Brain Cogn. **106**, 90–103 (2016)
8. Jung, T.P., Makeig, S., Humphries, C., Lee, T.W., Mckeown, M.J., Iragui, V., Sejnowski, T.J.: Removing electroencephalographic artifacts by blind source separation. Psychophysiology **37**(2), 163–178 (2000)
9. Shaffer, J.P.: Modified sequentially rejective multiple test procedures. J. Am. Stat. Assoc. **81**(395), 826–831 (1986)
10. Greenhouse, S.W., Geisser, S.: On methods in the analysis of profile data. Psychometrika **24**(2), 95–112 (1959)
11. Kaiser, V., Daly, I., Pichiorri, F., Mattia, D., Müller-Putz, G.R., Neuper, C.: Relationship between electrical brain responses to motor imagery and motor impairment in stroke. Stroke **43**(10), 2735–2740 (2012)
12. Derambure, P., Defebvre, L., Dujardin, K., Bourriez, J., Jacquesson, J., Destee, A., Guieu, J.: Effect of aging on the spatio-temporal pattern of event-related desynchronization during a voluntary movement. Electroencephalogr. Clin. Neurophysiol. Evoked Potentials Section **89**(3), 197–203 (1993)

Chapter 10
Behavioral and Brain Activity Modulation Through Neurofeedback Training Using Electroencephalography

Takuya Kimura and Jiro Okuda

Abstract Recent neurofeedback studies have revealed that mere self-induction of a specific brain activity state decoded from functional magnetic resonance imaging (fMRI) data suffices to cause plastic change in perceptual ability or cognitive state particular to that brain activity state. We tested such an induction effect by using frequency information of electroencephalography (EEG) data and further explored how the induction training affected EEG pattern itself in relation to behavioral changes. We found that repeated self-induction of an EEG power spectral pattern corresponding to motor preparation for gaining monetary reward caused shortening of reaction times to the motor task. Decoding of EEG data during the motor task after the induction training showed higher probability of appearance of the trained EEG spectral pattern during any task trials. These data reveal behavioral effectiveness of the EEG self-induction training, in parallel with overall enhancement effect of the trained brain activity state.

Keywords Decoding · EEG · Monetary reward · Causal relation · Plasticity

10.1 Introduction

A traditional approach in neuroscience studies has been to examine neurophysiological responses associated with experimental stimuli, cognitive tasks, and behavioral observations. Recently, a novel experimental approach using neurofeedback training has been developed to investigate cognitive and behavioral changes caused by induction of a specific brain activity pattern, i.e., an opposite cause-and-effect direction to the conventional approach, from experimental manipulation to brain responses. In the neurofeedback approach, a specific brain activity pattern corresponding to a

T. Kimura · J. Okuda (✉)
Department of Intelligent Systems, Faculty of Computer Science and Engineering, Kyoto Sangyo University, Kita-ku, Kyoto, Japan
e-mail: jokuda@cc.kyoto-su.ac.jp

J. M. Delgado-García et al. (eds.), *Advances in Cognitive Neurodynamics (VI)*, Advances in Cognitive Neurodynamics,
https://doi.org/10.1007/978-981-10-8854-4_10

particular cognitive state of experimental interest was modeled for each individual subject by machine learning technique, and the subject tried to induce the modeled brain activity pattern only through a feedback signal indicating how his/her brain activity pattern was similar to the modeled pattern. Importantly, the subject was unaware of exactly what the feedback signal means. So far, this approach has succeeded in improvement of visual discrimination ability [1], association of specific color to achromatic visual images [2], modulation of facial preferences [3] as well as a metacognitive state of self-confidence [4], and extinction of fear conditioning [5]. These studies used information decoded from spatial patterns of fMRI data as a feedback signal. Although intriguing, the fMRI neurofeedback technique had a limitation in temporal resolution of the training procedure. Each trial of the neurofeedback took around 6 s because of hemodynamic delay in the fMRI signal. Another limitation was that time-relevant information such as oscillatory neuronal activity changes in time was unavailable for the feedback signal. Moreover, how the feedback training affected the brain state itself has been less investigated.

In the present study, we developed a feedback training system that utilized information decoded from scalp-recorded EEG data. We explored effectiveness of EEG neurofeedback with better temporal resolution that used frequency information of the brain electrical activities as a feedback signal. We tried to show a behavioral effect on improvement of reaction times (RTs) to monetary incentive delay (MID) task [6] that was designed to induce activation of reward-related brain regions. We also explored how brain activity patterns in the MID task were modulated through the neurofeedback training.

10.2 Experimental Procedure

10.2.1 Subjects

Ten healthy male volunteers (21–23 years old, right-handed) participated in the study after giving written informed consent. The subjects were divided into two groups of five subjects. We assigned different EEG neurofeedback conditions to the two groups (see Feedback Training section below). The study was done in accordance with a protocol approved by our local ethics committee.

10.2.2 Experimental Schedule

The experiment took 5 days with three steps: a pre-training test in the first day, training sessions from the second to the fifth days, and a post-training test in the end of the final day. In the pre-training test, the subjects performed the MID task.

EEG data recorded during the task were used to construct a decoder model that classified across different trial types of the MID task. In the training, they underwent EEG self-induction training so that their EEG state became similar to one of the states modeled by the decoder (a target state). Finally, in the post-training test, the same MID task as in the pre-training was given to the subjects. We compared performances of the pre- and post-training MID tasks to examine the behavioral effect of the EEG training.

10.2.3 MID Task

The MID task is a simple RT task in which monetary gain or loss for each trial is determined according to the subject's RT [6]. The present MID task consisted of three task conditions: gain, loss, and neutral. In the gain condition, subjects were rewarded with 50 yen if their RT was faster than a time constraint that corresponded to 50% success rate for the subject. In the loss condition, they were able to save loss of 50 yen if their RT was faster than the constraint. In the neutral condition, neither monetary reward nor loss was imposed regardless of the RT. We indicated thc task condition by onc of thc thrcc cuc figurcs of a piggy bank after a fixation cross (Fig. 10.1). A delay period of random interval between 1000 and 1800 ms was inserted after the cue, followed by a go stimulus of small square dot for which the subjects made a button press response as soon as they could. Finally, we displayed a notification of success (○) or failure (×) of the trial and a monetary outcome. Each subject underwent 80 trials for each condition, resulting in a total of 240 trials.

The MID task was known to produce robust effects of RT improvement for the gain and the loss conditions, with reliable fMRI activations of reward-related regions during the delay period of the conditions [6]. Therefore, we adopted the MID task paradigm to induce reward-related brain activity and behavioral effect. We expected

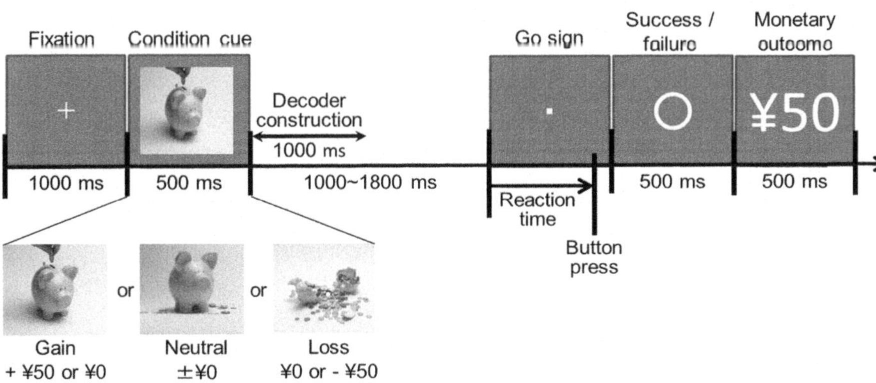

Fig. 10.1 Schematic drawing of a trial of MID task

that repeated induction of the EEG state corresponding to the delay period of the MID gain/loss trials might cause further improvement of RTs in the post-training MID task.

10.2.4 EEG Recording and Decoder Construction

We used an EEG amplifier (BrainAmp MR plus, Brain Products GmbH, Germany) with 32-channel active electrode system to record EEG signals from subjects' scalp. Four electrodes were used to measure vertical and horizontal electrooculogram (EOG). Sampling rate was 1000 Hz. Recorded EEG data were band-passed by a digital filter with 0.3 s time constant and 70 Hz high-frequency cutoff. Ocular artifacts were removed by using the EOG data [7]. We extracted data during initial 1000 ms of the delay period of each trial and calculated spectral power at 69 frequency bins from 3.6 to 70 Hz by a fast Fourier transform (FFT) algorithm.

We applied sparse logistic regression (SLR) algorithm [8] to construct a decoder model that classified three trial types of the MID task by using an input vector of 1932 spectral power (69 frequencies by 28 electrodes). We used multi-class SLR function in the SLR toolbox (http://www.cns.atr.jp/~oyamashi/SLR_WEB.html). For the first group of the subjects, we constructed a decoder that classified trials of the gain, loss, and neutral conditions irrespective of the subjects' RT performance. For the second group, on the other hand, to examine a possible effect according to subjects' response performance, we constructed a decoder that classified across trials with faster RTs in the gain and loss conditions (success trials), trials with slower RTs in the gain and loss conditions (failure trials), and trials of the neutral condition.

10.2.5 Feedback Training

In each trial of the EEG neurofeedback training, the subject was first presented with a fixation cross for 1000 ms, followed by a blank screen of 1500 ms. EEG data during the blank period was decoded by the SLR model for the subject. After a processing delay shorter than 500 ms, feedback of similarity of the subject's EEG spectral pattern to a target pattern was visually presented as a radius of a circle (grater for more similar). We told subjects that the radius reflected their EEG state during the blank period in some way and asked them to try to enlarge it via trial and error. Each subject underwent 10 sets of 80 trials of the EEG induction training for each day, resulting in a total of 3200 induction trials during the 4 days of training.

The target EEG state was different between two groups of the subjects. For the first group, the target state corresponded to all trials of the gain condition regardless of the subject's RT performance. We assumed that any RT improvement after the EEG training might be observed specifically in the gain condition of the post-training MID task. For the second group, the target state corresponded to trials with faster RT in the gain and the loss conditions (success trials). We tested whether the behavioral improvement would be more related to training of the EEG state corresponding to subject's performance state rather than the task condition per se.

10.3 Analyses

10.3.1 Behavioral Effects

We examined behavioral effects of the EEG training by comparing success rate (numbers of trials with faster RT than the constraint divided by total trial numbers) for each task condition (gain, loss, neutral) between the pre- and the post-training MID tasks for each subject.

10.3.2 EEG Induction Performance

We used likelihood value of the target state calculated by the SLR model (equivalent to the radius of the feedback circle presented to the subject) as an index of how well the subjects induced the target EEG state during the training. We examined whether the likelihood increased significantly during the course of the 4 training days.

10.3.3 Modulation of EEG State in MID Task

Finally, we explored possible changes in the EEG spectral pattern in the MID task as a result of the EEG induction training. We hypothesized that the EEG state repeatedly induced during the training would be strengthened and stabilized, which in turn makes the state more likely to occur in the corresponding MID task situations after the training. To test this, we classified EEG data during the delay period of the post-training MID task by using the SLR model defined by the data in the pre-training MID task. We examined if classification accuracy increased specifically for the trial category corresponding to the target state that was required to be induced during the training.

10.4 Results

We first confirmed significant increase in likelihood of induction of the target EEG state during the training for all subjects. Repeated measures analysis of variance (ANOVA) on the likelihood value with a factor of the training day (1–4) revealed significant main effect of the factor ($p < 0.05$) for all the subjects. Post hoc tests showed significantly higher likelihood in the third or fourth day as compared with the first day for all subjects. Six subjects (two in the first group and four in the second group) showed significantly higher likelihood than a chance level (33%) in the fourth day (Wilcoxon test, $p < 0.05$).

Behaviorally, on the other hand, RT improvement in the post-training MID task was observed only in the subjects in the second group. No subjects in the first group showed higher success rate of the gain trials in the post-training MID task (even worse than the pre-training task in three subjects), whereas four out of five subjects in the second group showed 105–142% increases in the success rate (the success rate of the post-training task divided by that of the pre-training task) for both the gain and the loss conditions. It is noteworthy that two subjects in the first group who showed significantly higher likelihood to the target EEG state than the chance level in the last training day even failed to perform better in the post-training MID task.

Decoding analyses of EEG data during the post-training MID task revealed higher probability of appearance of the trained EEG state during any MID trial types. Confusion matrices of the decoder's input-output relationship showed the highest probability of correct output class to each input data (e.g., "gain condition" output for the EEG input data during trials of the gain condition) for the pre-training MID task. For the post-training MID data, however, the classifier output of the trained EEG state was generally higher even for input data that did not belong to the state (e.g., "gain condition" output for the EEG input data during trials with the loss and the neutral conditions). This tendency of general enhancement of the trained EEG state was apparent in seven out of ten subjects (three in the first group and four in the second group).

10.5 Discussion

The results of the present study clearly indicated behavioral effectiveness of the neurofeedback training using EEG frequency information. In addition, the results also revealed relevance of to-be-induced EEG state to the behavioral effect. In the present results, induction training of the EEG state determined just by the task condition (gain condition for the first group) did not produce any improvement of the behavioral RTs, whereas induction of the state relevant to behavioral performance (success trials for the second group) reliably caused behavioral improvement. The negative result in the first group is not attributable to mere failure of enough induction of the EEG state since the subjects with higher induction likelihood even failed to show the behavioral effect. Rather, we suggest the possibility that selection of the to-be-induced target state may not be appropriate in the first group since the state included both trials of shorter and longer RTs. This must have promoted the state in which faster and slower behavioral responses were mixed. On the other hand, the state relevant only to faster RT was trained in the second group, which should have caused the effect of RT improvement.

Another important finding in the present study was the general trend of increases in occurrence of the trained EEG state for any conditions of the post-training task. We expected that probability of correct classification would increase for the category whose corresponding EEG state was repeatedly induced, under the hypothesis that repeated induction of a particular brain state would make the state more stable and likely to occur in that particular task situation. In a sense, the present results are consistent with this hypothesis but further suggest that the probability increase of appearance of the trained EEG state could extend to situations that were not originally associated with the state. Because this might be regarded as an unwanted side effect of the neurofeedback training in a certain case, we need to further explore appropriate training procedures that could cause the intended effect in more selective manner if we apply the neurofeedback method to possible physical and mental self-training.

In summary, the present study clarified that repeated induction of the brain state decoded from EEG frequency information could effectively cause a plastic change in behavior, with the suggestion that selection of to-be-induced EEG state should be important to cause the relevant behavioral effect. The present results further disclosed the general reinforcement effect of the induced EEG state. The mechanism by which the EEG neurofeedback with rich temporal information contributes to the behavioral change and the enhancement of the brain state should be addressed in future studies.

Acknowledgments This study was supported by KAKENHI Grant Numbers 15H05878 from MEXT and 26240037 from JSPS, Japan.

References

1. Shibata, K., Watanabe, T., Sasaki, Y., Kawato, M.: Perceptual learning incepted by decoded fMRI neurofeedback without stimulus presentation. Science. **334**, 1413–1415 (2011)
2. Amano, K., Shibata, K., Kawato, M., Sasaki, Y., Watanabe, T.: Learning to associate orientation with color in early visual areas by associative decoded fMRI neurofeedback. Curr. Biol. **26**, 1861–1866 (2016)
3. Shibata, K., Watanabe, T., Kawato, M., Sasaki, Y.: Differential activation patterns in the same brain region led to opposite emotional states. PLoS Biol. **14**, e1002546 (2016)
4. Cortese, A., Amano, K., Koizumi, A., Kawato, M., Lau, H.: Multivoxel neurofeedback selectively modulates confidence without changing perceptual performance. Nat. Comm. **7**, 13669 (2016)

5. Koizumi, A., Amano, K., Cortese, A., Shibata, K., Yoshida, W., Seymour, B., Kawato, M., Lau, H.: Fear reduction without fear through reinforcement of neural activity that bypasses conscious exposure. Nat. Hum. Behav. **1**, 0006 (2016)
6. Knutson, B., Westdorp, A., Kaiser, E., Hommer, D.: fMRI visualization of brain activity during a monetary incentive delay task. NeuroImage. **12**, 20–27 (2000)
7. Gratton, G., Donchin, E., Coles, G.H.M.: A new method for off-line removal of ocular artifact. Electroencephal. Clin. Neurophysiol. **55**, 468–484 (1983)
8. Yamashita, O., Sato, M., Yoshida, T., Tong, F., Kamitani, Y.: Sparse estimation automatically selects voxels relevant for the decoding of fMRI activity patterns. NeuroImage. **42**, 1414–1429 (2008)

Part II
Cognitive Network and Multi-scale Neural Network Dynamics

Chapter 11
Network Model for Dynamics of Perception with Reservoir Computing and Predictive Coding

Yuichi Katori

Abstract The sensory information processing in the brain is achieved by mutual interactions between externally given sensory signals and internally generated neural dynamics rather than by a one-directional bottom-up processing. However, the underlying mechanism of the dynamical properties of the sensory processing largely remains to be explored. Here, we propose a neural network model based on the predictive coding and reservoir computing as a model of the dynamical process of perception. The internal network dynamics of the proposed model is trained so that the network reproduces given multidimensional time courses of sensory signal and the prediction error is sent to higher-order network and is triggering the internal network dynamics. The proposed model may contribute to uncover the mechanism of higher-order cognitive function and can be a basis for the application of the neural dynamics for artificial intelligence.

Keywords Neural network · Reservoir computing · Predictive coding · Perception · Nonlinear dynamics

11.1 Introduction

The sensory information processing or perception is not mere a bottom-up processing of the sensory signal. Internal states of the brain have much influence on the process of perception with its top-down process. The internal network generates attentional modulation for the sensory processing and makes a prediction for the incoming sensory signals. Robust and flexible processing of the perception is achieved by the concert between the bottom-up and top-down processing. Indeed,

---

Y. Katori (✉)
The School of Systems Information Science, Future University Hakodate, Hakodate, Hokkaido, Japan

Institute of Industrial Science, The University of Tokyo, Meguro-ku, Tokyo, Japan
e-mail: katori@fun.ac.jp

J. M. Delgado-García et al. (eds.), *Advances in Cognitive Neurodynamics (VI)*, Advances in Cognitive Neurodynamics,
https://doi.org/10.1007/978-981-10-8854-4_11

some disorders of perception, e.g., visual hallucination, can be caused by the mismatch between the sensory information and the internal representation of the sensory information [1]. Mismatch negativity observed in human subjects with electroencephalography is reflecting a deviance of auditory stimuli from a prediction based on a learned regularity of stimuli [2]. Even without sensory stimuli, the internal neural network exhibits neural activity: this spontaneous activity is related to higher-order cognitive functions, e.g., planning of behavior and thought.

The process of the perception can be modeled as a dynamical system. In the associative memory network, the recurrent neural network rectifies a given sensory signal with its attractor dynamics [3]. Dynamic synapses improve the performance of the memory association with its transitive dynamics [4, 5]. In the present paper, we propose a hierarchical network model of dynamics of perception based on the concept of the reservoir computing and the predictive coding.

Reservoir computing or echo state network [6, 7] is a framework for training recurrent neural network with a simple learning strategy. The recurrent and feedback connections for the neural network can be fixed with sparsely and randomly generated weight values. The output from the network is trained so that the output time course matches the given time course.

Predictive coding [8] is a computational model of sensory information with a hierarchical structure of the network. Each level of the hierarchy employs a generative model to predict representation on the level below, and the network is trained so that the prediction error is minimized, and the prediction error sends to higher level. Several experimental studies justify the framework of the predictive coding (see a review paper [9]).

11.2 Model

We propose the hierarchical network model with the predictive coding and the reservoir computing. The module at each hierarchical level is composed of the dynamical reservoir, the predictive layer, the input layer, and the prediction error layer (Fig. 11.1a, b). The schematic of the phase space in Fig. 11.1c describes the dynamics of the module.

The connectivity and strength of the connections on the network are configured so that the dynamics of the internal network (the dynamical reservoir and the prediction layer) predicts the given input time course. Equivalently, the orbits of the input layer in the phase space (subspace of the internal network) match the orbit of prediction layer. We assume that the sensory information is given as spatially and temporally correlated multidimensional smooth time course and that a part of individual units of the input layer is mutually associated and exhibits repetition of certain spatiotemporal patterns.

If there is no input, the internal network stays in its resting state. When an input comes, firstly the prediction error (difference between the given time course and the internally generated time course) becomes larger, and then it triggers the

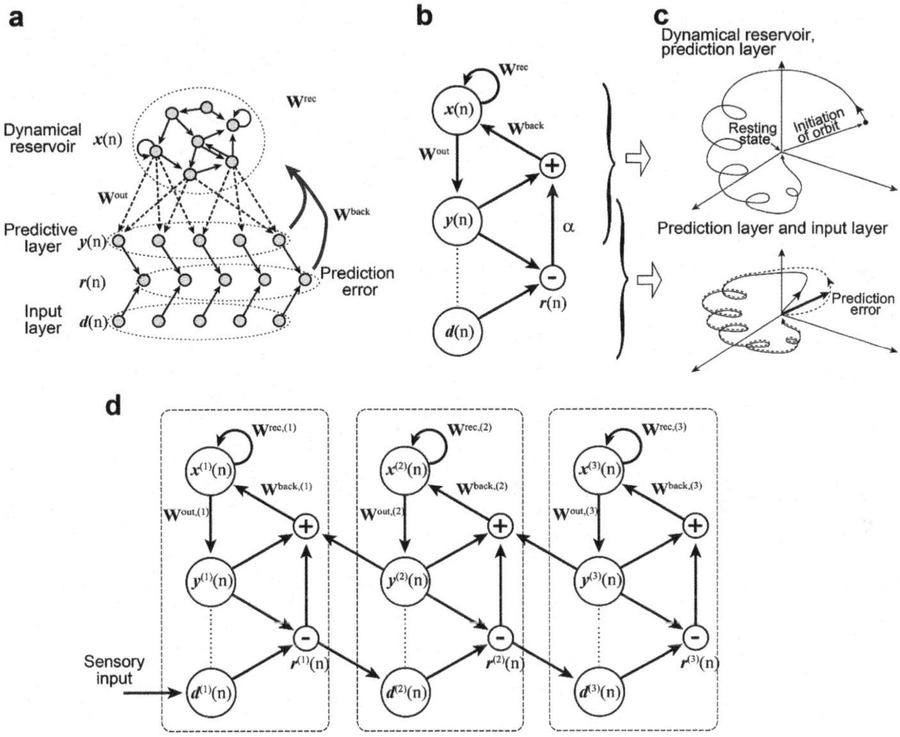

Fig. 11.1 Architecture of the network model. (**a**) The network structure of the module. (**b**) Simplified network diagram of (**a**). (**c**) Phase space of the dynamical reservoir (top) and that of the prediction and the input layer. (**d**) Hierarchical network model

internal network dynamics for a specific direction on the phase space. If the internal dynamics successfully reproduces the input time course, the prediction error fades away, and thus, the internal network is driven independently from the input layer.

In the hierarchical network model, the prediction error of each module sends to the higher-order module as the input time course. The higher-order module makes a prediction regarding the input (the prediction error in the lower module) and sends back the prediction to the lower module. The lower module modifies its internal dynamics according to the prediction.

In the following, we focus on the analysis of the properties of the single module. The dynamics of the module can be formulated as follows. The state of the N_x dimensional dynamical reservoir at time n is denoted by $\mathbf{x}(n)$. The states of the predictive layer, the input layer, and the prediction error layer are denoted by $\mathbf{y}(n)$, $\mathbf{d}(n)$, and $\mathbf{r}(n)$ (N_y dimension), respectively. The state of each layer is updated according to the following equations:

$$\mathbf{x}(n+1) = \mathbf{x}(n) + \{-\alpha_0 \mathbf{x}(n) + \mathbf{f}_x (\mathbf{W}^{\text{rec}} \mathbf{x}(n-k) \\ + \mathbf{W}^{\text{back}} (\mathbf{y}(n-k) + \mathbf{r}(n-k)))\} / \tau \tag{11.1}$$

$$\mathbf{y}(n) = \mathbf{f}_y \left(\mathbf{W}^{\text{out}} \mathbf{x}(n) \right), \tag{11.2}$$

$$\mathbf{r}(n) = \mathbf{f}_r (\mathbf{d}(n) - \mathbf{y}(n)), \tag{11.3}$$

In Eqs. (11.1, 11.2, and 11.3), $\mathbf{W}^{\text{rec}}$ is $N_x \times N_x$ weight matrix of recurrent connections, and $\mathbf{W}^{\text{back}}$ is $N_x \times N_y$ weight matrix of feedback connections. $f_x(x) = \tanh(x)$, $f_y(x) = \tanh(x)$, and $f_r(x) = \max(x, 0)$ are nonlinear response functions. k is propagation delay between each unit pair.

The configuration of the connectivity and the strength of the connections are achieved by the following two steps. Firstly, assign randomly and sparsely the recurrent connections $\mathbf{W}^{\mathbf{rec}}$ and the feedback connections $\mathbf{W}^{\text{back}}$ (see also [6]). Then the output connections (the connection from the dynamical reservoir to the prediction layer) $\mathbf{W}^{\text{out}}$ are trained with a given input time course.

To generate $\mathbf{W}^{\text{rec}}$, firstly, generate $N_x \times N_x$ matrix $\mathbf{W}_0$ by assigning 1 or -1 to randomly selected $\beta_r N_x \times N_x$ components and 0 to others. Then normalize and scale $\mathbf{W}_0$ with its spectral radius $\lambda_{\max}$ and coefficient α_r: $\mathbf{W}_1 = \alpha_r \mathbf{W}_0 / \mid \lambda_{\max} \mid$, where $\lambda_{\max}$ is maximum eigenvalue of $\mathbf{W}_0$. As to $\mathbf{W}^{\text{back}}$, firstly, assign 1 or -1 to randomly selected $\beta_b N_x \times N_y$ components and 0 to others. Then, normalize this matrix with the sum of columns of this matrix and scale with a coefficient α_b.

The output connection $\mathbf{W}^{\text{out}}$ is computed by a few iteration steps. As initial values of the matrix $\mathbf{W}^{\text{out},\,(0)}$, all the components of $\mathbf{W}^{\text{out},\,(0)}$ are set to 0. The m-th iteration step of output connection $W^{\text{out},\,(m)}$ is computed with $m-1$-th iteration step of the output connection $W^{\text{out},\,(m-1)}$. By computing the state of the network according to Eqs. (11.1)–(11.3) from n = 0 to $n = T_1$ with $\mathbf{W}^{\text{out},\,(m-1)}$ and the given time course $\mathbf{d}(n)$, the time course of dynamical reservoir $\mathbf{x}(n)$ is obtained. Then, put the time course in the range from n = T_0 to $n = T_1$ into state collecting matrix $\mathbf{M}$, where $\mathbf{M}$ is $(T_1 - T_0 - 1) \times N_x$ matrix. Similarly, put the sigmoid-inverted input time course $\tan^{-1}\mathbf{d}(n)$ into state collective matrix $\mathbf{G}$. Then, the $\mathbf{W}^{\text{out},\,(m)}$ is calculated using ridge regression as the following equation:

$$\left(\mathbf{W}^{\text{out},(m)}\right)^T = \left(\mathbf{M}^T \mathbf{M} + \lambda \mathbf{E}\right)^{-1} \mathbf{M}^T \mathbf{G} \tag{11.4}$$

In Eq. (11.4), λ is a coefficient for adjusting a sparseness, and $\mathbf{E}$ is a unit matrix. After the I iterations of above calculation, we obtain $\mathbf{W}^{\text{out}} = \mathbf{W}^{\text{out},\,(I)}$. After the iteration, weight values are fixed.

11.3 Results

As the input time course, we used 20-dimensional time courses, which show some repetition of specific patterns of time courses. Randomly selected four units in each pattern exhibit oscillatory waveform with different initiation time, and other units stay silence (see Fig. 11.2b, thin curve). We train and run the network with the following parameter values: $N_x = 200$, $N_y = 20$, $T_0 = 400$, $T_1 = 2000$, $\alpha_r = 0.6$, $\beta_r = 0.1$, $\alpha_b = 1.0$, $\beta_b = 0.1$, $\alpha_0 = 0.7$, $\tau = 2.5$, $k = 5$, $I = 10$, and $\lambda = 0.1$.

Figure 11.2 shows the network after the training with two patterns of oscillatory inputs reproduces the input time courses in the prediction layer. During the onset of the input patterns (arrows and thin curves in Fig. 11.2b), firstly, the prediction error increases (Fig. 11.2c), and it triggers the fluctuation on the dynamical reservoir (Fig. 11.2a, d). Then, the dynamical reservoir generates the time course in the prediction layer (Fig. 11.2b, thick curves). Orbits on the phase space (Fig. 11.2e) indicate that the difference between the prediction layer and the input layer increased just after the onset, then the orbit of the prediction layer follows the orbit of the input layer, and the prediction error is successfully shrunk. The prediction error exhibits larger response just after the onset of the input pattern composed of the four oscillatory waveforms. For the after-coming waveform in the pattern, the prediction error exhibits a relatively small response, because the first larger prediction error triggered the activities on the dynamical reservoir and generated the prediction, and thus the

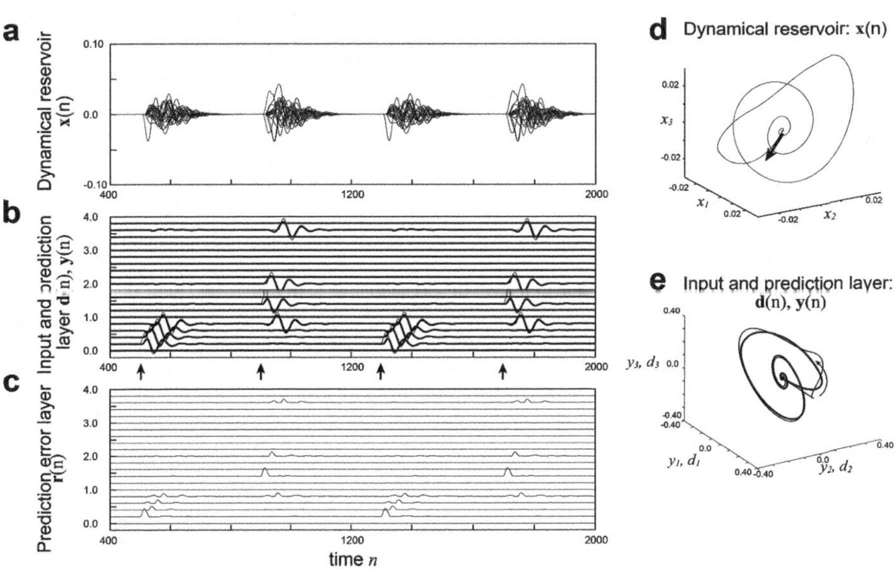

Fig. 11.2 Responses on the network after the training. (**a**) Dynamical reservoir. (**b**) The prediction and the input time courses. (**c**) The prediction error. (**d**) A trajectory of the dynamical reservoir in the phase space. Time courses of x_1, x_2, x_3 are shown. (**e**) Phase space of the prediction layer (thick curve) and the input layer (thin curve). Each first three components of $\mathbf{y}(n)$ and $\mathbf{d}(n)$ are shown

prediction error became small. The larger responses on the prediction error reflect these unpredictable input time courses. If unlearned patterns are applied to the network, the network cannot produce appropriate time courses, but its amplitude of activities is relatively small.

The prediction error depends on the several configurations of the network and the parameter values (not shown). The prediction error increases as the number of patterns increases. This indicates the limitation of a capacity of the dynamical reservoir. The capacity is increased with the size of the dynamical reservoir.

11.4 Discussion

We proposed a network model for dynamics of the perception with the framework of reservoir computing and the predictive coding. The network is driven by the prediction error to generate a prediction of the given time courses. We show that the given high-dimensional time courses can be reproduced by the internal network dynamics.

The larger peaks on the prediction error reflect on the onset of the input pattern; these responses trigger activities of the internal network and reproduce the given input time courses. In other words, this process can be understood as a process of encoding high-dimensional and complex time courses into the low-dimensional and sparse time courses. This encoding process might be necessary for the sensory information processing in the brain that transforms the detailed sensory information into abstract information.

The mechanism we analyzed here can be extended to the hierarchical network model. The encoded low-dimensional time course is carried into the higher-order module as a bottom-up signal. Then, the higher-order module makes a prediction and further encodes the time course into more abstract information. This predicted time course is sent back to the lower network for the prediction in the lower module as a top-down signal. The dynamical process with cycle of the bottom-up and top-down signaling is crucial for the perception, prediction, and generation of sensory signals.

Acknowledgment This research was supported by Grant-in-Aid for Scientific Research (C) Grant Number 16K00246.

References

1. Naatanen, R., Tervaniemi, M., Sussman, E., Paavilainen, P., Winkler, I.: Primitive intelligence in the auditory cortex. Trends Neurosci. **24**, 283–288 (2001)
2. Collerton, D., Taylor, J., Tsuda, I., Fujii, H., Nara, S., Aihara, K., Katori, Y.: How can we see things that are not there? Current insights into complex visual hallucinations. J. Conscious. Stud. **23**, 195–227 (2016)

3. Hopfield, J.J.: Neural networks and physical system with emergent collective computational abilities. Proc. Natl. Acad. Sci. U. S. A. **79**, 2554–2558 (1982)
4. Katori, Y., Otsubo, Y., Okada, M., Aihara, K.: Stability analysis of associative memory network composed of stochastic neurons and dynamic synapses. Front. Comput. Neurosci. **7**, 1–12 (2013)
5. Katori, Y.: Enhanced memory association on neural network with dynamic synapses. In: Wang, R., Pan, X. (eds.) Advances in cognitive neurodynamics (V), pp. 741–748. Springer, Singapore (2015)
6. Jaeger, H.A.: Tutorial on training recurrent neural network, Covering BPPT, RTRL, EKF, and the 'echo state network' approach. GMD Report. **159**, 1–46 (2005)
7. Jaeger, H.: Harnessing nonlinearity: predicting chaotic systems and saving energy in wireless communication. Science. **304**, 78–80 (2004)
8. Rao, R.P.N., Ballard, D.H.: Predictive coding in the visual cortex: a functional interpretation of some extra-classical receptive-field effects. Nat. Neurosci. **2**, 79–87 (1999)
9. Kilner, J.M., Friston, K.J., Frith, C.D.: Predictive coding: an account of the mirror neuron system. Cogn. Process. **8**, 159–166 (2007)

Chapter 12
Analysis of Structure-Function Relationship Using a Whole-Brain Dynamic Model Based on MRI Images of the Common Marmoset

Hiromichi Tsukada, Hiroaki Hamada, Ken Nakae, Shin Ishii, Junichi Hata, Hideyuki Okano, and Kenji Doya

Abstract How brain functions emerge from anatomical networks of the brain is still an open fundamental question. One approach for understanding the structure-function relationship is computational modeling of structural and functional connectivity data from MRI and to explore the behavior of the dynamic model by computer simulation. Here we constructed a whole-brain model based on the structural connectivity between 96 anatomical regions of the marmoset brain estimated from diffusion MRI data. We compared the brain activity simulated by the model and the brain activity observed by resting state functional MRI. We found that the correlation between the simulated and empirical functional connectivities increases within balanced parameter regions of excitatory-inhibitory connections, although the models with shuffled weights break the correlation. This result suggests that these parameters are crucial factors for the relationship between anatomical and functional network in the resting state MRI.

Keywords Marmoset · MRI · Structural connectivity · Functional connectivity · Neural network model

---

H. Tsukada (✉) · H. Hamada
Okinawa Institute of Science and Technology Graduate University, Okinawa, Japan
e-mail: hiromichi.tsukada@oist.jp

K. Nakae · S. Ishii
Integrated Systems Biology Laboratory, Graduate School of Informatics, Kyoto University, Kyoto, Japan

J. Hata · H. Okano
Laboratory for Marmoset Neural Architecture, Brain Science Institute RIKEN, Saitama, Japan

Department of Physiology, Keio University School of Medicine, Tokyo, Japan

K. Doya
Neural Computation Unit, Okinawa Institute of Science and Technology Graduate University, Okinawa, Japan

J. M. Delgado-García et al. (eds.), *Advances in Cognitive Neurodynamics (VI)*,
Advances in Cognitive Neurodynamics,
https://doi.org/10.1007/978-981-10-8854-4_12

12.1 Introduction

Multiple brain regions with different functions exchange information with each other. These brain regions are highly organized both anatomically and functionally, making it possible to respond flexibly to the environment. Each of anatomical and functional network can be measured by diffusion and functional MRI, respectively. Therefore, the research to investigate the relationship between the anatomical and functional networks of the brain using MRI has become an active area of interest.

Functional networks are thought to be driven by an anatomical network. However, most of the functional networks cannot be explained with an anatomical network alone. One factor to connect these relations is neuromodulators such as acetylcholine, dopamine, and serotonin, which tune the connection strength of functional networks [1–3]. Simulation studies suggest that excitatory-inhibitory balance impacts the dynamic state of the neuronal population activity [4] and gating multiple signals [5].

In recent years, methods to investigate the relationship between anatomical and functional networks using mathematical models incorporating anatomical connections have been proposed in human studies [6, 7]. These works are very powerful to understand how the anatomical network structure can shape functional networks.

In this study, we apply these methods to the common marmoset MRI data in order to allow investigation of neural dynamics using invasive microscopic approaches in the future. Furthermore, we extend the Wilson-Cowan model [8] and examine how the balance of excitatory-inhibitory connections affect the relationship between anatomical and functional networks.

12.2 Methods

12.2.1 Structural (Anatomical) Connectivity (SC)

Structural connectivity data was obtained using ex vivo high-angular resolution diffusion imaging (HARDI) data and global tractography [9] from two healthy marmosets (Fig. 12.1a). Brain regions were subdivided into 452 regions of interest (ROIs) defined according to the Hashikawa ATLAS [10], which are grouped into 96 areas in total.

The connection of a region to itself was set to 0 in the connectivity matrix for the simulations. For the HARDI data acquisition, the following parameters were used: $TR = 4000$ ms, $TE = 21.8$ ms, flip angle $= 90°$, reconstructed matrix size of $128 \times 128 \times 36$, voxel size of $0.35 \times 0.35 \times 0.7$ mm with slice thickness of 0.7 mm. Furthermore, the data were collected with 128 optimal nonlinear diffusion gradient directions at $b = 1500$ s/mm^2.

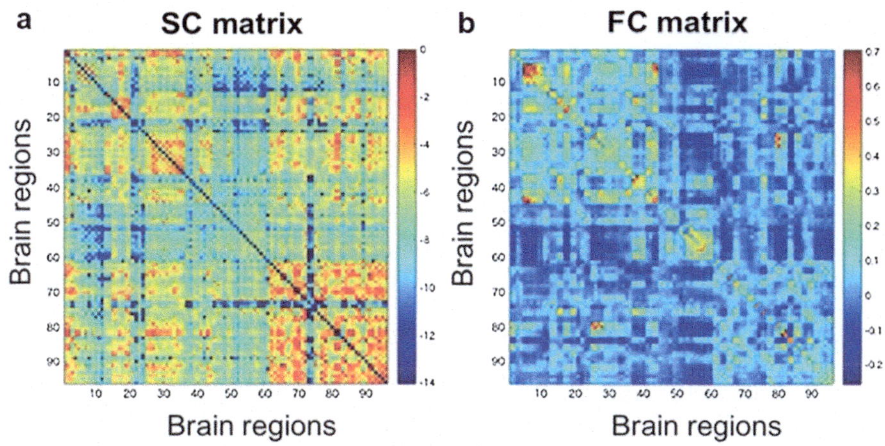

Fig. 12.1 Structural and functional connectivity matrices. (**a**) SC matrix averaged across two marmosets. (**b**) FC matrix averaged across three marmosets under resting state awake condition

12.2.2 *Functional Connectivity (FC)*

The empirical FC was calculated for three healthy awake common marmosets in rest, and averaged across the three marmosets (Fig. 12.1b). For each scanning run, initial five volumes were discarded, and the next 555 functional volumes were used for the analysis. Data was acquired using a 9.4 T small animal MRI scanner (Bruker Biospin, Ettlingen, Germany) with parameters: $TR = 2000$ ms, volumes = 560, TE = 16 ms, and flip angle = 70°. Each functional volume comprised of 35 slices with a resolution of 0.5 × 0.5 × 1 mm. *T2*-weighted scan was also acquired before the functional scans with parameters: $TR = 8000$ ms, $TE = 36.7$ ms, and flip angle = 180°.

Each functional volume comprised of 80 slices with a resolution of 0.15 × 0.15 × 0.3 mm. Data preprocessing was performed using the SPM12 software package (Wellcome Department of Cognitive Neurology, London, UK) running under MATLAB (MathWorks). We subsequently performed denoising steps with functional connectivity toolbox (CONN). FC was determined using Pearson correlation between the average time courses of the brain regions.

12.2.3 *Whole-Brain Model*

We employed Wilson-Cowan model [8] as a base whole-brain model, which consists of 96 coupled brain regions. To evaluate the influence of excitatory-inhibitory connection balance, we modified the model as follows:

$$\tau \frac{dE_i}{dt} = -E_i(t) + (k_E - r_E E_i(t)) \\ S_E \left(L \left(c_E E_i(t) - c_I I_i(t) \right) + G \sum_j C_{ij} E_j(t) + P_i(t) \right) + \sigma w_i(t) \quad (12.1)$$
$$\tau \frac{dI_i}{dt} = -I_i(t) + (k_I - r_I I_i(t)) \, S_I \left(L \left(c_E E_i(t) - c_I I_i(t) \right) \right) + \sigma v_i(t),$$

In Eq. (12.1) $E_i(t)$ and $I_i(t)$ stand for the firing rate of the excitatory and inhibitory population of i-th brain area at time t, respectively, $P_i(t) = 1.25$ is an external stimulus parameter, and $w_i(t)$ and $v_i(t)$ are additive noise with a standard normal distribution.

Other parameters are set to $\tau = 8$ ms, $\sigma = 0.0001$, $Ke = 1$, $k_i = 1$, $r_e = 1$, and $r_i = 1$. C_E and C_I stand for excitatory and inhibitory coupling parameters; L and G represent local and global scaling parameters, respectively. Regions are coupled through the excitatory population with a connectivity matrix C_{ij} derived from the average of two marmoset's tractography data.

The transfer function is given by the sigmoidal function:

$$S_Y(x) = \frac{1}{1 + e^{(-a_Y(x - \theta_Y))}} - \frac{1}{1 + e^{a_Y \theta_Y}}. \quad (12.2)$$

Here parameters are $a_E = 1.3$, $a_I = 2$, $\theta_E = 4$, and $\theta_I = 3.7$ ($Y = E, I$) [8].

12.3 Results

We examined how the balance of excitatory-inhibitory connection affects the relationship between the anatomical and functional networks. To evaluate the relationship, we computed the extended Wilson-Cowan model based on the structural connectivity matrix and compared the brain activity simulated by the model and the brain activity observed by resting state functional MRI. We generated 5 runs of 10 s of simulated neural activity for each of coupling parameters, c_E, c_I, L, and G. The output data corresponding to the first 1 s of the simulations were discarded from the analysis to avoid initial transient dynamics, resulting 10s length of simulated neural activity in total. Simulated FC was calculated by Pearson's correlation coefficient between the time series of 96 brain regions. The model was simulated by the Euler-Maruyama method at a sampling frequency of 20 kHz.

We compared the similarity of simulated and empirical brain activities by calculating the pairwise Pearson's correlation coefficient between 96 brain regions. The correlation between the simulated and empirical FC (r_{SE}) in the space of coupling parameters c_E and c_I, where L and G are adjusted so that r_{SE} is maximized, is shown in Fig. 12.2a.

Notice that r_{SE} increased in the region where c_E and c_I parameters were balanced (the highest r_{SE} was 0.33). We investigated the influence of each parameter (c_E, c_I, L, and G) on r_{SE} in Fig. 12.2b. We selected the parameters, c_E, c_I, L, and G, in

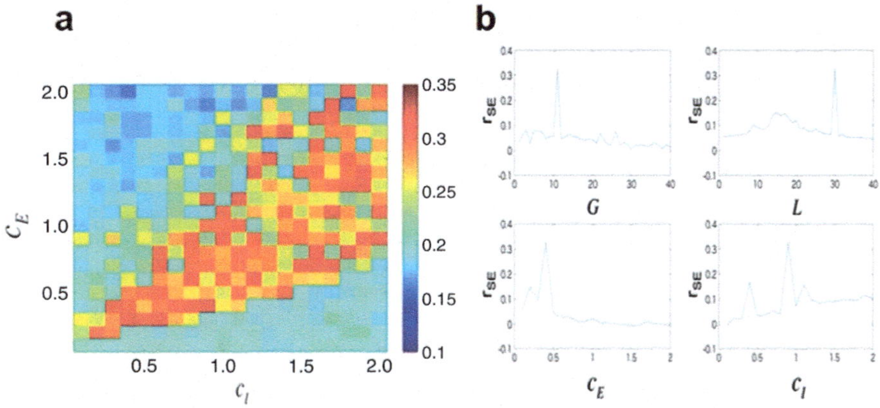

Fig. 12.2 Correlation between simulated and empirical FC (r_{SE}) with model parameters. (**a**) r_{SE} in the parameter space of excitatory and inhibitory coupling strength. (**b**) Relationship among the parameters (c_E, c_I, L, and G) and r_{SE}

order to maximize r_{SE} and changed one of the parameters. Then, a sharp peak was appeared with all the parameters.

12.4 Conclusions

In this study, we constructed a whole-brain network model based on the structural connectivity matrix of marmosets and compared the brain activity simulated by the model and the brain activity observed by resting state functional MRI. We found that the correlation between simulated FC and empirical FC (r_{SE}) increased within the balanced region of excitatory-inhibitory connections. This result indicates that excitatory-inhibitory connections within a brain region are well tuned to fit the empirical resting state functional network.

We also examined the influence of each parameter (c_E, c_I, L, and G) on r_{SE} and found a sharp peak was appeared for all the parameters. This result suggests that these parameters become one of crucial factors for the relationship between anatomical and functional networks in the resting state MRI.

As a future study, we consider a role of excitatory and inhibitory balance on neural dynamics by comparing resting state, attention task, and disease marmoset fMRI data. We also improve the accuracy of structural connectivity by comparison with tracer data and functional connectivity by developing fMRI data preprocessing.

Acknowledgments This work was supported by the program for Brain Mapping by Integrated Neurotechnologies for Disease Studies (Brain/MINDS) from Japan Agency for Medical Research and Development, AMED.

References

1. Kruglikov, I., Rudy, B.: Perisomatic GABA release and thalamocortical integration onto neocortical excitatory cells are regulated by neuromodulators. Neuron. **58**, 911–924 (2008)
2. Seamans, J.K., Yang, C.R.: The principal features and mechanisms of dopamine modulation in the prefrontal cortex. Prog. Neurobiol. **74**, 1–58 (2004)
3. Moreau, A.W., Amar, M., Le-Roux, N., Morel, N., Fossier, P.: Serotoninergic fine-tuning of the excitation-inhibition balance in rat visual cortical networks. Cereb. Cortex. **20**, 456–467 (2010)
4. Tsukada, H., Yamaguti, Y., Tsuda, I.: Transitory memory retrieval in a biologically plausible neural network model. Cogn. Neurodyn. **7**, 409–416 (2013)
5. Vogels, T.P., Abbott, L.F.: Gating multiple signals through detailed balance of excitation and inhibition in spiking networks. Nat. Neurosci. **12**, 483–491 (2009)
6. Honey, C., Kotter, R., Breakspear, M., Sporns, O.: Network structure of cerebral cortex shapes functional connectivity on multiple time scales. Proc. Natl. Acad. Sci. U. S. A. **104**, 10240–10245 (2007)
7. Deco, G., Ponce-Alvarez, A., Mantini, D., Romani, G.L., Hagmann, P., Corbetta, M.: Resting-state functional connectivity emerges from structurally and dynamically shaped slow linear fluctuations. J. Neurosci. **33**, 11239–11252 (2013)
8. Wilson, H.R., Cowan, J.D.: Excitatory and inhibitory interactions in localized populations of model neurons. Biophys. J. **12**, 1–24 (1972)
9. Reisert, M., Mader, I., Anastasopoulos, C., Weigel, M., Schnell, S., Kiselev, V.: Global fiber reconstruction becomes practical. NeuroImages. **54**, 955–962 (2011)
10. Hashikawa, T., Nakatomi, R., Iriki, A.: Current models of the marmoset brain. Neurosci. Res. **93**, 116–127 (2015)

Chapter 13
A Structure and Function of Hippocampal Memory Networks in Consolidating Spatiotemporal Contexts

Hiromichi Tsukada, Minoru Tsukada, and Yoshikazu Isomura

Abstract A feature of information processing in the brain is to have the ability of integrating novelty information into past experience. In other words, this integration ability is to manipulate spatiotemporal context information in learning and memory. In this paper, we propose a neural network model, which accounts for this feature, and show the possibility to acquire a spatiotemporal context information by the coding method of spatial clustering and its self-similarity in learning and memory based on the hippocampal experimental data. We also consider the physiological validity of this model.

Keywords Spatiotemporal context · Spatiotemporal learning rule (STLR) · Learning and memory · Recurrent network · Hippocampus

13.1 Introduction

In research on neural network model for memory, several models have been proposed so far. Associative memory models are widely known as a neural network memory model, and these models can produce a stable attractor state in the memory space [1, 2]. Chaotic itinerancy model has been proposed, and this model can generate successive retrieval of memory (transition between quasi-attractors), whose trajectory is chaotic and history-dependent [3, 4]. An appropriate balance of excitatory-inhibitory (E/I) connection strengths yields the stable attractor in associative memory, and imbalance of E/I produces successive retrieval of memory [5].

H. Tsukada
Okinawa Institute of Science and Technology Graduate University, Okinawa, Japan

M. Tsukada (✉) · Y. Isomura
Brain Science Institute, Tamagawa University, Tokyo, Japan
e-mail: tsukada@eng.tamagawa.ac.jp

J. M. Delgado-García et al. (eds.), *Advances in Cognitive Neurodynamics (VI)*,
Advances in Cognitive Neurodynamics,
https://doi.org/10.1007/978-981-10-8854-4_13

These studies contribute to manipulate the state of the memory; however, how to manipulate this state to acquire spatiotemporal context information in the memory is still not cleared.

On the other hand, the learning rule also plays an important role to acquire the spatiotemporal context information. In particular, to identify the difference in contextual information and to stabilize the identified context information are essential process. For storing a spatiotemporal sequence, physiologically, the magnitude of LTP in hippocampal CA1 depends on spatiotemporal stimuli [6], and based on the results, Tsukada et al. proposed the spatiotemporal learning rule (STLR) [7]. The STLR is the synaptic weight change which depends on both spatial coincidence and its temporal summation of the source neuron activity, unrelated to the destination neuron activity, which is different from Hebbian learning (HEB) depending on both the source and destination neuron activity. It was identified by model simulation that the STLR plays an important role in the pattern separation and Hebb in pattern completion [8, 9]. Tsukada et al. [10] clarified in neurophysiological experiment that both rules coexist in the hippocampal CA1 neuron [10].

Theoretically, Tsuda et al. showed the possibility of Cantor coding in hippocampal network [11–14]. Experimentally, Cantor-like coding in membrane potentials of hippocampal CA1 pyramidal neurons was identified [15].

Based on the studies mentioned above, we propose a neural network model, which consists of a single-layered neural network with a feed-forward excitatory connection, two types of feedback connections, and three types of learning rules, to achieve spatiotemporal context learning. We also consider the structural and functional mechanisms for spatial clustering and its self-similarity in hippocampal networks.

13.2 Models

13.2.1 Neuron Model

Any McCulloch-Pitts of artificial neuron is an example of which called a finite automaton. However, there has been no mention about spatiotemporal pattern discrimination and attractor in memory neural networks. We adopt integrate-and-fire neuron model, which have internal state (membrane potential). The model is defined by the following equation:

$$s_j(t) = \sum_i x_{ij}(t) w_{ij}(t) + \sum_m s_j\,(t-m)\,e^{-m\lambda_1}, \tag{13.1}$$

In Eq. (13.1), $x_{ij}(t)$ is an input from neuron i to neuron j at time instant t, $w_{ij}(t)$ is a synaptic weight from neuron i to neuron j at time instant t, and λ_1 is the decay constant of membrane potential. t is a discrete time. Output of neuron j is given by

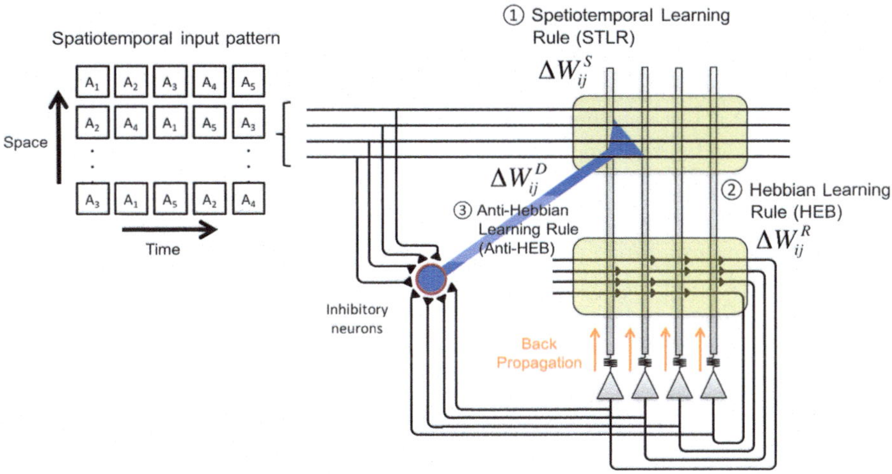

Fig. 13.1 A schematic diagram of the network architecture

$$
\begin{aligned}
y_j(t) &= F\left\{s_j(t) - T\right\}, \\
F(x) &= \begin{cases} 1 & \text{if } x \geq 0 \\ 0 & \text{if } x < 0 \end{cases},
\end{aligned}
\tag{13.2}
$$

In Eq. (13.2), T is the threshold.

13.2.2 The Network Architecture

The structure of the network is illustrated in Fig. 13.1. The network is a single-layered network, which consists of N neurons, with feed-forward excitatory connection and two types (excitatory and inhibitory) of feedback connections. Each neuron connects to all source neurons $(x_1, x_2, \ldots, x_i, \ldots, x_N)$ via an excitatory synapse.

An input neuron i $(i = 1, 2, \ldots, N)$ connects an output neuron j $(j = 1, 2, \ldots, N)$ through a synaptic weight w_{ij} and consists of the spatiotemporal pattern, whose spatial snap at one moment corresponds to a spatial frame of N-dimensional binary elements. The weight w_{ij} is modified by three types of learning rules as follows:

- Δw_{ij}^{S} by spatiotemporal learning rule (STLR) in the feed-forward excitatory connection
- Δw_{kj}^{R} by Hebb learning rule (HEB) in the recurrent connection
- Δw_{ij}^{D} by anti-Hebb learning rule (anti-HEB) and the inhibitory effect (LTD) induced by a coincidence between input and output.

13.2.3 The Learning Rules

Training the network requires the steps as follows:

First step: Δw_{ij}^{S} is modified by STRL depending on input pattern *x*(*t*) under the subthreshold condition of membrane potentials *s*(*t*) in each neuron. Second step: Δw_{kj}^{R} is modified by Hebb learning rule (HEB) in the recurrent network depending on *y*(*t*) · *y*(*t* − 1). Third step: Δw_{ij}^{D} is modified by anti-Hebb learning rule (anti-HEB). The inhibitory effect (LTD) is induced by a coincidence between input *x*(*t*) and output *y*(*t*). After enough repetitions of these three steps, the outputs become constant (attractor).

13.2.3.1 Spatiotemporal Learning Rule (STLR)

The synaptic weight change depends on both spatial coincidence and its temporal summation of the source neuron activity, unrelated to the destination neuron activity [7]. The equation is given by

$$\Delta w_{ij}^{s}(t) = \eta h \left\{ \sum_{m} I_{ij}\,(t-m)\, e^{-m\lambda_2} - \theta \right\}, \tag{13.3}$$

In Eq. (13.3), λ_2 is the time constant of NMDA channel, η is the learning rate coefficient, *h*(*u*) is a sigmoid output function of the potentiation force, and θ is the thresholds. $I_{ij}(t)$ is the coincidence among the source neuron activities at instant *t* and given by

$$I_{ij}(t) = x_{ij}(t) s_j(t). \tag{13.4}$$

13.2.3.2 Hebb Learning Rule (HEB) in the Recurrent Connection

The weight modification depends on the coincidence between previous output and current output and is given by

$$\Delta w_{kj}^{R}(t) = \alpha y_k\,(t-1)\, y_j(t), \tag{13.5}$$

In Eq. (13.5), α is the learning rate coefficient. This modification can exhibit properties similar to short-term memory in humans.

13.2.3.3 Anti-HEB in the Recurrent Inhibitory Connection

The weight modification induces the LTD in each recurrent inhibitory connection, which works as a coincident detector between source and target neuron, and the equation is given by

$$\Delta w_{ij}^{D}(t) = -\beta x_i(t) y_j(t), \tag{13.6}$$

In Eq. (13.6), β is the learning rate coefficient. The weight at instant $w_{ij}(t)$ is given by

$$w_{ij}(t) = w_{ij}\,(t-1) + \left\{\Delta w_{ij}^{s}(t) + \Delta w_{ij}^{D}(t)\right\} + \Delta w_{kj}^{R}(t). \tag{13.7}$$

13.3 Characteristic of the Model

Physiologically, it is believed that contextual information is temporary stored in the hippocampus in the form of short-term memory. How information is coded and represented in the hippocampus is an important topic in the investigation of memory systems. In the CA1 area of the hippocampus, the magnitude of LTP depends not only on the frequency of the applied stimuli, but also on the time sequence [6, 7]. Fukushima et al. showed that the membrane potentials of CA1 are hierarchically clustered in a self-similar manner to the input sequences [15].

Based on these physiological results, we propose a neural network consisting of the three types of synaptic modifications (Fig. 13.1). First is the modification by STRL, second is that by HEB in recurrent network, and third is that by anti-HEB based on the input-output coincidence of pyramidal cells.

STLR, depending on spatial coincidence and its time history, has high ability in clustering spatiotemporal pattern. It plays an important role in pattern separation. On the other hand, HEB in the recurrent network has feedback paths from their outputs to their inputs; the response of such networks is dynamic; that is, after applying a new input, the output is calculated and fed back to modify the input. The successive iterations of this process produce smaller output changes until the output becomes constant (attractor).

For the spatial clustering and its self-similarity (Cantor-like coding) in learning and memory, the outputs are hierarchically clustered in a self-similar manner to the input sequences. In the process, STLR plays an important role in pattern clustering in the neural network depending on input pattern, while HEB in recurrent network forms the dynamics depending on the input pattern and leads to pattern completion. However, the increased weight by STLR influences the next process, so that the next pattern clustering is strongly influenced by the previous input pattern; that is, the hierarchical clustering becomes out of shape. In order to avoid this influence, anti-HEB based on the input-output coincidence of pyramidal cells adjusts the increased weights related to output.

The cascade weight modification described above is an important algorithm for the spatial clustering and its self-similarity (Cantor-like coding), which could be the spatiotemporal attractor for memory. Physiologically, parvalbumin (PV) basket cell and oriens-lacunosum moleculare (O-LM) cell seem to play an important role to modify w_{ij} as an inhibitory effect (Δw_{ij}^D) in the hippocampus [16].

Acknowledgments This work was supported by Grants-in-Aid for Scientific Research on Innovative Areas (JP26112005; JP15K21715) and KAKENHI C (15K00325; 17K00322).

References

1. Amari, S.: Learning patterns and pattern sequences by self-organizing nets of threshold elements. Inst. Electr. Electro. Eng. Trans. **C-21**, 1197–1206 (1972)
2. Hopfield, J.J.: Neural networks and physical systems with emergent collective computational properties. Proc. Natl. Acad. Sci. U. S. A. **79**, 2554–2558 (1982)
3. Tsuda, I.: Dynamic link of memories - chaotic memory map in nonequilibrium neural networks. Neural Netw. **5**, 313–326 (1992)
4. Tsuda, I.: Chaotic itinerancy and its roles in cognitive neurodynamics. Curr. Opin. Neurobiol. **31**, 67–71 (2015)
5. Tsukada, H., Yamaguti, Y., Tsuda, I.: Transitory memory retrieval in a biologically plausible neural network model. Cogn. Neurodyn. **7**, 409–416 (2013)
6. Tsukada, M., Aihara, T., Mizuno, M., Kato, H., Ito, K.: Temporal pattern sensitivity of long-term potentiation in hippocampal CA1 neurons. Biol. Cybern. **70**, 495–503 (1994)
7. Tsukada, M., Aihara, T., Saito, H., Kato, H.: Hippocampal LTP depends on spatial and temporal correlation of inputs. Neural Netw. **9**, 1357–1365 (1996)
8. Tsukada, M., Pan, X.: The spatiotemporal learning rule and its efficiency in separating spatiotemporal patterns. Biol. Cybern. **92**, 139–146 (2005)
9. Pan, X., Tsukada, M.: A model of the hippocampal-cortical memory system. Biol. Cybern. **95**, 159–167 (2006)
10. Tsukada, M., Yamazaki, Y., Kojima, H.: Interaction between the spatio-temporal learning rule (STLR) and Hebb type (HEBB) in single pyramidal cells in the hippocampal CA1 area. Cogn. Neurodyn. **1**, 1157–1167 (2007)
11. Tsuda, I.: Toward an interpretation of dynamic neural activity in terms of chaotic dynamical systems. Behav. Brain Sci. **24**, 793–810 (2001)
12. Tsuda, I., Kuroda, S.: Cantor coding in the hippocampus. Jpn. J. Ind. Appl. Math. **18**, 249–258 (2001)
13. Tsuda, I., Kuroda, S.: A complex systems approach to an interpretation of dynamic brain activity II: does cantor coding provide a dynamic model for the formation of episodic memory. In: Erdi, P., et al. (eds.) Cortical Dynamics, LNCS 3146, pp. 129–139. Springer, Heidelberg (2004)
14. Yamaguti, Y., Kuroda, S., Fukushima, Y., Tsukada, M., Tsuda, I.: A mathematical model for Cantor coding in the hippocampus. Neural Netw. **24**, 43–53 (2011)
15. Fukushima, Y., Tsukada, M., Tsuda, I., Yamaguti, Y., Kuroda, S.: Spatial clustering property and its self-similarity in membrane potential of hippocampal CA1 pyramidal neurons for a spatiotemporal input sequence. Cogn. Neurodyn. **1**, 305–316 (2007)
16. Klausberger, T., Somogyi, P.: Neuronal diversity and temporal dynamics: the unity of hippocampal circuit operations neuronal diversity and temporal dynamics. Science. **321**, 53–57 (2008)

Chapter 14
A Pseudo-neuron Device and Firing Dynamics of Their Networks Similar to Neural Synchronizing Phenomena Between Far Local Fields in the Brain

Tomoyuki Yano, Yoshitomo Goto, Tomoyuki Nagaya, Ichiro Tsuda, and Shigetoshi Nara

Abstract A pseudo-neuron hardware device called dynamic self-electrooptic effect device (DSEED) is proposed and considered. First, its neuron-like pulsed oscillations ((i) fast spiking type, (ii) threshold spiking type) are theoretically predicted, and their oscillating properties are discussed using computer experiments. Second, firing (pulsing) pattern dynamics of their diffusion-coupled networks are considered, and the computer experiments indicate that there are three kinds of firing dynamics, (a) transient chaotic firing state, (b) entirely synchronized (coherent) firing state, and (c) partly synchronized firing state, between far local fields in the network. In particular, the case (c) shows that the firing patterns are quite similar to neural synchronization phenomena between far local fields as observed in the brain. So, the present device and their networks are, if successfully realized in hardware, quite useful to verify and develop theories of neural (coupled oscillator) networks to consider or to understand the mechanisms of brain functions based on the viewpoint of nonlinear dynamics.

Keywords Pseudo-neuron device · DSEED · Neural networks · Diffusion-coupled networks · Hardware implementation · Chaos · Nonlinear oscillator · Firing synchronization · Brain functions

---

T. Yano · S. Nara (✉)
Graduate School of Natural Science and Technology, Okayama University, Okayama, Japan
e-mail: nara@ec.okayama-u.ac.jp

Y. Goto
Junior College, Beppu University, Beppu, Japan

T. Nagaya
Division of Natural Sciences, Oita University, Oita, Japan

I. Tsuda
Chubu University Academy of Emerging Sciences, Kasugai, Aichi, Japan

J. M. Delgado-García et al. (eds.), *Advances in Cognitive Neurodynamics (VI)*, Advances in Cognitive Neurodynamics,
https://doi.org/10.1007/978-981-10-8854-4_14

14.1 Introduction

Since these several decades, neuroscience and brain science have been greatly progressing not only with the background of rapid and big development of computer science and technologies but also by the development of measurement technologies to observe their activities. However, we still have no detailed understanding about mechanisms of advanced brain functions as well as neural information or control processing, except rather small number of cases in certain restricted contexts. An approach based on the viewpoint of nonlinear dynamics has been expected to shed a light to them, but huge complex dynamics observed in systems having very large but finite degrees of freedom have been preventing us from easy investigation of them.

Even with the use of a few simplified neuron models, such as binary state neuron model or oscillator neuron model, certain networks consisting of a large number of coupled neurons reveal huge complex dynamics and do not easily allow us to understand them with conventional methodologies. So, in these situations, it would be useful that we propose a new device indicating neuron-like behaviors utilizing optoelectronic semiconductor technology and their integrated and monolithically designed network.

The neuron device and their network could enable us to make easy hardware experiments and computer analyses as well. Thus, using a pseudo-neuron hardware device (see Fig. 14.1) called dynamic self-electrooptic effect device (DSEED) proposed in the previous paper [1], we consider first the two kinds of neuron-like pulsed oscillations, (i) fast spiking type and (ii) threshold spiking type, that are theoretically predicted, and their nonlinear oscillation properties are discussed. The case (i) was partly reported in our previous papers [2], so this paper is the first report of the case (ii). Next, firing (pulsing) pattern dynamics of their diffusion-coupled DSEED networks of both spiking types are considered and discussed.

14.2 Dynamic Self-Electrooptic Effect Device (DSEED) and Pulse Operation (Neural Firing)

In our previous works [1, 2], a pseudo-neuron device called DSEED was proposed (see Fig. 14.1), where the device consists of series-coupled *pin* diode (1st layer, *p*-type semiconductor; 2nd layer, 10 nm intrinsic semiconductor layer made from multi-quantum wells, abbreviated as i-MQW; 3rd layer, *n*-type semiconductor).

As shown in Fig. 14.1, when incident light having appropriately chosen power (P_{in}) and wave length (frequency ω), with feedback light from lower device as well, is given, then pulse oscillation of photocurrent occurs in the circuit, as shown in Fig. 14.2.

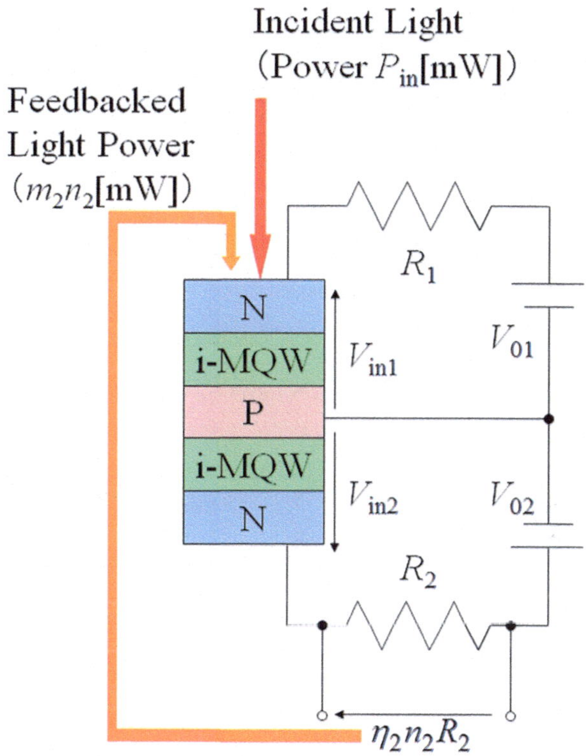

Fig. 14.1 Schematic representation of the proposed pseudo-neuron device. N means n-type semiconductor and P, p-type semiconductor, respectively. i-MQW is a thin film consisting of intrinsic semiconductor multi-quantum well made from GaAs and AlGaAs, for instance

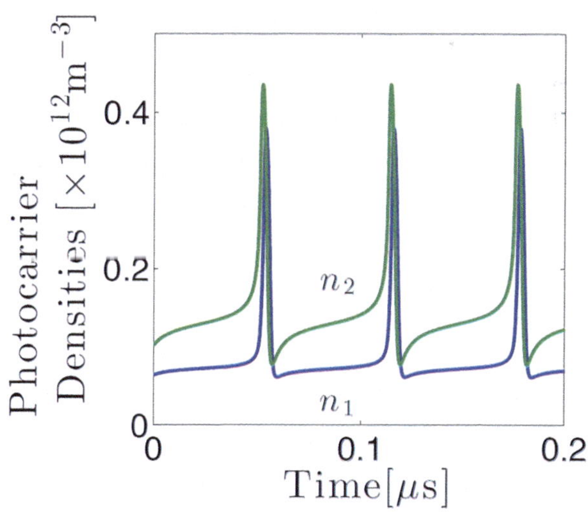

Fig. 14.2 An example of pulsing oscillation of DSEED

The two rate equations of photo-carriers in the upper circuit and lower circuit are written in Eq. (14.1), which are coupled with two nonlinear differential equations written below.

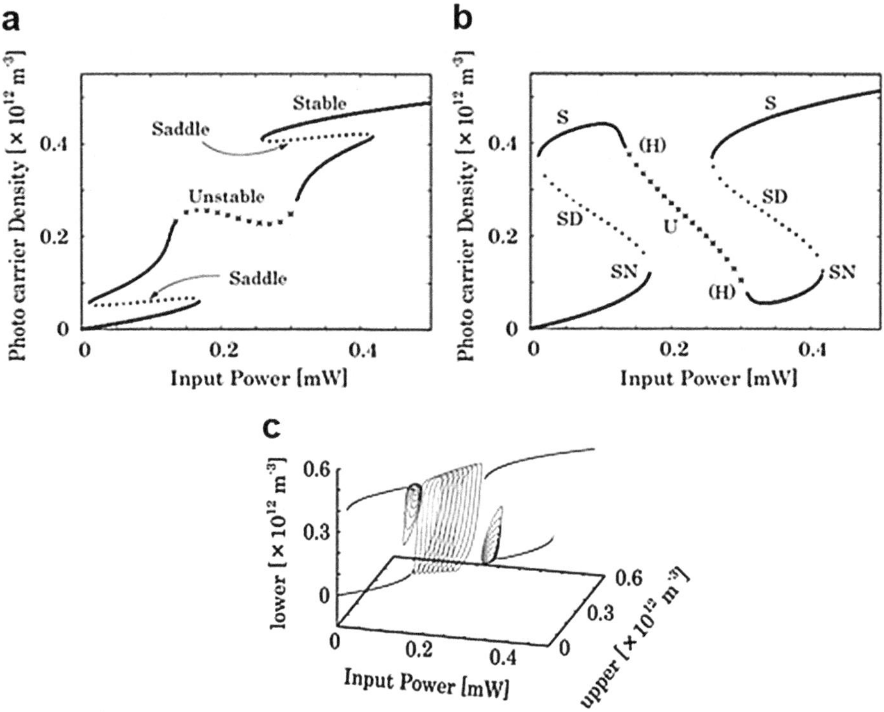

Fig. 14.3 (**a**) and (**b**) are the bifurcation diagrams of n_1 and n_2, with respect to P_{in}. (**c**) indicates three-dimensional plots of them, simultaneously with showing the limit cycles after Hopf bifurcation and the existence of crisis (sudden vanish of limit cycles)

$$\begin{aligned}
\frac{dn_1}{dt} &= -\frac{n_1}{\tau_1} + \frac{\alpha_1 \Omega_{01}\left(P_{\text{in}} + m_2 n_2\right)}{\left[\omega - \omega_{01} + \beta_1\left(V_{01} - \eta_1 R_1 n_1\right)\right]^2 + \left(\Omega_{01}/2\right)^2} \\
\frac{dn_2}{dt} &= -\frac{n_2}{\tau_2} + \frac{\alpha_2 \Omega_{02}\left(P_{\text{in}} - m_1 n_1 + m_2 n_2\right)}{\left[\omega - \omega_{02} + \beta_2\left(V_{02} - \eta_2 R_2 n_2\right)\right]^2 + \left(\Omega_{02}/2\right)^2}
\end{aligned} \tag{14.1}$$

When we give appropriate parameter values, $\boldsymbol{\alpha}_l$, $\boldsymbol{\beta}_l$, $\boldsymbol{\tau}_l$, $\boldsymbol{\eta}_l$, $\boldsymbol{\Omega}_l$, m_l, and R_l ($l = 1$ or 2), we can make the bifurcation diagram with respect to the primary parameter, the power of incident light power, $P_{\text{in},}$ under the fixed frequency ω.

The result is shown in Fig. 14.3. There are two important bifurcation points. The first is a *saddle node bifurcation* on limit cycle. The second is a *crisis* of limit cycle after Hopf bifurcation. The former gives continuous pulse train, which is called "fast spiking (firing)." The latter gives only one pulse (spike) oscillation when an initial state is set at rather small values of n_1 and n_2, in just outside of the *destabilized limit cycle*, where the limit cycle ceases to exist by touching the basin boundary of stable fixed point in the state space, n_1 vs. n_2.

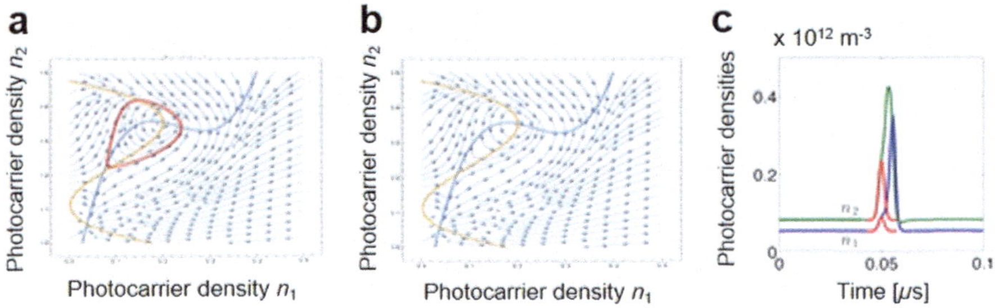

Fig. 14.4 The null cline of Eq. (14.1) and outlines of orbits. (**a**) is near "crisis" and (**b**) is after crisis. (**c**) is an example of one spike firing after an excitation bigger than threshold (red line) is applied

The null cline and outlines of orbits near the crisis are shown in Fig. 14.4 in the state space of n_1 vs. n_2, where Fig. 14.4a is just before the crisis and Fig. 14.4b is just after the crisis. One can easily know that a single pulse is generated by applying an appropriate external excitation bigger than threshold value to n_1 and n_2 when they stay at the stable fixed point, where this single pulse excitation is equivalent to neural firing having threshold in biological neuron cell.

14.3 Diffusion-Coupled DSEED Network of Both Spike-Type Elements

As noted in the previous section, we consider dynamic states of pulsing (firing) operation of diffusion-coupled DSEED network consisting of the two types of elements, fast spiking type and threshold spiking type, in terminology of neuroscience.

14.3.1 Nearest Neighbor Diffusion-Coupled Network and Spiking (Firing) Pattern Dynamics

$$D_\alpha\left(-4{n_\alpha}^{(i,j)} + {n_\alpha}^{(i-1,j)} + {n_\alpha}^{(i+1,j)} + {n_\alpha}^{(i,j-1)} + {n_\alpha}^{(i,j+1)}\right)$$

When we employ a square lattice network and introduce diffusion coupling between four neighboring sites as to be added to the right side of Eq. (14.1) in the previous section, where D is the diffusion constant (parameter), the suffix (i,j) denotes the site position in the square lattice, α is the suffix to denote that carrier density of the upper layer or lower layer shown in Figs. 14.1 and 14.2,

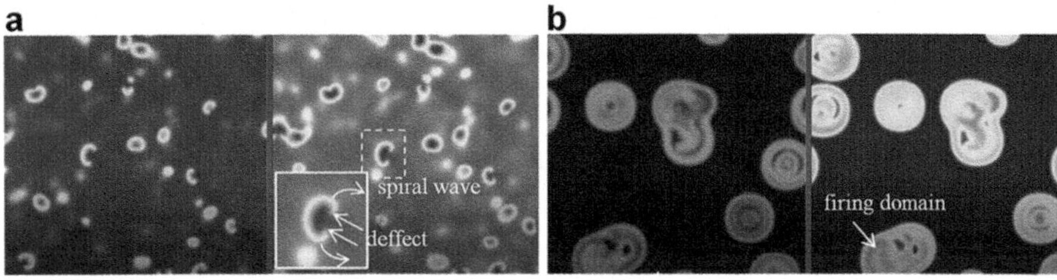

Fig. 14.5 (**a**) A snapshot of spiking pattern dynamics of the $n_\alpha{}^{(i,j)}$ (left, $\alpha = 1$; right, $\alpha = 2$) in fast spiking case. (**b**) The same with (**a**) in threshold spiking case

so $\alpha = 1$ or 2. In continuum limit, this term converges to usual diffusion term: $D(\partial^2 n/\partial x^2 + \partial^2 n/\partial y^2)$.

Including the diffusion effects, the computer experiments indicate that there are two kinds of firing dynamics, (a) transient chaotic firing state and (b) entirely synchronized (coherent) firing state. The results with employing the fast spiking case were partly reported in ICCN2011 briefly with taking rather less neuron number (400 and 900) for simplicity. The new results in both cases reported in this paper are the experiments in larger neuron number case (up to $512 \times 512 = 262.144$) which results in the more realistic and new phenomena (see Fig. 14.5), where Fig. 14.5a shows the fast spiking case and Fig. 14.5b, the threshold spiking case, respectively.

In the case of fast spiking (Fig. 14.5a), pairs of spiral waves (white curve arrows) and topological defects (white straight arrows) emerge in the network. The defects make pairs and connect the wave arm extending from each defect and survive long time in spatial-temporal chaotic field. Then the distance between a pair of defect decreases as neurons firing and the pair of defects annihilate finally. In contrast to the proceeding argument, there are no topological defects in the threshold spiking case (Fig. 14.5b), and the wide spread dark area in the figure is stable state that is depressed under the threshold. In addition the, neurons in the bright area, are not allowed to settle down to the stable state because of the stimulation by diffusion coupling four neighboring neurons and firing continuously in spite of global stable (Fig. 14.4b). In the firing domains, one can find some dark areas that are stable state under the threshold. Such a stable region nucleates in continuous firing area and remains while being eroded by the surrounding bright area. In this situation, the neurons irregularly alternate stable state and chaotic firing state. These various pattern dynamics and self-organizations show that the DSEED network has ability to investigate the behavior of threshold-type neurons and their corporative dynamics.

14.3.2 Distant Coupling and Synchronization via Chaos

Now, we extend our consideration in the previous section to more artificially designed network, where in this paper we employed a randomly connected network that loses the concept of Euclidean distance in the configuration space. In this

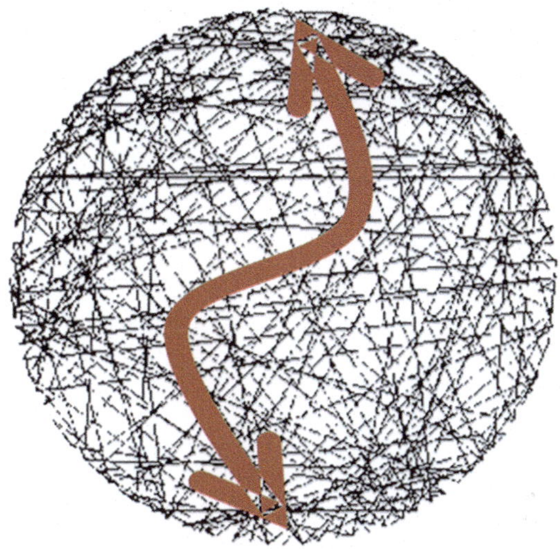

Fig. 14.6 A randomly connected network of 400 fast spiking neuros put on circumference and neuron groups, a and b (20 neurons), are far separated in the measure of "the averaged shortest path length"

network, when we take the coupling as

$$D_\alpha \sum_j \left(n_\alpha{}^{(j)} - n_\alpha{}^{(i)} \right) \tag{14.2}$$

That is, the coupling between arbitrary i-th and j-th sites and then chaotic dynamics continues within enough longtime intervals in comparison with spiking time, say several 10 microsecond order that is several three or four order of magnitude longer than typical spiking time in this device, provided we employ appropriate incident optic power, P_{in}, that gives fast spiking oscillation.

In this setting, we tried to realize synchronization phenomena via chaos between certain far distant partial element groups, where the incident power of light to the two small groups is kept strong so as to result quickly in synchronized state. The key point is that the two neuron groups are separated each other by chaos caused by the majority of the other elements, where the measure of "distance" is defined by the "averaged shortest path length" between the neurons of the two groups in a given configuration of elements.

As shown in Fig. 14.6, the neurons are put on circumference, and connection existence is shown by lines, where in this case the total number of neurons is 400 and the number of the neuron groups, a and b, is about 20, respectively.

In this setting of computer experiments, we evaluate the cross correlation functions between the two neuron groups, a and b, as shown in Fig. 14.7, which indicates that the firing of both groups is *synchronizing* each other via *chaos*

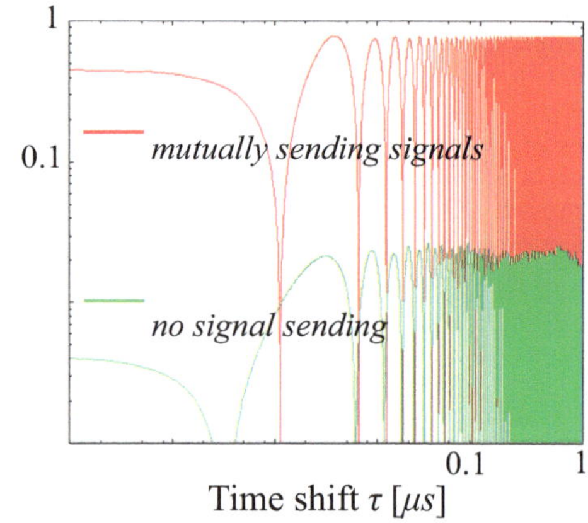

Fig. 14.7 Normalized cross correlation function $<n_a^{(i)}(t)n_b^{(j)}(t+\tau)>_t$ between neuron groups, a and b, as a function of time shift τ, with and without signal sending, respectively

generated by the other neurons and it is quite similar to neural synchronization phenomena derived from the other neural systems [3] or actually observed in the brain [4]. Chaos in the brain has been attracting much interest of many people related with the question, *what is the role of chaos in brain functions* [4–6]. Besides them, on the viewpoint of information dynamics in chaos, there have been many works by the pioneering people from rather long ago, for instance [7–9], while the detailed descriptions and discussions are discarded due to the given space restriction.

14.4 Concluding Remarks

1. A pseudo-neuron hardware device called dynamic self-electrooptic effect device (DSEED) is proposed, and the two kinds of neuron-like pulsed oscillations ((i) fast spiking type, (ii) threshold spiking type) are theoretically predicted.
2. The firing (spiking) pattern dynamics of their nearest neighbor diffusion-coupled DSEED networks of both types in a square lattice are considered, and the computer experiments indicate that, in fast spiking neurons, firing patterns have a tendency to form circle-like chain or one *spiral with phase defects* on both edges, which behave as nonlinear wave propagation of synchronously firing groups in the network with changing of spatial shapes and extension.
3. In threshold neurons, a certain number of survived firing neurons play a role of *pacemaker* of multiconcentric rings of firing pattern, where nearest neighbor diffusion coupling corresponds to gap junction coupling between GABA (*g*amma *a*mino*b*utyric *a*cid) neurons that has tendency to make inhibitory neuron network in the brain.
4. In a network with diffusion coupling between arbitrary elements, there are three kinds of firing dynamics, (a) transient chaotic firing state, (b) entirely synchronized (coherent) firing state, and (c) *partly synchronized firing state*

between far distant fields via *chaos*, in the network depending on the network design and setting of the parameters.

5. In particular, the case (c) reveals that the firing patterns are quite similar to neural synchronization phenomena observed in the brain and in theoretical models as well.

Acknowledgments This work is supported in part both by MEXT KAKENHI Grant Number JP24120707 and by the Cooperative Research Program of the "Network Joint Research Center for Materials and Devices."

References

1. Ohkawa, T., Yamamoto, Y., Nagaya, T., Nara, S.: Dynamic behaviors in coupled self-electro-optic effect devices. Appl. Phys. Lett. **86**, 111107 (2005)
2. Yamamoto, T., Ohkawa, Y., Kitamoto, T., Nagaya, T., Nara, S.: Bifurcation phenomena in coupled dynamic self-electro-optic-effect devices. Int. J. Bifurcation Chaos. **16**, 3717–3725 (2006)
3. Soma, K., Mori, R., Sato, R., Furumai, N., Nara, S.: Simultaneous multichannel signal transfers via chaos in a recurrent neural network. Neural Comput. **27**, 1083–1101 (2015)
4. Skarda, C.A., Freeman, W.J.: How brains make chaos in order to make sense of the world. Behav. Brain Sci. **10**, 161–195 (1987)
5. Tsuda, I.: Chaotic itinerancy as a dynamical basis of hermeneutics in brain and mind. World Future. **32**, 167–184 (1991)
6. Nara, S., Davis, P.: Chaotic wandering and search in a cycle-memory neural network. Prog. Theor. Phys. **88**, 845–855 (1992)
7. Shaw, R.: Strange attractors, chaotic behavior, and information flow. Z. Naturforsch. **36a**, 80–112 (1981)
8. Matsumoto, K., Tsuda, I.: Calculation of information flow rate from mutual information. J. Phys. A. **21**, 1405–1414 (1988)
9. Arhem, P., Blomberg, C., Liljenström, H. (eds.): Disorder Versus Order in Brain Function - Essays in Theoretical Neurobiology. World Scientific Pub, London (2000)

Chapter 15
Neurodynamics on Up and Down Transitions of Membrane Potential: From Single Neuron to Network

Xuying Xu, Rubin Wang, and Jianting Cao

Abstract The phenomenon of up and down transition is an important characteristic of spontaneous brain activity, which happens in various levels of the nervous system. In level of membrane potentials, it shows spontaneous periodic transitions between two subthreshold preferred stable states. Here, we have studied the mechanisms and characteristics of up and down transitions of membrane potentials from single neuron to network model. Furthermore, we have developed the model by considering both excitatory and inhibitory neurons and introducing synaptic dynamics into network model. Based on this model, we studied the influence of intrinsic characteristics and network parameters on up and down activities. The main output of this study is that the network parameters have little impact on these spontaneous periodic up and down transitions. However, the intrinsic currents were found to play a leading role in the process. In this regard, we expect to explain the dynamics of up and down transitions and to lay the foundation for future work on the role of these transitions in cortex activity.

Keywords Up and down transitions · Spontaneous activity · Ionic channel · Dynamical characteristics

---

X. Xu (✉) · R. Wang
Institute of Cognitive Neurodynamics, East China University of Science and Technology, Shanghai, China

J. Cao
Saitama Institute of Technology, Fukaya, Japan

Brain Science Institute, RIKEN, Wako, Japan

J. M. Delgado-García et al. (eds.), *Advances in Cognitive Neurodynamics (VI)*,
Advances in Cognitive Neurodynamics,
https://doi.org/10.1007/978-981-10-8854-4_15

15.1 Introduction

The phenomenon of up and down transition is an important characteristic of spontaneous brain activity, which happens in various levels of the nervous system. Neural electrophysiology experiments show that during slow-wave sleep in the primary visual cortex of anesthetized animals [1–3] and during quiet wakefulness in the somatosensory cortex of unanesthetized animals [4], the membrane potentials make spontaneous transitions between two different levels called up and down states [5].

Why these transitions occur and whether this spontaneous activity plays a role in brain functions are still not clear. In fact, we know little about the induction of neuron membrane potentials and interactions among neural networks, especially the relationship between neural coding and cognitive behaviors.

So in this paper, we study on the neurodynamics on up and down transitions of membrane potential from single neuron to network, to figure out the mechanisms and characteristics of these transitions.

15.2 Up and Down Transitions of Single Neuron

We start our research on up and down transition from one single neuron, which is undoubtedly an effective method to study neural information processing mechanism. On the other hand, resent electrophysiology experiment can only record neural activities of a single or a few neurons, rather than large neuron clusters. However, data of one single neuron is easy to measure. So it is obvious to clean up the dynamics of a single neural up and down transitions before we study the neurodynamics of the network, which is a basic for us to learn cognitive behavior and network activity.

15.2.1 Biophysical Model of Single Neuron

The single neural dynamic model based on H-H equations consists of the following three ionic currents [6]: an instantaneous, inward current (sodium current), a slow h-like current, and two outward currents (a potassium current and a leak current). Specific equations are described in our previous work [7]. According to the numerical simulation of this model, we observe the results that have been observed in electrophysiology experiments.

15.2.2 Characteristics of Up and Down Transitions in Single Neuron

Here, we indicate two characteristics, bistability and spontaneity, of up and down transitions, using the dynamic model of one single neuron.

The calculation results are shown in Fig. 15.1. The top one shows that membrane potential presents spontaneous periodic transitions between two preferred stable states without any external stimulation. And the bottom one is the distribution of membrane potential, from which we can also find that the membrane potential almost stays in one of the up state and down state.

This dynamic model can describe the bistability of up and down transitions of neural membrane potential. The neuron can stay any one of the two without input. When the neuron is stimulated, it can switch its state from one to another to adjust itself to a new balance. These two states are called up state and down state, respectively. That is to say, the up and down transitions can be modulated by external stimulations.

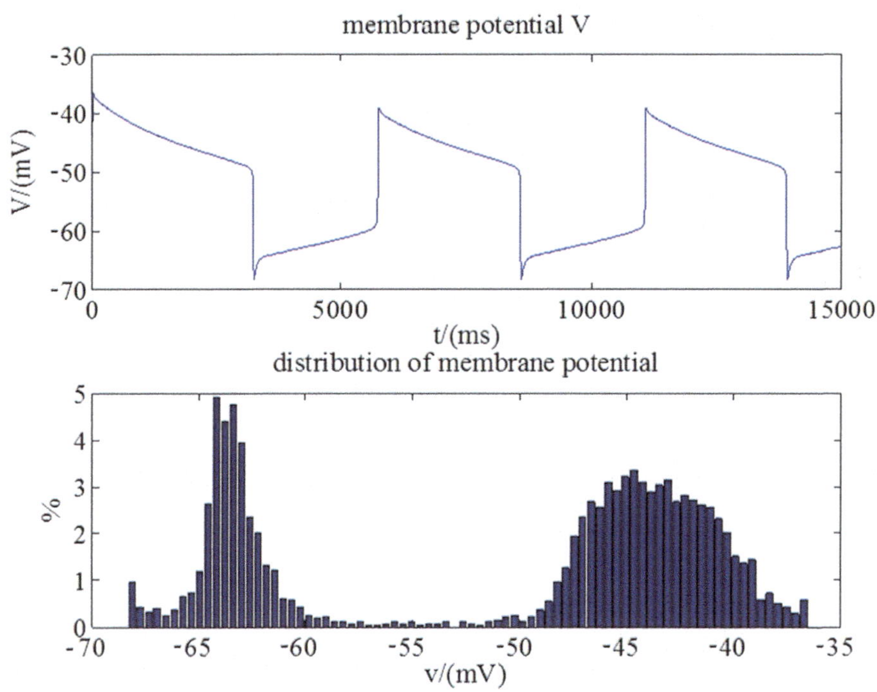

Fig. 15.1 Membrane potentials show spontaneous periodic transitions between two stable states – up and down states – and the corresponding distribution

The three dynamic variable models can also describe periodic spontaneous transitions between the up and down states in the absence of synaptic input. And in this case, the neuron's response to stimulation has a certain relationship with which state the neuron stays. If a stimulus is applied when the neuron is in a down state, responses are stronger than if it is applied during an up state.

15.3 Network Dynamics of Up and Down Transitions

In this section, we expand the model by clarifying the mechanism in two types of neurons, namely, excitatory and inhibitory neurons, and by substituting the constant connection state with a changing connecting function state involving two types of neurons so that the model reflects the in vivo mechanism better. Using this model, we explore the factors that influence spontaneous periodic up and down transitions.

15.3.1 Intrinsic and Synaptic Dynamics

The intrinsic current is adopted as introduced in the above section.

For synaptic dynamics, three types of receptors, namely, AMPA receptors, NMDA receptors, and $GABA_A$ receptors, are significant for the transmission of information between neurons. The AMPA and NMDA receptors are excitatory in nature, while the $GABA_A$ receptors are inhibitory. Furthermore, the AMPA receptors mediate the prototypical fast excitatory synaptic currents in the brain, while the NMDA receptors mediate synaptic currents that are substantially slower than the AMPA current.

The excitatory neurons excite other connecting neurons, including excitatory and inhibitory neurons, via AMPA receptors and NMDA receptors. However, the inhibitory ones only inhibit one type of neuron, the excitatory neurons, through the $GABA_A$ receptors. The three types of currents are represented by equations in our work [8, 9].

15.3.2 Network Structure

The following pattern is true for all the networks described in this paper, regardless of size: One excitatory neuron excites other connecting neurons including excitatory and inhibitory neurons. At the same time, it is also excited by other excitatory neurons and inhibited by other inhibitory ones. However, an inhibitory neuron only inhibits one other type of neuron, the excitatory neuron, and is excited by connecting excitatory neurons. All the excitatory connections are mediated by a fast AMPA component as well as a slow voltage-dependent NMDA component. The inhibitory connections are modulated by fast $GABA_A$ synaptic receptors.

15.3.3 Mechanism of Up and Down Transitions

In this section, we try to understand the mechanism underlying the effect of ionic currents. At the ionic level, the ion channel properties and conditions may play a role. Therefore, we examine different channels under different conductance conditions.

The results of the quantitative analysis are illustrated in Fig. 15.2 [9]. Besides the average up state duration of the membrane potential (top row), we also study the average cycle of the transitions (middle row) and the average ratio of the up state in the cycle (bottom row) of the three types of channels – potassium channel (left column), sodium channel (center column), and h channel (right column).

In most observations, the up state duration increases with decrease in the potassium channel conductance and increase in sodium channel conductance and h channel conductance, during appropriate ranges in which the membrane potential shows up and down transitions. However, the cycle of transitions does not change much under most conditions, except for cases in which the membrane potential is close to but does not reach that of a permanent down state. The up state ratio in the cycle showed a similar tendency to the up state duration.

In view of ionic channel, the mechanism of up and down transitions can be briefly described as follows. Depolarization activates the sodium channel letting the sodium ions move into cells, and then the potassium channel was activated slowly making the potassium ions move out of the cells which leads to the result of hyperpolarization. And the hyperpolarization then activates the h channel which allows sodium and potassium ions [10] going through.

Based on this point, the ionic channel interpretation for Fig. 15.2 would be obvious. The bigger value of potassium conductance makes the faster movement of potassium ions going out of the cells, leading to shorter up state duration, while the smaller value of sodium conductance gives rise to much more sodium ions moving into the cells which results in longer up state duration.

As to the conductance of h channel, the results are similar with the case of sodium conductance. Therefore, we tend to believe that the sodium ions largely contribute to the h current.

15.4 Discussion

Here, we have studied the mechanisms and characteristics of up and down transitions of membrane potentials from single neuron to network model, including bistability and spontaneity.

Furthermore, we have developed the model by considering both excitatory and inhibitory neurons and introducing synaptic dynamics into network model. Based on our model, we studied the influence of intrinsic characteristics and network parameters on up and down activities. The main output of this study is that the

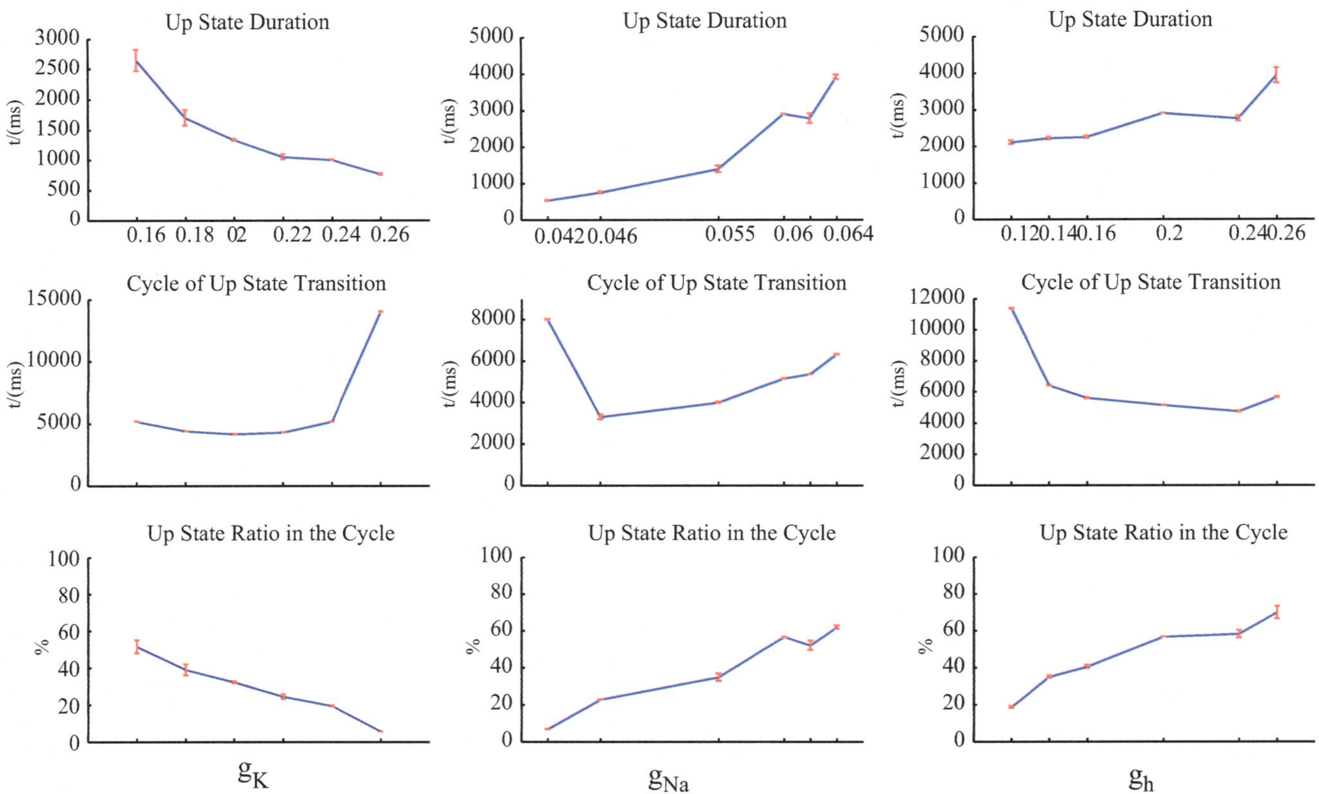

Fig. 15.2 Average duration of the up state (top row), average cycle of the up and down transitions (middle row), and average ratio of the up state in one cycle (bottom row) under different potassium channel conductance values (left column), sodium conductance (center column), and conductance of the h current (right column)

network parameters have little impact on these spontaneous periodic up and down transitions. However, the intrinsic currents were found to play a leading role in the process.

In this regard, we expect to explain the dynamics of up and down transitions and to lay foundation for future work on the role of these transitions in cortex activity [11].

Acknowledgments This study is supported by the National Natural Science Foundation of China (No. 11232005) and the Fundamental Research Funds for the Central Universities (No. 222201714020).

References

1. Anderson, J., Lampl, I., Reichova, I., Carandini, M., Ferster, D.: Stimulus dependence of two-state fluctuations of membrane potential in cat visual cortex. Nat. Neurosci. **3**, 617–621 (2000)
2. Lampl, I., Reichova, I., Ferster, D.: Synchronous membrane potential fluctuations in neurons of the cat visual cortex. Neuron. **22**, 361–374 (1999)
3. Steriade, M., Nunez, A., Amzica, F.: Intracellular analysis of relations between the slow (< 1 Hz) neocortical oscillation and other sleep rhythms of the electroencephalogram. J. Neurosci. **13**, 3266–3283 (1993)
4. Petersen, C.C.H., Hahn, T.T.G., Mehta, M., Grin-Vald, A., Sakmann, B.: Interaction of sensory responses with spontaneous depolarization in layer 2/3 barrel cortex. Proc. Natl. Acad. Sci. U. S. A. **100**, 13638–13643 (2003)
5. Parga, N., Abbott, L.F.: Network model of spontaneous activity exhibiting synchronous transitions between up and down states. Neuroscience. **1**, 57–66 (2007)
6. Loewenstein, Y., Mahon, S., Chadderton, P., Kitamura, K., Sompolinsky, H., Yarom, Y., Häusser, M.: Bistability of cerebellar Purkinje cells modulated by sensory stimulation. Neuroscience. **8**, 202–211 (2005)
7. Xu, X., Wang, R.: Neurodynamics of up and down transitions in a single neuron. Cogn. Neurodyn. **8**, 509–515 (2014)
8. Xu, X., Wang, R.: Neurodynamics of up and down transitions in network model. Abstr. Appl. Anal. **2013**, 486178 (2013)
9. Xu, X., Ni, L., Wang, R.: A neural network model of spontaneous up and down transitions. Nonlinear Neurodynamics. **84**, 1541–1551 (2016)
10. Ermentrout, G.B., Terman, D.H.: Mathematical Foundations of Neuroscience. Springer, New York (2010)
11. Xu, X., Ni, L., Wang, R.: Synchronous transitions of up and down states in a network model based on stimulations. J. Theor. Biol. **412**, 130–137 (2017)

Chapter 16
Effects of Temporal Integration on Computational Performance of Spiking Neural Network

Fangzheng Xue, Yang Zhang, Hongjun Zhou, and Xiumin Li

Abstract In spiking neural networks (SNN), information is considered to be encoded mainly in the temporal patterns of their firing activity. Temporal integration of information plays a crucial role in a variety of cognitive processes, such as sensory discrimination, decision-making, or interval timing. However, it is rarely considered in traditional computational SNN models. In this paper, we investigate the influence of temporal integration on the computational performance of liquid state machine (LSM) from two aspects: the synaptic decay constant and time delay from presynaptic neurons to the output neurons. LSM is a biologically spiking neural network model for real-time computing on time-varying inputs, where the high dimensionality of dynamical spikes is transformed into smoothly changing states through synaptic integration into the readout neuron. Our experimental results show that increasing the decay constant of synapses from SNN to the output neuron can remarkably improve the computational performance due to the enhancement of temporal integration. Moreover, transmission delays have an even larger impact on the richness of dynamical states, which in turn significantly increase the computational accuracy of SNN. These results may have important implications for the modeling of spiking neural networks with excellent computational performance.

Keywords Spiking neural network · Time constant · Time delay · Computational performance

F. Xue · Y. Zhang · X. Li (✉)
College of Automation, Chongqing University, Chongqing, China
e-mail: xmli@cqu.edu.cn

H. Zhou
School of Economics and Business Administration, Chongqing University, Chongqing, China

J. M. Delgado-García et al. (eds.), *Advances in Cognitive Neurodynamics (VI)*,
Advances in Cognitive Neurodynamics,
https://doi.org/10.1007/978-981-10-8854-4_16

16.1 Introduction

With the rapid development of theories and applications in brain science, spiking neural network (SNN) has very solid theoretical and experimental basis and shows more powerful calculation and associative memory ability than traditional artificial neural network. Liquid state machine (LSM) represents one of the most efficient computing models of SNNs proposed by Maass et al. [1]. It exploits the recurrent SNNs without training the entire network online, which remarkably reduces computational complexities. LSM has been applied to many applications, including word recognition [2], real-time speech recognition [3], and robotics [4], and it is shown to have universal computational power for time varying inputs and is good at dealing with temporal pattern recognition task. The key problem of liquid computing is the design of spiking neural network.

So far, most optimizations for LSM focus on the structure of SNN in LSM; very few studies pay attention to the readout component where readout neurons are trained simply by linear regression for different targets. However, the input stream to readout is not a vector of discrete spike trains but integrated synaptic signals from all of the presynaptic neurons in LC. Therefore, how the temporal information of the firing activity of the liquid network is integrated and transformed into the readout is indeed quite crucial for the computational performance.

In this paper, we investigate the influence of temporal integration on the computational performance of LSM from two aspects: the synaptic decay constant and time delay from recurrent networks to the output neurons. LSM is a biologically spiking neural network model for real-time computing on time-varying inputs, where the high dimensionality of dynamical spikes are transformed into smoothly changing states through synaptic integration into the readout neuron. Our experimental results show that increasing the decay constant of synapses from SNN to the output neuron can remarkably improve the computational performance due to the enhancement of temporal integration. Moreover, transmission delays have an even larger impact on the richness of dynamical states, which in turn significantly increase the computational accuracy of SNN. These results may have important implications for the modeling of spiking neural networks with excellent computational performance.

16.2 Methods

LSM mainly consists of three components: input component (IC), liquid component (LC), and readout component (RC) (as shown in Fig. 16.1). The liquid component in LSM uses spiking neurons connected by synapses to project inputs into a high-dimensional feature space called "liquid state." For specific tasks, the liquid state can be mapped onto a target signal through a readout component, which acts as a memoryless readout function [1]. Instead of updating all of the weights online, like in the abovementioned recurrent neural networks, synaptic connections in the liquid

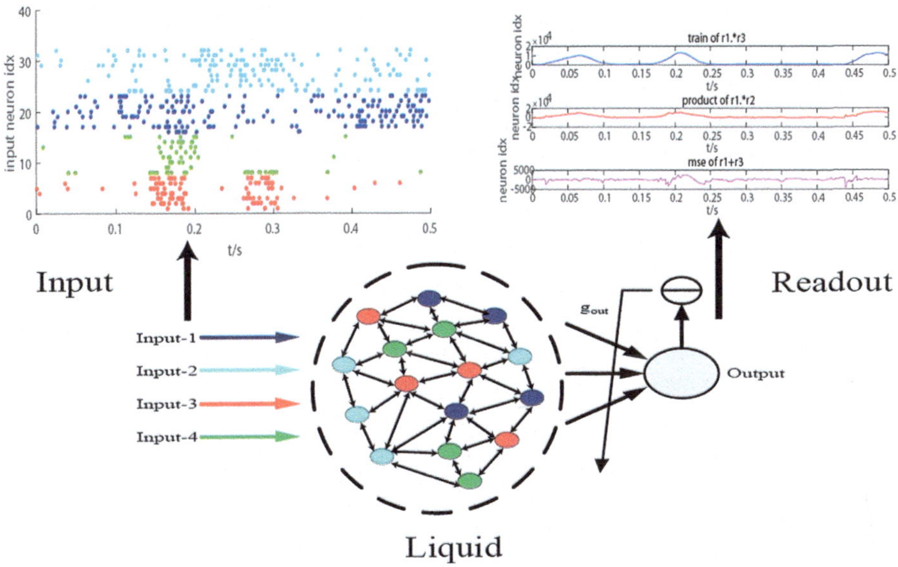

Fig. 16.1 Neural network structure. In this model, 400 neurons in LC are equally connected to four input streams independently. Neurons applied with different inputs are marked with different colors. The readout neuron is connected to all of the neurons in the liquid with synaptic weights g_{out}, which are trained by linear regression for different tasks (three examples are shown in the top right corner)

network are usually selected randomly and fixed during the training process; only the readout is trained using a simple classification/regression technique according to specific tasks. Detailed configurations of LSM in this paper are:

Input component: In this paper, four different inputs are independently added to four equivalently divided groups in the liquid network, as shown in Fig. 16.1. Note that inputs in the first layer (IC) are temporal sequences of spikes rather than firing rate, which are generated randomly by Poisson process. The inputs are four independent signal streams generated by the Poisson process with randomly varying rates $r_i(t), i = 1, \ldots 4$, where each input stream consists of eight spike trains (Fig. 16.1). The time-varying firing rates $r_i(t)$ of the eight Poisson spike trains were chosen as follows [1, 5]. The baseline firing rates for streams 1 and 2 are chosen to be 5 Hz, with randomly distributed bursts of 120 Hz for 50 ms. The rates for Poisson processes that generated the spike trains for input stream 3 and 4 are periodically updated, by randomly drawing from the two options 30 and 90 Hz.

Liquid component: In this paper the liquid network consists of 400 neurons which are all-to-all bidirectionally connected with synaptic weights randomly distributed in the range of $[0, g_{max}]$. The two-variable Izhikevich neuron model [6] is used in the network. It satisfies the following equations:

$$\begin{aligned}\dot{v_i} &= 0.04v_i^2 + 5v_i + 140 - u_i + I + I_i^{\text{syn}} \\ \dot{u_i} &= a(bv_i - u_i) + D\xi_i\end{aligned} \tag{16.1}$$

$$\text{if } v_i > 30\text{ mV, then } \begin{cases} v_i \leftarrow c \\ u_i \leftarrow u_i + d \end{cases} \tag{16.2}$$

where $i = 1, 2, \ldots, N$, i represents the total number of neurons; v_i represents the number of neurons membrane voltage; u_i is the number of neurons to restore variables; the variable ξ_i is subject to the mean to 0; the density D represents Gaussian white noise; I stands for the injection of the synaptic current of neurons; and I_i^{syn} is the total synaptic current through neuron i and is governed by the dynamics of the synaptic variable s_j. The synaptic model is:

$$\begin{aligned}I_i^{\text{syn}} &= -\textstyle\sum_{1(j\neq i)}^{N} g_{ji}s_j(v_i - v_{\text{syn}}) \\ \dot{s_j} &= \alpha(v_j)(1 - s_j) - s_j/\tau \\ \alpha(v_j) &= \alpha_0/(1 + e^{-v_j/v_{\text{shp}}})\end{aligned} \tag{16.3}$$

Here the excitatory synaptic reversal potential v_{syn} is set to be 0. Other parameters used in this paper are $\alpha_0 = 3$, $\tau = 2$, $V_{\text{shp}} = 5$, and $D = 0.1$. In particular, the other parameters are $a = 0.02$, $b = 0.2$, $c = -65$, and $d = 6$ for regular spiking neuron and for bursting neuron $a = 0.02$, $b = 0.2$, $c = -50$, and $d = 2$ [6].

Readout component: To investigate the computational capability of the LSM, usually learning tasks can be the approximation or functional combination of several real-time independent inputs. For example, if there are three independent inputs r_1, r_2, and r_3, the teaching signal could be $r_1+r_2+r_3$, r_1*r_2, or r_3^2 as considered in [7]. During the training process, a linear regression is employed to obtain the optimal readout weights that can generate the least mean squared error (MSE) according to the teaching signal. The synaptic transformation from LC to RC is achieved through the application of a filter with an exponentially decaying kernel [7]. The exponentially decaying kernel is calculated by $e^{\delta/\tau}$, where δ is the time lag between spike time and sample time and τ is the time constant which is set to 30 ms in this paper.

16.3 Results

In this paper, we investigate the influence of temporal integration on the computational performance of liquid state machine (LSM) from two aspects: the synaptic decay constant and time delay from presynaptic neurons of LC to the RC output neurons.

Firstly, we investigated how the synaptic integration of a single readout neuron affects the accuracy of network computation. In other words, the decay time constant of the readout filter in the original LSM model is investigated. The synaptic current

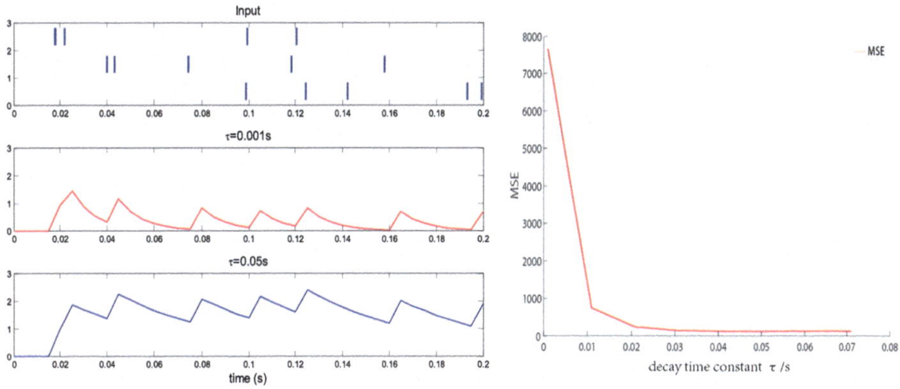

(a) (b)

Fig. 16.2 Effects of the decay time constant of output synapse. (**a**) Real-time synaptic current targeting to the readout neuron with different decay time constant. (**b**) The computational performance of LSM measured by mean square error (MSE) with the increase of decay time constant. Here the target signal for the output curve is r1 + r3

from LC to RC with different time constant is shown in Fig. 16.2a. It demonstrates that individual synaptic signals generated from each presynaptic spiking stimulus can be much more efficiently integrated into smooth currents with larger decay time constant, which indicates LSM contains more memory of previous information. Such enhanced temporal integration is beneficial for improving the computational performance of LSM as shown in Fig. 16.2b. By increasing the decay constant of synapses from SNN to the output neuron, the mean square error (MSE) can be remarkably reduced due to the enhancement of temporal integration.

In the brain cortex, a wide range of axonal transmission delays have been observed and suggested to be beneficial for computational power [8, 9] and learnability [10–12]. Therefore, synaptic delays linking the network to the readout neurons are considered in this paper. Spike trains received from the liquid network are added delays which are generated from a Gaussian distribution with the average value of 0 and standard deviation of 0.02 s. From Fig. 16.3 it can be seen that the complexity of discharge or network activity is greatly increased with the involvement of time delay, which also effectively improve the calculation accuracy of the network. In Fig. 16.4a, we can see that with the increase of delay the computational performance is significantly improved when the delay value reaches to about 0.02 s. The histogram of interspike interval (ISI) of the spike sequence of liquid network with or without output delay is shown in Fig. 16.4b. Without synaptic delay, the high-degree global intrinsic synchronization impairs computational performance. Adding synaptic delay can increase the information entropy, which reduces the likelihood of pathological synchronization. The enhancement of transmitting time-varying signals is beneficial for the training of output weight coefficients through the linear regression and ultimately improves the computational power of neural network.

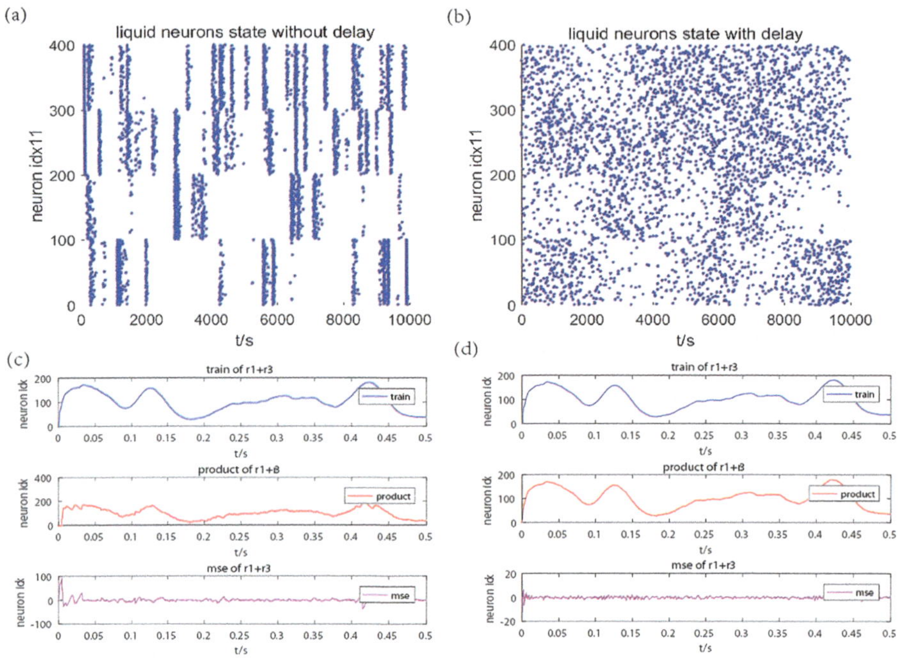

Fig. 16.3 (**a**, **b**) Liquid network discharge before and after the application of delay disturbance in the output synapse. (**c**, **d**) Corresponding computational performance of the linear readouts (red dashed line) which were trained according to the teaching signal $r_1 + r_3$ (blue line)

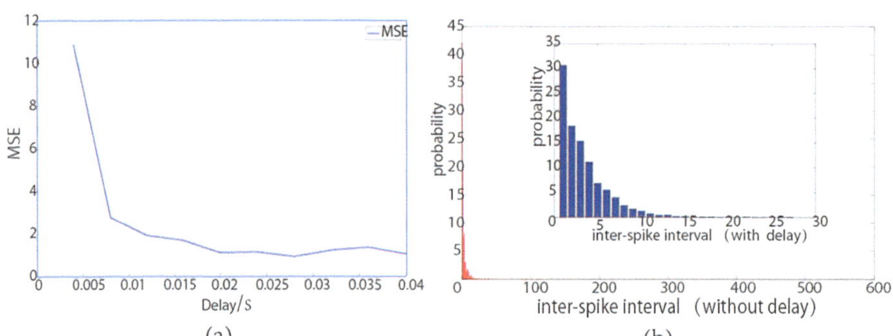

Fig. 16.4 Effects of the delay disturbance of output synapse. (**a**) The computational performance of LSM measured by mean square error (MSE) with the increase of delay constant. Here the target signal for the output curve is r1 + r3. (**b**) The histogram of interspike interval (ISI) of the spike sequence of liquid network with or without output delay

16.4 Conclusion

In this paper, we investigate the influence of temporal integration on the computational performance of liquid state machine from two aspects: the synaptic decay constant and time delay from liquid network to the output neurons. Our experimental results show that increasing the decay constant of synapses from SNN to the output neuron can remarkably improve the computational performance due to the enhancement of temporal integration. Moreover, transmission delays have an even larger impact on the richness of dynamical states, which in turn significantly increase the computational accuracy of SNN. These results may have important implications for the modeling of spiking neural networks with excellent computational performance.

Acknowledgements This paper is supported by the National Natural Science Foundation of China (No. 61304165 and No. 61473051) and Natural Science Foundation of Chongqing (No. cstc2016jcyjA0015).

References

1. Maass, W., Natschläger, T., et al.: Real-time computing without stable states: a new framework for neural computation based on perturbations. Neural Comput. **14**(11), 2531–2560 (2002)
2. Verstraeten, D., Schrauwen, B., Stroobandt, D., et al.: Isolated word recognition with the liquid state machine: a case study. Inf. Process. Lett. **95**(6), 521–528 (2005)
3. Schrauwen, B., D'Haene, M., Verstraeten, D., et al.: Compact hardware liquid state machines on FPGA for real-time speech recognition. Neural Netw. Off. J. Int. Neural Netw. Soc. **21**(2–3), 511–523 (2008)
4. Werner, M., Genta, V.M., Williams, D.: Movement generation and control with generic neural microcircuits. In: Biologically Inspired Approaches to Advanced Information Technology, First International Workshop, BioADIT 2004, Lausanne, 29–30 Jan 2004. Revised Selected Papers, pp. 258–273. DBLP (2004)
5. Maass, W., Natschläger, T., Markram, H.: A model for Real-Time computation in generic neural Microcircuits[C]// DBLP. pp. 213–220 (2002)
6. Izhikevich, E.M.: Simple model of spiking neurons. IEEE Trans. Neural Netw. **14**(6), 1569 (2003)
7. Maass, W., Joshi, P., Sontag, E.D.: Computational aspects of feedback in neural circuits. PLoS Comput. Biol. **3**(1), e165 (2007)
8. Maass, W.: Networks of spiking neurons: the third generation of neural network models. Neural Netw. **10**(9), 1659–1671 (1997)
9. Schmitt, M.: On Computing Boolean Functions by a Spiking Neuron. Kluwer Academic, Boston (1998)
10. Maass, W., Schmitt, M.: On the complexity of learning for a spiking neuron. In: Proceedings of the Tenth Annual Conference on Computational Learning Theory, pp. 54–61. ACM (1997)
11. Maass, W., Schmitt, M.: On the complexity of learning for spiking neurons with temporal coding[M]. Academic Press, Inc (1999)
12. Maass, W.: On the relevance of time in neural computation and learning. Theor. Comput. Sci. **261**(1), 157–178 (2001)

Chapter 17
Anticipatory Top-Down Interactive Neural Dynamics

Steven L. Bressler

Abstract Engagement in a cognitive task typically involves configuration of the mental resources needed to perform the task, and switching from one task to another involves the reconfiguration of those resources. It is believed that this configuration and reconfiguration require the prefrontal cortex of the brain to facilitate the activation of distributed brain areas that will be involved in task execution and defacilitate others. Although facilitation and defacilitation may occur during task execution, I consider here that they may also be initiated in advance of, and as preparation for, task execution. This point of view comes from a large-scale, distributed neurocognitive network understanding of brain function that emphasizes processes of task preparation and expectation (or task set) in the brain in addition to processes of perception and action. In short, it requires that a distinction be made between task configuration processes and task execution processes. This paper presents and discusses some convergent lines of evidence suggesting that the prefrontal cortex exerts top-down configuration of sensory and motor brain areas to construct task set, that this configuration depends on top-down brain processes in the beta-frequency range, and that resource configuration may occur when the brain is at rest prior to task execution.

Keywords Prefrontal cortex · Distributed neurocognitive network · Large-scale brain circuits · Top-down configuration · Cognitive dynamics

---

S. L. Bressler (✉)
Center for Complex Systems and Brain Sciences and Department of Psychology, Florida Atlantic University, Boca Raton, FL, USA
e-mail: bressler@fau.edu

J. M. Delgado-García et al. (eds.), *Advances in Cognitive Neurodynamics (VI)*,
Advances in Cognitive Neurodynamics,
https://doi.org/10.1007/978-981-10-8854-4_17

17.1 Introduction

Cognitive resources are allocated as needed to perform a cognitive task, and each task that is performed utilizes a unique set of resources [1, 2]. Switching from one task to another requires the allocation of a different set of resources as appropriate for the new task. This paper is based on the proposal that the prefrontal cortex (pFC) is the primary brain region supporting resource allocation [3]. In top-down control by the pFC, different areas within the pFC, which occupies a vast expanse in the frontal lobes, have been proposed to be involved in determining which resources are needed to perform a particular task [4–6] and then orchestrating the participation in task processing of posterior neocortical regions (in frontal, parietal, temporal, and occipital lobes) when those areas are needed to perform a task [7, 8].

The topic of top-down control by the pFC fits naturally within the theory of neural coordination dynamics [9–11], because the pFC is viewed as coordinating the participation in task processing of posterior neocortical areas [12]. That role has been proposed to be fundamentally domain-nonspecific: rather than representing a specific domain of knowledge, pFC activity is thought to coordinate the activities of other, domain-specific, regions.

A number of experimental and analytic techniques are available in cognitive neuroscience to assess the influence exerted by the pFC on other areas. One prominent method is to stimulate the pFC and record the sensory effects, either by electrophysiological recording in sensory neocortex, by fMRI recording, or by psychophysics. The pFC stimulus has been delivered by microstimulation in monkeys [13, 14] and transcranially in humans [15–19]. Another technique is to apply time series analysis to simultaneously recorded neural processes. Time series analytic techniques have been developed that can reveal statistical causal influences [20, 21] between time series. In addition to Wiener-Granger causality, these techniques include the directed transfer function [22], the partial directed coherence [23], and the transfer entropy [24]. A spectral Wiener-Granger causality has also been created [25, 26]. These techniques have been used to measure causal influences from the pFC to other neocortical areas [12, 27–29], and, although the problem of determining the pFC's influence on other brain areas inherently requires the measurement of top-down interareal interactions, the experimental and analytic techniques themselves may also be applied within, as well as between, individual brain areas and to bottom-up and lateral, as well as top-down, interactions.

One cognitive function where the allocation of cognitive resources by the pFC has been observed is visuospatial attention. When areas of pFC and visual occipital cortex (VOC) are simultaneously recorded during performance of a visuospatial attention task, the pFC is seen to interact with the VOC [12, 27, 28, 30]. In this interaction, top-down directed influences (from pFC to VOC) are stronger than bottom-up directed influences (in the other direction). Top-down influences not only are found to affect visuomotor processing following visual target presentation but also prior to it, following an auditory cue and before visual target presentation, when visuomotor processing has not yet occurred, but is expected.

17.2 Methods

This report summarizes a body of work employing the Wiener-Granger causality time series analysis method. This method follows the approach of Wiener (1956) [20], in being aimed at statistical time series predictability, and Granger (1969) [21], in determining this predictability by autoregressive (AR) modeling. When applied to monkey local field potential (LFP) data, spectral AR methods [25] were employed, and the model order was high (over 10). The high model order and spectral analysis allowed Wiener-Granger causality to measure the interactive dynamics of oscillatory neural processes in monkey LFPs [29, 31]. When applied to human fMRI BOLD data, time-domain Wiener-Granger causality was employed, and the model order was one. Statistical causal influence in the time domain was thus equivalent to the predictability of the random variable of one voxel's time series from that of a second voxel's time series one time step earlier. The relation between the fast interactive dynamics of oscillatory neural processes and the slow interactive dynamics of the fMRI BOLD signal has yet to be determined.

Whether applied in the spectral (frequency) or time domain, Wiener-Granger causality from one random variable to another measures the predictability between them. When the random variables are physiological processes at different brain sites (either LFP electrodes or fMRI BOLD voxels), Wiener-Granger causality measures the neural influence exerted by the neurons at one brain site on those at the other. Since one random variable is always selected as the sender, and the other as the receiver, Wiener-Granger causality can be measured in both directions between two sites, thus matching the bidirectional axonal pathways known to exist between neurons in most connected areas. Of course, the measurement of Wiener-Granger causality from one electrode or voxel to another does not mean that a direct, monosynaptic projection exists, since the measurement may also reflect indirect multisynaptic pathways.

When the Wiener-Granger causality value measured between brain sites during a cognitive task is significantly above chance levels, the statistically significant connection between those two sites is considered to represent a functional edge in a large-scale neurocognitive network [32]. Brain networks are biophysical brain systems that can be represented by mathematical entities, called graphs, that consist of collections of nodes (or vertices) and edges. A common technique for computing the statistical significance of functional connections is by resampling [33]. In this approach, a surrogate analysis is performed by randomly permuting the channel labels on each of the many iterations and computing the connectivity as in the veridical analysis for each iteration. The result is a permutation distribution (over iterations), which is used to determine the statistical significance of the veridical results. For example, the 95% confidence limit for veridical results is the value of the permutation distribution that just exceeds 95% of the values in that distribution. This resampling approach is a powerful method for determining the statistical significance of an interdependency test statistic because it is derived directly from the same data set as are the veridical results and does not require an analytic expression for the test statistic's distribution.

17.3 Results

To uncover brain networks in the human brain, directed edges were measured by directed functional connectivity analysis of fMRI BOLD data. In the first experiment, directed edges were computed from human fMRI BOLD recordings from the dorsal attention network (DAN), including the frontal eye field (FEF), a part of the pFC, and multiple areas of the VOC, taken as subjects performed a visuospatial attention task [30]. It was found that the DAN➔VOC (top-down) directed influence is consistently greater than the VOC➔DAN (bottom-up) directed influence during the task [27]. The same asymmetry is seen in subjects with eyes open at rest [12]. Task engagement enhances DAN➔VOC top-down influences, but not bottom-up influences, and bidirectional within-DAN influences (between FEF and the inferior parietal sulcus – iPS – another DAN region) [12]. Furthermore, top-down influences are greater than bottom-up following the auditory cue stimulus and before presentation of the visual target stimulus, as well as after visual target presentation [12]. These directed influences were confirmed by also measuring undirected functional connectivity. Overall, these results suggest that top-down influences from the DAN not only coordinate posterior visual neocortical areas for task processing but do so also in anticipation of task processing.

In a second fMRI analysis, this time of human visuomotor coordination behavior, bidirectional influences were measured between fMRI BOLD time series recorded during behavioral coordination from the dorsal anterior cingulate cortex (dACC), in the medial prefrontal neocortex, and in the supplementary motor area (SMA) and between those from the dACC and the primary motor cortex (M1) [34, 35]. An asymmetric functional relation was found for the dACC-SMA pair, but not for the dACC-M1 pair, with the dACC➔SMA directed influence being consistently greater than the SMA➔dACC influence. In a follow-up study [36], visuomotor coordination task blocks, alternating with rest blocks, were contrasted with working memory control task blocks, which also alternated with rest blocks in separate scans. It was found that the previously known asymmetric relation between the dACC and SMA was significantly larger in the visuomotor coordination task than the working memory control task indicating that this asymmetric relation was specific to motor control. Moreover the asymmetry was reversed during the rest condition of the motor coordination task, but not that of the working memory task. Thus, the SMA➔dACC influence was significantly greater in the motor rest condition than in the working memory control rest condition. These results suggest that top-down influences from prefrontal neocortical areas (the dACC in this case) configure sensory and motor areas (the SMA in this case) in relation to task set (instructions to perform the task) and that preparatory processes work to accentuate top-down resource configuration when the brain is at rest in advance of task execution.

In functional connectivity studies of macaque monkeys performing a visual pattern discrimination task, with a GO/NO-GO motor response, distributed LFPs were simultaneously recorded from sensorimotor and visual neocortical regions. Crucially, each trial of this task was self-initiated by the monkey pressing and holding a hand lever, and LFPs were recorded in a prestimulus window when the lever was held down and the monkey anticipated the subsequent appearance of a visual stimulus on the display screen. Spectral directed functional connectivity analysis of prestimulus LFPs from the sensorimotor neocortex (with electrodes in primary and secondary motor and somatosensory areas) revealed interareal-coupled beta-frequency LFP oscillations [37]. Their pattern of directed influence was consistent with a sensorimotor loop that includes motor outflow from the primary motor cortex, ultimately exciting the hand and arm muscles, and proprioceptive feedback inflow from hand and arm muscles and joints to the somatosensory cortex [38]. The pattern of directed influence from LFPs also reflected the visual anticipatory aspect of task performance. Spectral directed functional connectivity analysis applied to LFPs from the visual neocortex (with electrodes in area V1 and extrastriate areas V4 and TEO) showed that top-down (extrastriate➔V1), but not bottom-up (V1➔extrastriate), directed influences in the beta-frequency range occurred significantly above chance [29]. Most importantly, linear support vector machine pattern classification of the pattern of top-down beta-frequency directed influence within the visual cortex significantly discriminated two task rules that determined the correct motor response to a given visual stimulus [39]. The results thus not only indicate that top-down cortico-cortical functional influences are carried by beta-frequency oscillations but that these influences convey behavioral context to primary visual cortex (V1) in advance of task processing.

Finally, spectral directed functional connectivity analysis was also applied to LFPs recorded from the pFC and posterior parietal cortex (pPC) of macaque monkeys during the delay period of a delayed match-to-sample working memory task with an oculomotor response [40]. Single-cell spike trains recorded from the same electrodes were used to exclude white-matter LFPs. Significant beta-frequency cortico-cortical directed influences were observed in both directions between pFC and pPC, again suggesting that cortico-cortical functional influences are carried by beta-frequency oscillations, here in anticipation of match stimulus and oculomotor response processing, i.e., task execution.

17.4 Discussion

Experimental evidence has been presented showing that the prefrontal cortex is involved in the anticipation as well as the execution of task processing. In a set of human fMRI experiments, the FEF, a part of the pFC that is coupled with other areas of the dorsal attention network (DAN), was observed to exert top-down control of visual occipital cortex (VOC) following the auditory cue stimulus and preceding the visual target stimulus in a visuospatial attention task, indicating that the pFC coordinates posterior visual neocortical areas in anticipation of task processing. Since the asymmetry between top-down (DAN➔VOC) and bottom-up (VOC➔DAN) directed influences was observed both following and before visual target stimulus presentation, the top-down control of VOC in those experiments may have been a tonic effect related to task set. In a second set of human fMRI experiments, the dorsal anterior cingulate cortex (dACC), a part of the medial pFC that is structurally connected with motor system areas [41, 42], was observed to exert top-down control of the supplementary motor area (SMA) in a visuomotor coordination task, suggesting that the pFC coordinates motor control areas in anticipation of task execution. In fact, one of these studies [36] revealed asymmetric directed functional connectivity between the dACC and SMA, but with bottom-up (SMA➔dACC) greater than top-down (dACC➔SMA) connectivity, during a block-interleaved rest condition. This result suggests that top-down resource configuration by the pFC may be accentuated by sensory and motor areas in advance of task processing.

Macaque monkey experiments allow the recording of local field potentials (LFPs) from neuronal populations, and the analysis of spectral directed functional connectivity, in animals performing cognitive tasks. In one such study of macaque monkeys performing a visuomotor pattern discrimination task [37], beta-frequency directed functional connectivity was observed among sensorimotor neocortical areas before visual stimulus presentation, in conformance with the maintenance of sustained motor output by the monkey. In the visual cortex [29], top-down (extrastriate➔V1), directed functional connectivity was observed in the beta-frequency band but was absent at all frequencies in the bottom-up (V1➔extrastriate) direction. This study did not find cortico-cortical directed functional connectivity from the pFC to VOC. However, Richter et al. (2016) [39] recently reported that the pattern of top-down beta-frequency directed influence within the visual cortex before the visual stimulus significantly discriminates task rules determining the correct motor response to the stimulus. This result implies that top-down influences convey behavioral context to the visual cortex in anticipation of task execution. Presumably, these top-down influences are conveyed from a network of association cortical areas that includes pFC areas [31]. In a second experiment, LFPs were recorded from the pFC and pPC during the delay period of a delayed match-to-sample task performed by macaque monkeys. Cortico-cortical beta-frequency directed influences were observed in both directions between pFC and pPC, implying that the pFC and pPC interact in the beta-frequency range in anticipation of task execution (match stimulus and oculomotor response processing).

Mental resources need to be configured in order to perform a cognitive task, and substantial evidence exists that task resource configuration is performed by the pFC [3]. The present paper presents experimental evidence in support of this claim, implying that the pFC performs this function using top-down influences in the beta-frequency range [31]. Also evident are the findings that the configuration of mental resources may occur in anticipation of, as well as during, task execution and that resource configuration may occur at rest prior to task processing.

References

1. Allport, A., Styles, E.A., Hsieh, S.: Shifting intentional set: exploring the dynamic control of tasks. In: Umilta, C., Moscovitch, M. (eds.) Attention and Performance XV, pp. 421–452. MIT Press, Cambridge, MA (1994)
2. Monsell, S.: Task switching. Trends Cogn. Sci. **7**, 134–140 (2003)
3. Sakai, K.: Task set and prefrontal cortex. Ann. Rev. Neurosci. **31**, 219–245 (2008)
4. Allport, A., Wylie, G.: Task switching: positive and negative priming of task-set. In: Humphreys, G.W., Duncan, J., Treisman, A. (eds.) Attention, Space, and Action: Studies in Cognitive Neuroscience, pp. 273–296. Oxford University Press, Oxford (1999)
5. Koechlin, E., Ody, C., Kouneiher, F.: The architecture of cognitive control in the human prefrontal cortex. Science. **302**, 1181–1185 (2003)
6. Ridderinkhof, K.R., Ullsperger, M., Crone, E.A., Nieuwenhuis, S.: The role of the medial frontal cortex in cognitive control. Science. **306**, 443–447 (2004)
7. Fuster, J.M.: Cortex and Mind: Unifying Cognition. Oxford University Press, New York (2003)
8. Fuster, J.M., Bressler, S.L.: Cognit activation: a mechanism enabling temporal integration in working memory. Trends Cogn. Sci. **16**, 207–218 (2012)
9. Bressler, S.L., Kelso, J.A.: Cortical coordination dynamics and cognition. Trends Cogn. Sci. **5**, 26–36 (2001)
10. Bressler, S.L., Kelso, J.A.: Coordination dynamics in cognitive neuroscience. Front. Neurosci. **10**, 397 (2016)
11. Miconi, T.: Biologically plausible learning in recurrent neural networks reproduces neural dynamics observed during cognitive tasks. elife. **6**, e20899 (2017)
12. Meehan, T.P., Bressler, S.L., Tang, W., et al.: Top-down cortical interactions in visuospatial attention. Brain Struct. Funct. (2017). https://doi.org/10.1007/s00429-017-1390-6
13. Moore, T., Armstrong, K.M.: Selective gating of visual signals by microstimulation of frontal cortex. Nature. **421**, 370–373 (2003)
14. Ebitz, R.B., Moore, T.: Selective modulation of the pupil light reflex by prefrontal cortex microstimulation. J. Neurosci. **37**, 5008–5018 (2017)
15. Ruff, C.C., Blankenburg, F., Bjoertomt, O., et al.: Concurrent TMS-fMRI and psychophysics reveal frontal influences on human retinotopic visual cortex. Curr. Biol. **16**, 1479–1488 (2006)
16. Ruff, C.C., Bestmann, S., Blankenburg, F., et al.: Distinct causal influences of parietal versus frontal areas on human visual cortex: evidence from concurrent TMS-fMRI. Cereb. Cortex. **18**, 817–827 (2008)
17. Ruff, C.C., Blankenburg, F., Bjoertomt, O., et al.: Hemispheric differences in frontal and parietal influences on human occipital cortex: direct confirmation with concurrent TMS-fMRI. J. Cogn. Neurosci. **21**, 1146–1161 (2009)
18. Bachinger, M., Zerbi, V., Moisa, M., et al.: Concurrent tACS-fMRI reveals causal influence of power synchronized neural activity on resting state fMRI connectivity. J. Neurosci. **37**, 4766–4777 (2017)
19. Bogdanov, M., Ruff, C., Schwabe, L.: Transcranial stimulation over the dorsolateral prefrontal cortex increases the impact of past expenses on decision-making. Cereb. Cortex. **27**, 1094–1102 (2017)
20. Wiener, N.: The theory of prediction. In: Beckenbach, E.F. (ed.) Modern Mathematics for the Engineer, pp. 165–190. McGraw-Hill, New York (1956)
21. Granger, C.W.J.: Investigating causal relations by econometric models and cross-spectral methods. Econometrica. **37**, 424–438 (1969)
22. Blinowska, K.: Review of the methods of determination of directed connectivity from multichannel data. Med. Biol. Eng. Comput. **49**, 521–529 (2011)
23. Baccala, L.A., Sameshima, K.: Overcoming the limitations of correlation analysis for many simultaneously processed neural structures. Prog. Brain Res. **130**, 33–47 (2001)
24. Barnett, L., Bossomaier, T.: Transfer entropy as a log-likelihood ratio. Phys. Rev. Lett. **109**, 138105 (2012)

25. Geweke, J.F.: Measures of conditional linear dependence and feedback between time series. J. Am. Stat. Assoc. **79**, 907–915 (1984)
26. Chicharro, D.: On the spectral formulation of Granger causality. Biol. Cybern. **105**, 331–347 (2011)
27. Bressler, S.L., Tang, W., Sylvester, C.M., et al.: Top-down control of human visual cortex by frontal and parietal cortex in anticipatory visual spatial attention. J. Neurosci. **28**, 10056–10061 (2008)
28. Tang, W., Bressler, S.L., Sylvester, C.M., et al.: Measuring Granger causality between cortical regions from voxelwise fMRI BOLD signals with LASSO. PLoS Comput. Biol. **8**, e1002513 (2012)
29. Bressler, S.L., Richter, C.G.: Interareal oscillatory synchronization in top-down neocortical processing. Curr. Opin. Neurobiol. **31**, 62–66 (2015)
30. Corbetta, M., Shulman, G.L.: Control of goal-directed and stimulus-driven attention in the brain. Nat. Rev. Neurosci. **3**, 201–215 (2002)
31. Bressler, S.L., Richter, C.G., Chen, Y., Ding, M.: Cortical functional network organization from autoregressive modeling of local field potential oscillations. Stat. Med. **26**, 3875–3885 (2007)
32. Bressler, S.L., Menon, V.: Large-scale brain networks in cognition: Emerging methods and principles. Trends Cogn. Sci. **14**, 277–290 (2010)
33. Efron, B.: The Jackknife, the Bootstrap and Other Resampling Plans. Society for Industrial & Applied Mathematics U.S, New York (1982)
34. Asemi, A., Ramaseshan, K., Burgess, A., et al.: Dorsal anterior cingulate cortex modulates supplementary motor area in coordinated unimanual motor behavior. Front. Hum. Neurosci. **9**, 309 (2015)
35. Silverstein, B.H., Bressler, S.L., Diwadkar, V.A.: Inferring the dysconnection syndrome in schizophrenia: interpretational considerations on methods for the network analyses of fMRI data. Front. Psychiat. **7**, 132 (2016)
36. Diwadkar, V.A., Asemi, A., Burgess, A., Chowdury, A., Bressler, S.L.: Potentiation of motor sub-networks for motor control but not working memory: interaction of dACC and SMA revealed by resting-state directed functional connectivity. PLoS One. **12**, e0172531 (2017)
37. Brovelli, A., Ding, M., Ledberg, A., et al.: Beta oscillations in a large-scale sensorimotor cortical network: directional influences revealed by Granger causality. Proc. Natl. Acad. Sci U. S. A. **101**, 9849–9854 (2004)
38. MacKay, W.A.: Synchronized neuronal oscillations and their role in motor processes. Trends Cogn. Sci. **1**, 176–183 (1997)
39. Richer, C.G., Coppola, R., Bressler, S.L.: Top-down beta oscillatory signaling conveys behavioral context to primary visual cortex. bioRxiv. 074609 (2016). https://doi.org/10.1101/074609
40. Salazar, R.F., Dotson, N.M., Bressler, S.L., Gray, C.M.: Content-specific fronto-parietal synchronization during visual working memory. Science. **338**, 1097–1100 (2012)
41. Picard, N., Strick, P.L.: Motor areas of the medial wall: a review of their location and functional activation. Cereb. Cortex. **6**, 342–353 (1996)
42. Dum, R.P., Strick, P.L.: Motor areas in the frontal lobe of the primate. Physiol. Behav. **77**, 677–682 (2002)

Chapter 18
Coherence-Based Coding in Spiking Neural Network with Global Inhibitory Feedback

Jinli Xie, Qinjun Zhao, and Jianyu Zhao

Abstract It is widely assumed that the amplitudes of neural response depend on presynaptic history, and signaling between neurons can be regulated through synaptic connections. We focus on the coherence-based coding properties in this study and investigate how global inhibitory feedback shapes information transmission in spiking neural networks. Numerical simulations and computed input-output transfer functions are used to determine the coding properties and show the changes in network response amplitude resulting from feedback interaction with the related effects of inhibitory synapses. The coherence is decreased with increasing frequency, suggesting that the inhibitory feedback is capable of low pass filter characteristics. More importantly, however, the overall coherence drops with enhancing feedback gain with respect to the feedback modulation of the network. Our results further indicate that the transmission time of inhibitory synapses also plays a key role in modulating the coherence. The monotonic coherence improves as the transmission time increases. Thus, inhibitory feedback that controls the firing state of postsynaptic neurons can be significant in altering the network coding property.

Keywords Inhibitory feedback · Firing rate · Coherence · Information coding

18.1 Introduction

The spiking activity of neuronal populations shows highly complex temporal dynamics. It is well known from electrosensory system studies that correlated activities between action potentials play a substantial role in solving crucial computational problems [1–3]. Remarkably, the realization of neural coding depends on the features of neuronal firing or the correlated spiking activity [4–6]. Similarly,

J. Xie (✉) · Q. Zhao · J. Zhao
School of Electrical Engineering, University of Jinan, Jinan, China
e-mail: cse_xiejl@ujn.edu.cn

J. M. Delgado-García et al. (eds.), *Advances in Cognitive Neurodynamics (VI)*, Advances in Cognitive Neurodynamics,
https://doi.org/10.1007/978-981-10-8854-4_18

the fluctuations in membrane potential, reflecting the postsynaptic spike trains, can be controlled by modulatory synaptic response amplitude, and it suggests that the changes in synaptic gain and temporal dynamics have distinct roles in information transmission.

Recent studies revealed the functional contributions of neuronal and synaptic interactions to the correlated firing activities by theoretical and experimental techniques [7–10]. What is important, however, is that shifts in connection strength result in changes in correlations [11, 12]. Considering the significant influence of network connectivity on neural correlations, the inhibitory synapse is integrated into the feedback loop of a spiking neural network. Numerical simulations based on this model may lead insight into the potential roles of inhibitory feedback in coding and processing information.

Our computational model, enlightening by the available architecture of electrosensory system [13, 14], consists of two populations of neurons employing global inhibitory feedback. The strategy here is to construct the projection to pyramidal cells from P-unit cells. We first characterized the neuronal firing activity in the context of low-pass Gaussian white noise and then quantified the ability of information transmission by frequency-dependent coherence. The effects of inhibitory synapse on coding efficacy were further examined in detail. The transmission of information could be dependent of the intensity of these modulations of inhibitory feedback. Our studies thus reveal more general properties of information coding with inhibitory feedback in sensory networks.

18.2 Model and Methods

We consider a network consisting of two interacting layers of M excitatory neurons and N inhibitory neurons with all-to-all connectivity and without local connections among themselves [11, 12, 15]. The neurons are described by leaky integrate-and-fire (LIF) model, the membrane potential of which submits to a spike-and-reset rule: the neuron fires immediately once the membrane potential labeled V crosses the spike threshold V_T, after that resets to the reset value V_R, and keeps for an absolute refractory period τ_R. The spike train of each LIF neuron follows:

$$y(t) = \sum_f \delta\left(t - t_f\right) \tag{18.1}$$

Here t_f represents the instants of threshold crossing. Between every two spikes, the dynamics are governed by the equation:

$$\tau_m \frac{dV}{dt} = V_R - V + R_m I(t) \tag{18.2}$$

where τ_m is the membrane time constant and R_m is the membrane resistance. $I(t)$ represents the input current to each neuron, which consists of the following components for excitatory neurons:

$$I_i(t) = \mu + \xi_i(t) + S_i(t)\text{-}g \int_{\tau_D}^{\infty} d\tau \frac{(\tau - \tau_D)\, e^{-(\tau-\tau_D)/\tau_I}}{\tau_I^2} \sum_{i=1}^{N} y_i\,(t - \tau) \quad (18.3)$$

The bias current μ and the internal Gaussian white noise $\xi_i(t)$ with intensity D describe the dynamics of the neuron itself. $S_i(t)$ represents the external stimuli, which is low-pass filtered Gaussian white noise (0–120 Hz) with zero mean and variance W. The last term in Eq. (18.3), named postsynaptic conductance, defines the inhibitory feedback selecting g as the gain projecting to all the excitatory neurons after a time delay τ_D, where a delayed α function convolutes the sum of the spike trains of the inhibitory neurons. This term is subtracted, indicating an inhibitory feedback. The decay time τ_I here regards to the inhibitory synaptic transmission time.

Besides the autonomous stochastic dynamics, the inhibitory neurons also accept input from the excitatory neurons. According to Eq. (18.3), we obtain the input current I_{EI} to the inhibitory neurons as follows:

$$I_{EI} = \int_{0}^{\infty} d\tau \frac{\tau e^{-\tau/\tau_E}}{\tau_E^2} \sum_{i=1}^{M} y_i\,(t - \tau) \quad (18.4)$$

where τ_E is the synaptic transmission time of the feedforward pathway.

The information coding ability of the network can be quantified by the coherence [7, 8]. The frequency-dependent coherence $C_{S,\,y}$ between external stimuli S and output spike train y is measured by:

$$C_{S,y}(f) = \frac{\left|P_{S,y}(f)\right|^2}{P_S(f)P_y(f)} \quad (18.5)$$

where $P_{S,\,y}(f)$ is related to the cross power spectrum between S and y. $P_S(f)$ and $P_y(f)$ are, respectively, the power spectrum of the external stimuli and the output spike train.

The model parameters are $M = 80$, $N = 20$, $V_T = 1$, $V_R = 0$, $R_m = 1$, $W = 0.5$ $\tau_R = 0.5\,\text{ms}$, $\tau_m = 1\,\text{ms}$, $\tau_D = 1\,\text{ms}$, $\tau_E = 0.5\,\text{ms}$, and $\mu = 0.8$, $D = 0.08$. Euler-Maruyama integration method is used to Eq. (18.1) with a time step of 0.05 ms.

18.3 Results

The mean firing rate of the output LIF neurons is first presented as a function of the feedback gain and the transmission time of inhibitory synaptic. Remarkably, the variations across both variables are monotonic in Fig. 18.1. While increasing feedback gain, the mean firing rate decreases obviously. Further, the mean firing rate $\overline{\lambda}$ slightly increases with transmission time at shorter τ_I. However, there is little variation of $\overline{\lambda}$ across transmission time at larger τ_I values.

These changes are expected since variations of the feedback gain and the synaptic time constant change the mean synaptic conductance, and consequently the overall output firing activities of the network. We then quantify the information coding behaviors by analyzing the frequency-dependent coherence function. The coherence measures the linearity of the transmission of the stimulus at each frequency, which ranges between 0 (nonlinearity) and 1 (linearity).

The coherence of the network is compared as a function of the feedback gain and the synaptic transmission time constant in Fig. 18.2. For both cases, monotonically decreasing coherence function is observed. The coherence $C_{S,y}$ decreases with increasing frequency, which indicates low-pass filtering, owing to the similar low-pass cross-spectrum. The postsynaptic conductance due to the total synaptic inputs from the inhibitory neurons is related to the gain and the time constant of the feedback pathway; the resulting variabilities of which in these two cases increase the nonlinearity of the network, leading a drop in coherence in the high-frequency

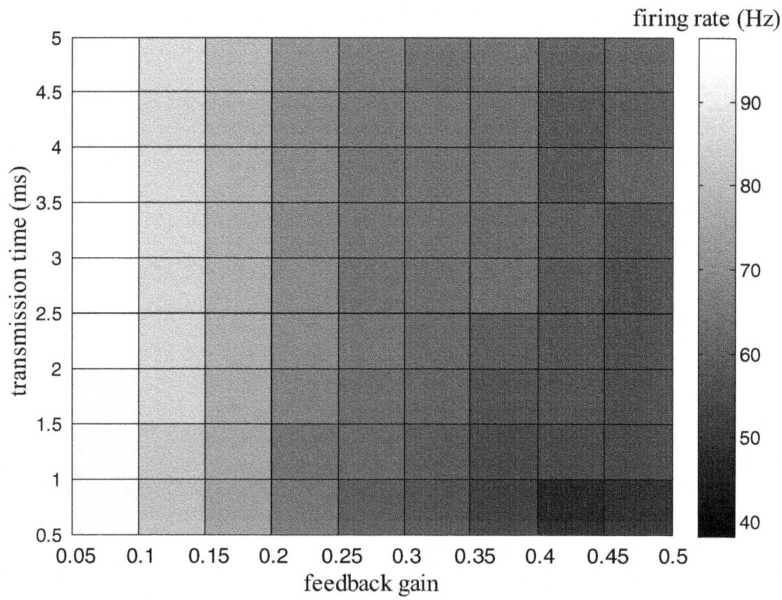

Fig. 18.1 Mean firing rate of the network vs. feedback gain and transmission time

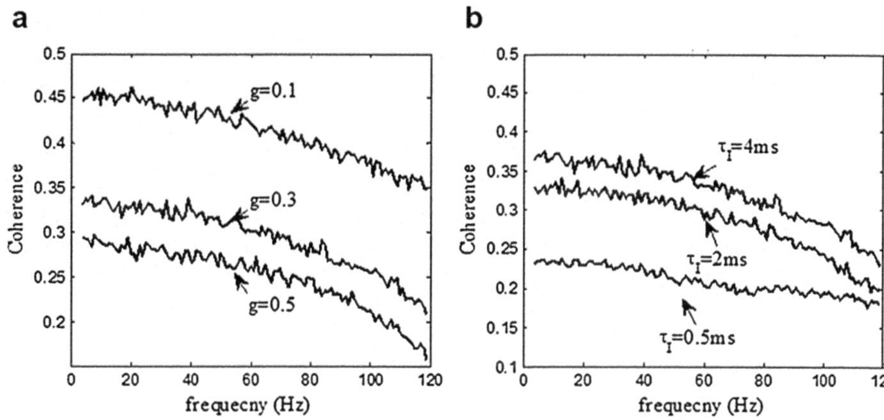

Fig. 18.2 Coherence of the network. (**a**) Coherence vs. frequency for different values of transmission time. (**b**) Coherence vs. frequency for different values of feedback gain

range. Our simulation results thus suggest that the inhibitory feedback is capable of low-pass filter characteristics. Moreover, in Fig. 18.2a, growing feedback gain g results in low coherence in overall frequency ranges. We also find that the monotonic coherence function reaches higher values with longer transmission time τ_I, as shown in Fig. 18.2b. Trends are similar for mean firing rate and spectral coherence with varying feedback gain and transmission time. The latter reflecting the linearity of the information transmission seems to be closely related to the amplitude of network response. Remarkably, the variation in coherence with respect to varying feedback gain turns to be greater than that with shifted transmission time, conforming to the results obtained in Fig. 18.1. This spectrum of coding behaviors thus gives us clues to the coding characteristics of inhibitory feedback, combined with consideration of the network response strength.

18.4 Conclusions

Simultaneous recording from neurons in the nervous system has revealed information coding within the activity of a neuronal ensemble, the firing patterns of which involve abundant structures. Thus the temporal character of the spike trains is an important information carrier. Yet many studies have focused on measuring correlated firings, particularly on the portion of correlations that can be extracted with simple realizable algorithms. Evidence from numerical simulations based on multi-neuron firing activities reveals that overall network correlations can be modulated by the inhibitory feedback pathway. Therefore, it is crucial for studying the neural information coding in network involving inhibitory feedback.

The neuronal firing dynamics modeled in this paper aimed to quantify the contribution of inhibitory feedback to the coding capabilities. We used conductance power spectra to identify the information transmission between the stimuli and the output spike trains. We found that the power spectra coherence was relatively suppressed in the high-frequency range. The feedback conductance dynamics was proved to act as a nonlinear filter on the total synaptic input. Giving consideration to the depression in combination with the conductance dynamics, the reduction of power in high-frequency range could be explained. However, static synapse is generally thought to be no frequency-dependent filter. Indeed, our results showed that the filtering effects on information could be modulated by network feedback and synaptic dynamics via their influence on nonlinearity of postsynaptic firing activities. The reliability of such a low-pass coherence coding property need further extends to other dynamics.

The properties of coding were then examined by modulating the gain and the transmission time of inhibitory feedback. Specifically, the integral level of coherence was decreased with increasing feedback gain, but improved with larger transmission time. These findings could be traced to the effects these parameters had on the mean firing rate. Consequently, all the coding results in this paper suggested that the inhibitory feedback might modulate the coding strategies of the network.

Acknowledgments This work is supported by the National Natural Science Foundation of China (No. 61203375).

References

1. Wallace, E., Benayoun, M., van-Drongelen, W., Cowan, J.D.: Emergent oscillations in networks of stochastic spiking neurons. PLoS One. **6**, e14804 (2011)
2. Doiron, B., Litwin-Kumar, A., Rosenbaum, R., Ocker, G.K., Josić, K.: The mechanics of state-dependent neural correlations. Nat. Neurosci. **19**, 383–393 (2016)
3. Arandia-Romero, I., Tanabe, S., Drugowitsch, J., Kohn, A., Moreno-Bote, R.: Multiplicative and additive modulation of neuronal tuning with population activity affects encoded information. Neuron. **89**, 1305–1316 (2016)
4. Puzerey, P.A., Galan, R.F.: On how correlations between excitatory and inhibitory synaptic inputs maximize the information rate of neuronal firing. Front. Comput. Neurosci. **8**, 59–59 (2014)
5. Vijay, S., Martin, T., Ilya, N.: Extrinsic and intrinsic correlations in molecular information transmission. Phys. Rev. E. **94**, 022425 (2016)
6. Elliot, M., Hlinka, J., Jorn, D.: Pairwise network information and nonlinear correlations. Phys. Rev. E. **94**, 0403301 (2016)
7. Krisztina, S., Andre, L., Leonard, M.: Effect of synaptic plasticity on sensory coding and steady-state filtering properties in the electric sense. BioSystems. **92**, 16–28 (2008)
8. Benjamin, L., Dorian, G., Andre, L., John, E.L.: Broadband coding with dynamics synapses. J. Neurosci. **29**, 2076–2088 (2009)
9. Gregory, D., Georg, N., Andre, L.: Finite size effect induces stochastic gamma oscillation in inhibitory network with conduction delay. BMC Neurosci. **15**, 115 (2014)
10. Adam, K., Ruben, C.-C., Ingmar, K., Alexandre, P.: Correlations and neuronal population information. Annu. Rev. Neurosci. **39**, 237–256 (2016)

11. Jinli, X., Zhijie, W., Andre, L.: Correlated firing and oscillations in spiking networks with global delayed inhibition. Neurocomputing. **83**, 146–157 (2012)
12. Jinli, X., Zhijie, W.: Effect of inhibitory feedback on correlated firing of spiking neural network. Cogni. Neurodyam. **7**, 325–331 (2013)
13. Maurice, J.C., Joseph, B.: Population coding by electrosensory neurons. J. Neurophy. **99**, 1825–1835 (2008)
14. Babadi, B., Sompolinsky, H.: Sparseness and expansion in sensory representations. Neuron. **83**, 1213–1226 (2014)
15. Daniele, M., Hilbert, J.K., Stan, C.A.M.G.: Input-driven oscillations in networks with excitatory and inhibitory neurons with dynamic synapses. Neural Comput. **19**, 1739–1765 (2007)

Chapter 19
Time-Varying Scalp EEG Network Patterns for Music Tempo Perception

Wei Xu, Yin Tian, Haiyong Zhang, Huiling Zhang, Zhongyan Wang, Li Yang, Shuxing Zheng, Yupan Shi, Xing Zhao, Dechun Zhao, Xiuxing Wang, Yu Pang, and Zhangyong Li

Abstract In the present study, we used the time-varying scalp network analysis method of electroencephalography (EEG) to investigate information flows among different tempos perception in the alpha band. The results showed the network hubs of different tempos existed variously. Only during listening to normal tempo (52 bpm), the strongest out-degree of hubs is distributed in the right hemisphere and the information flow transferred from the left to the right hemisphere. Based on these findings, we proposed that the left hemisphere did not prime the processing until the necessary information had been transferred from the right hemisphere. This study was the first to use time-varying network method based on adaptive directed transfer function (ADTF) to investigate music-related EEG activities and proposed a novel method to reveal the neural mechanisms on music tempo.

Keywords EEG · Time-varying network · Network pattern · Alpha · Music tempo

19.1 Introduction

Music tempo represents a rate of periodic events that individuals perceived to occur at regular intervals and reflects emotional experience [1]. Previous studies found that EEG signal elicited by music tempo was the most representative related to attention in the alpha band ranging from 8 to 12 Hz [2]. Researchers found that music impacted drivers' attention and normal tempo was optimal to focus attention [3]. Alpha power decreased when the tempo deviates from the original

W. Xu · Y. Tian (✉) · H. Zhang · H. Zhang · Z. Wang · L. Yang · S. Zheng · Y. Shi · X. Zhao · D. Zhao · X. Wang · Y. Pang · Z. Li
Bio-information College, Chongqing Key Laboratory of Photoelectronic Information Sensing and Transmitting Technology, Chongqing High School Innovation Team of Architecture and Core Technologies of Smart Medical System, Chongqing University of Posts and Telecommunications, Chongqing, China

J. M. Delgado-García et al. (eds.), *Advances in Cognitive Neurodynamics (VI)*, Advances in Cognitive Neurodynamics,
https://doi.org/10.1007/978-981-10-8854-4_19

tempo [4]. Moreover, our previous study revealed that alpha networks with the original tempo showed better network parameters than that of modulated tempos [5]. Network analysis method is an effective way to assess interactions between brain areas activated by music [6–8]. However, traditional EEG connectivity methods are stationary and invalid to insight dynamic neural activity [9]. In the present study, we further investigated the relationship between music tempo and dynamic brain activity by using the ADTF method [9]. To our best knowledge, few studies have examined the time-varying characteristic about music tempo; the information transfer variation of hub nodes over time could help us to understand the potential mechanism of music tempo.

19.2 Materials and Methods

19.2.1 Participants

Twelve right-handed males (22.4 ± 4.5 years) consented to participate in the EEG study. None reported any formal musical training or had prior exposure to the tested music. Nor had they a history of mental or neurological disorders. Informed consent was obtained prior to the study, and the participants were paid. All experiments were approved by the ethical committee of the University of Electronic Science and Technology of China.

19.2.2 EEG Recording and Data Analysis

Scalp EEG recordings of 12 subjects were performed by serially presenting them with Mozart sonata (piano) of different tempos: normal (52 bpm), fast (138 bpm), moderately fast (78 bpm), and slow (26 bpm). EEG was recorded with a 128-channel EEG system (sampling rate 500 Hz, impedance <40 kΩ), and the data was re-referenced to the infinity reference (IR) using the software REST (www.neuro.uestc.edu.cn/rest) [10, 11]. For each of the three conditions, 48 s EEG of each subject with artifacts removed was used for further analysis. Half of the participants heard the slower music before the faster music, and the other half vice versa (see details in our previous work [5]). Using fast Fourier transform, the alpha activity (8–12 Hz) was generated. Eighteen regions of interest (ROIs), Fp1, Fp2, Fz, F3, F4, F7, F8, C3, C4, T7, T8, Pz, P3, P4, P7, P8, O1, and O2, defined in the 10–20 international electrode system, were used to construct the time-varying network by ADTF, which can be described as:

$$X(f,t) = A^{-1}(f,t)\,\varepsilon(f,t) \tag{19.1}$$

where $A(f,t) = \sum_{m=0}^{k} A_m(t)e^{-j2\pi f \Delta t m}$, A_m is the matrix of time-varying multivariate adaptive autoregressive model coefficients, and $X(f,t)$ and $\varepsilon(f,t)$ are, respectively, the transformations of data vector and white noise in the frequency domain.

Then the cost threshold of 0.1 was applied to every network [12], and one sample t-test was used to obtain the statistical time-varying network with significant threshold $p < 0.05$. The threshold was corrected by Bonferroni correction. Based on time-varying network, the out-degree distribution over time was calculated.

19.3 Results

As shown in Fig. 19.1a, the location of hubs and connectivity pattern changed over time. For 26 bpm, the location of hubs varied from the left temporal region (T7) to the frontal region and occipital region (O1 and O2). For 52 bpm, the hubs were mainly located in the bilateral temporal regions (T7 and T8) and left temporal-parietal region (P7).

For 78 bpm, the right temporal-parietal region (P8) was the primary location of hubs and the variation of hub node from the temporal region (T8) to the posterior parietal area (Pz). For the 138 bpm, the location of hubs changed from the frontal region to the temporal-parietal region (P7, P8). The great out-degrees of P7 and P8 were separately observed for 52 bpm and 78 bpm (Fig. 19.1b).

One-way repeated measures ANOVA with channel were used to test the network properties on different tempos. For 52 and 78 bpm, significant differences of mean out-degree among all time points of different channels were found (52 bpm, $F = 2.821$, $p < 0.05$; 78 bpm, $F = 2.009$, $p < 0.05$). For 26 and 138 bpm, nonsignificant effects were found (all $ps > 0.05$). Post hoc tests were further conducted between two channels and revealed that the greatest values of out-degree at P7 and P8 were separately observed for 52 bpm and 78 bpm, respectively.

Through the above analysis, we found that the P7 and the P8 were important locations for tempo perception. Thus, we further separately measure the out-degree at the P7 and the P8 using one-way repeated measures ANOVA with tempo. The results showed that the out-degree of P7 had a significant effect in different tempos ($F = 6.053$, $p < 0.05$), while nonsignificant effect of the P8 was observed ($F = 0.554$, $p > 0.05$).

19.4 Discussion

Previous studies found that the left hemisphere could play an important role in the processing for music perception [4, 13]. In the present alpha-network study, we found that the out-degree of the original tempo (52 bpm) was stronger in the left

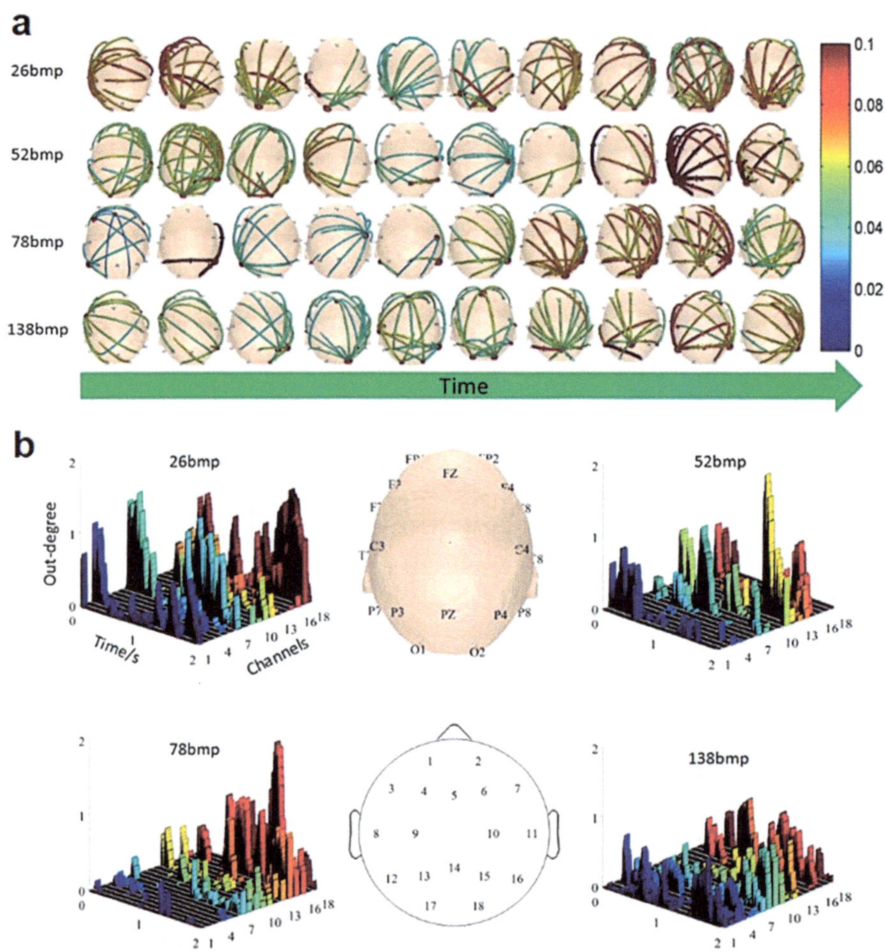

Fig. 19.1 Time-varying scalp EEG network patterns with four music tempos. (**a**) Time-varying networks. (**b**) Dynamic out-degrees

temporal-parietal than that of other tempos, indicating that only the original tempo induced the hub distribution on dynamic networks consistent with previous music perception [4]. We further observed the information flow from the right hemisphere to the left hemisphere during listening to 58 bpm (Fig. 19.1b).

Furthermore, previous studies found that music and language overlapped in the inferior temporal gyrus [14], and harmonious melody would promote the brain processing of lyric meaning [15]. Furthermore, the information transfer from the right hemisphere to the left hemisphere was found in the processing of language [16]. Thus, music and language may show a similar direction of information transfer to promote perception.

Based on these findings, we proposed that the left hemisphere does not prime the processing until the necessary information has been transferred from the right hemisphere during listening of the normal tempo (58 bpm). The frontal and occipital hubs for 26 bpm in the present study indicated that modulated-slow tempo could induce the enhancement of attentional consumption [7, 17]. The variation of hub nodes showed confusion during listening to 138 bpm, illustrating that modulated-fast tempo made individuals hard to focus their attention. The hub node distributed at the right temporal-parietal region during listening to 78 bpm could result from music perception of non-harmonious beat and thus enhanced activation at the right temporal-parietal regions [8]. As shown in Fig. 19.1b, information transfer for 52 bpm was presented from the right temporal-parietal to the left.

In summary, the present study is the first to use time-varying network method to investigate music tempo. Activations elicited by tempos showed different in the dynamic alpha networks. These findings provide a novel tool from the time-varying network level to investigate music tempo.

Acknowledgments This research was supported by the National Natural Science Foundation of China (#61671097); the Chongqing advanced and applied basic research project cstc2015jcyjA10024.

References

1. Knight, W.E., Rickard, N.S.: Relaxing music prevents stress-induced increases in subjective anxiety, systolic blood pressure, and heart rate in healthy males and females. J. Music. Ther. **38**, 254–272 (2001)
2. Schaefer, R.S., Vlek, R.J., Desain, P.: Music perception and imagery in EEG: alpha band effects of task and stimulus. Int. J. Psychophysiol. **82**, 254–259 (2011)
3. Dalton, B.H., Behm, D.G.: Effects of noise and music on human and task performance: a systematic review. Occup. Ergon. **7**, 143–152 (2007)
4. Ma, W., Lai, Y., Yuan, Y., Wu, D., Yao, D.: Electroencephalogram variations in the α-band during tempo-specific perception. Neuroreport. **23**, 125–128 (2012)
5. Tian, Y., Ma, W., Tian, C., Xu, P., Yao, D.: Brain oscillations and electroencephalography scalp networks during tempo perception. Neurosci. Bull. **29**, 731–736 (2013)
6. Tillmann, B., Janata, P., Bharucha, J.J.: Activation of the inferior frontal cortex in musical priming. Cogn. Brain Res. **999**, 145–161 (2003)
7. Simpson, G.V., Weber, D.L., Dale, C.L., Pantazis, D., Bressler, S.L., Leahy, R.M., Luks, T.L.: Dynamic activation of frontal, parietal, and sensory regions underlying anticipatory visual spatial attention. J. Neurosci. **31**, 13880–13889 (2011)
8. Schmithorst, V.J.: Separate cortical networks involved in music perception: preliminary functional MRI evidence for modularity of music processing. NeuroImages. **25**, 444–451 (2005)
9. Wilke, C., Ding, L., He, B.: Estimation of time-varying connectivity patterns through the use of an adaptive directed transfer function. IEEE Trans. Biomed. Eng. **55**, 2557–2564 (2008)
10. Tian, Y., Yao, D.: Why do we need to use a zero reference? Reference influences on the ERPs of audiovisual effects. Psychophysiology. **50**, 1282–1290 (2013)

11. Yao, D., Wang, L., Arendt-Nielsen, L., Chen, A.C.N.: The effect of reference choices on the spatio-temporal analysis of brain evoked potentials: the use of infinite reference. Comput. Biol. Med. **37**, 1529–1538 (2007)
12. Ding, J.R., An, D., Liao, W., Li, J., Wu, G.R., Xu, Q., Long, Z., Gong, Q., Zhou, D., Sporns, O.: Altered functional and structural connectivity networks in psychogenic non-epileptic seizures. PLoS One. **8**, e63850 (2013)
13. Pietro, M.D., Laganaro, M., Leemann, B., Schnider, A.: Receptive amusia: temporal auditory processing deficit in a professional musician following a left temporo-parietal lesion. Neuropsychologia. **42**, 868–877 (2004)
14. Sammler, D., Koelsch, S., Ball, T., Brandt, A., Elger, C.E., et al.: Overlap of musical and linguistic syntax processing: intracranial ERP evidence. Ann. N. Y. Acad. Sci. **1169**, 494–498 (2009)
15. Poulin-Charronnat, B., Bigand, E., Madurell, F., Peereman, R.: Musical structure modulates semantic priming in vocal music. Cognition. **94**, B67 (2005)
16. Wei, J., Peng, D., Yang, Z., Chan, T., Luo, Y.: The hemispheric difference of semantic processing of Chinese characters in two dimensions as revealed by ERPs. Neuroreport. **12**, 3697–3701 (2002)
17. Stuss, D.T.: Frontal lobes and attention: processes and networks, fractionation and integration. J. Int. Neuropsychol. Soc. **12**, 261–271 (2006)

Chapter 20
Serotonin 5-HT1A Receptors Modulate Neural Rhythms in Prefrontal Cortex and Hippocampus and Prefronto-Hippocampal Connectivity in Alert Mice

Thomas Gener, Adrià Tauste-Campo, Maria Alemany-González, Cristina Delgado-Sallent, and Maria Victoria Puig

Abstract The serotonergic system plays a crucial role in cognition and is a target of many psychiatric treatments. In particular, serotonin 5-HT_{1A} receptors (5-HT_{1A}R) in the prefrontal cortex and hippocampus play key roles in learning, memory, behavioral flexibility, and response inhibition. Here, we investigated how 5-HT_{1A} receptors influence neural network dynamics in the prefrontal cortex and hippocampus and prefronto-hippocampal functional connectivity in alert mice. We found that pharmacological stimulation of 5-HT_{1A}R with 8-OH-DPAT markedly reduces theta, beta, and high gamma oscillations in both areas and weakens prefronto-hippocampal phase synchronization at theta and beta frequencies. Pharmacological inhibition of 5-HT1A receptors with WAY-100635 reduces theta and high gamma oscillatory activity but increases beta and delta oscillations. It also weakens prefronto-hippocampal phase synchronization at theta frequencies. These results reveal that prefronto-hippocampal neurodynamics are highly sensitive to 5-HT_{1A} manipulation and may be relevant for understanding the actions of psychiatric medication targeting the serotonergic system.

Thomas Gener and Adrià Tauste-Campo have contributed equally.

T. Gener · M. V. Puig (✉)
Hospital del Mar Medical Research Institute, Barcelona Biomedical Research Park, Barcelona, Spain
e-mail: mpuig3@imim.es

A. Tauste-Campo · M. Alemany-González · C. Delgado-Sallent
Hospital del Mar Medical Research Institute, Barcelona Biomedical Research Park, Barcelona, Spain

University Pompeu Fabra, Barcelona, Spain

J. M. Delgado-García et al. (eds.), *Advances in Cognitive Neurodynamics (VI)*, Advances in Cognitive Neurodynamics,
https://doi.org/10.1007/978-981-10-8854-4_20

Keywords Neural circuits · Neural networks · Neuropharmacology

20.1 Introduction

Serotonin (5-hydroxytryptamine, 5-HT) is synthesized by serotonergic neurons of the midbrain raphe nuclei whose axons reach almost every brain structure. This widespread innervation allows a powerful modulation of brain activity and function, including cognition. The prefrontal cortex (PFC) and hippocampus (HPC) are two core structures for cognition and express densely $5\text{-HT}_{1A}R$ [1, 2]. Serotonin $5\text{-HT}_{1A}R$ are $G_{i/0}$-protein-coupled receptors that hyperpolarize neuronal membranes inhibiting neural spiking. They are expressed by both excitatory pyramidal neurons and fast-spiking inhibitory interneurons where they influence network activity [2]. Pharmacological manipulations of 5-HT transmission in the PFC and HPC have highlighted a crucial role of 5-HT in cognition [3, 4], $5\text{-HT}_{1A}R$ being the receptors more thoroughly investigated. Excessive or insufficient $5\text{-HT}_{1A}R$ activation in PFC increases impulsivity and cognitive inflexibility [5], whereas abnormal hippocampal $5\text{-HT}_{1A}R$ activation causes learning and memory deficits [6, 7]. Due to its anatomical and functional organization, the serotonergic system has become the target of many pharmacological interventions to treat brain disorders. For example, many antipsychotic drugs are $5\text{-HT}_{1A}R$ agonists [8]; $5\text{-HT}_{1A}R$ agonists display anxiolytic/antidepressant activity in animal models [9], whereas $5\text{-HT}_{1A}R$ antagonists reverse drug-induced cognitive deficits [7]. Here, we investigated the influences of a selective activation or inhibition of $5\text{-HT}_{1A}R$ with 8-OH-DPAT and WAY-100635, respectively, on network dynamics of the PFC and HPC and fronto-hippocampal functional connectivity in alert mice.

20.2 Materials and Methods

20.2.1 Experimental Subjects

C57BL/6 male mice ($n = 10$) were obtained from the local colony at the Barcelona Biomedical Research Park (PRBB) Animal Facility. Mice were 2–3 months old and weighed 25–30 g at the beginning of the experiment. Animals were housed in individual cages over the course of the experiments to avoid damage to their implant. Cages were maintained under a controlled atmosphere (humidity of 55 $\pm$ 7% and temperature of 22 $\pm$ 1 °C) and with a 12:12-h light-dark cycle. Food and water were available ad libitum. All procedures outlined in this work had authorization granted by the PRBB Animal Research Ethics Committee and were carried out in accordance with the guidelines of the European Union Council (2003/65/CE) and Spanish regulations (BOE 252/34367-91, 2005).

20.2.2 Surgical Procedure

Mice were anesthetized (induction with ketamine/xylazine and maintenance with isoflurane 0.5–4%) and placed in a stereotaxic apparatus. Three tungsten electrodes 25 μm wide were lowered down into the medial PFC (AP, 1.5–2.0 mm; ML, ±0.4 mm; DV, −1.7 mm from bregma) and three into the dorsal HPC (AP, −1.8–2.5 mm; ML, 1.3–2.3 mm; DV, −1.5 mm). Several micro-screws were screwed into the skull to stabilize the implant, and the one on top of the cerebellum was used as a general ground. Electrodes were implanted with dental cement. During the 7–10 day recovery period, animals were extensively monitored and received both analgesia (buprenorphine) and anti-inflammatory (meloxicam) treatments.

20.2.3 Electrophysiological Recordings and Pharmacology in Alert Mice

Following postsurgical recovery, we recorded local field potentials (LFPs) in freely moving mice exploring their own home cage. Recordings were implemented with the Open Ephys system (http://www.open-ephys.org/) at 0.1–6000 Hz and a sampling rate of 30 kHz. 8-Hydroxy-2-(di-*n*-propylamino)tetralin (8-OH-DPAT; 5-HT_{1A}R agonist) and WAY-100635 (5-HT_{1A}R antagonist) were obtained from Sigma/Aldrich. Drugs were diluted in saline and the pH corrected to be between 6 and 8. Drugs were first prepared in a concentrated solution and frozen at −20 °C. The day of the experiment, the solution was thawed and diluted before intraperitoneal administration.

20.2.4 Histology

Animals were sacrificed and the brains immediately extracted and frozen at −80°. Serial coronal sections of 30 μm thickness were cut through the entire brain using a cryostat at −20 °C. The sections were stained with cresyl violet for the reconstruction of the electrode tracks.

20.2.5 Data Analyses

LFP signals were detrended, Notch filtered to remove 50 Hz artifacts, and downsampled to 1000 Hz with custom-written scripts in Python. Signals were then bandpass filtered at 0.1–600 Hz. Power spectra and spectrograms were constructed using the multitaper method in MATLAB with the Chronux toolbox (www.chronux.org).

Band-limited (2 Hz) functional connectivity analysis of each experiment was performed in MATLAB using three coupling measures (Pearson correlation, phase-locking value, and phase-lag index) independently. Specifically, each coupling measure was computed across all PFC-HPC contact pairs over non-overlapping time windows (1 s) [10]. In essence, Pearson correlation is a linear coupling measure that captures instantaneous (zero-lag) amplitude fluctuations of both areas. In contrast, phase-locking values quantify the average phase coupling between pairs of signals, thus including zero-phase (zero-lag) and nonzero-phase contributions. Finally, the phase-lag index can be regarded as refinement of the phase-locking measure, where zero-phase contributions are discarded for the average computation. The frequency bands considered include delta (3–5 Hz), theta (9–11 Hz), beta (15–25 Hz), low gamma (30–50 Hz), and high gamma (50–80 Hz). We used Wilcoxon ranked test and Cohen's D [11] to test for statistical significance and corresponding effect sizes between baseline and drug-period samples (power and connectivity).

20.3 Results

Selective activation of 5-HT_{1A}R with 8-OH-DPAT at 1 mg/kg ($n = 7$ mice) exerted complex spectral changes in the neural signals of the PFC and HPC that were not observed after a previous injection of saline. Overall changes were larger in HPC compared to PFC ($p < 0.05$). In both areas, 8-OH-DPAT produced sharp decreases of theta waves accompanied by reductions in beta and gamma oscillations ($p < 0.001$). The gamma decrease was particularly pronounced at high frequencies (>60 Hz) (Fig. 20.1). The 5-HT_{1A}R antagonist WAY-100635 at 0.5 mg/kg completely reversed 8-OH-DPAT's effects in PFC ($p < 0.001$). In HPC, it restored theta and beta power but only partially reversed the decrease of high gamma oscillations ($p < 0.001$). Interestingly, it also produced a rebound increase in delta and beta oscillations in both areas ($p < 0.001$). By contrast, the effects of WAY-100635 when administered alone were smaller than those after 8-OH-DPAT ($p < 0.001$). Theta and high gamma oscillations were also reduced ($p < 0.001$), especially in HPC, but less than after 8-OH-DPAT. WAY-100635 alone also increased delta and beta oscillations in both areas ($p < 0.001$) (Fig. 20.1). These effects were not observed in the control group where only saline was injected ($n = 5$) (data not shown).

We further investigated whether selective activation or inhibition of 5-HT_{1A}R influenced PFC-HPC functional connectivity. We analyzed three complementary connectivity measures that reflect different aspects of circuit dynamics related to phase synchronization: global, zero-lag effects of the drugs that occur simultaneously in both areas (Pearson correlation), nonzero-lag effects that quantify constant lags between areas and likely reflect direct PFC-HPC connectivity (PLI), and a mix of zero-lag and nonzero-lag effects (PLV). Both 8-OH-DPAT and WAY-100635 decreased functional connectivity at theta and beta frequencies,

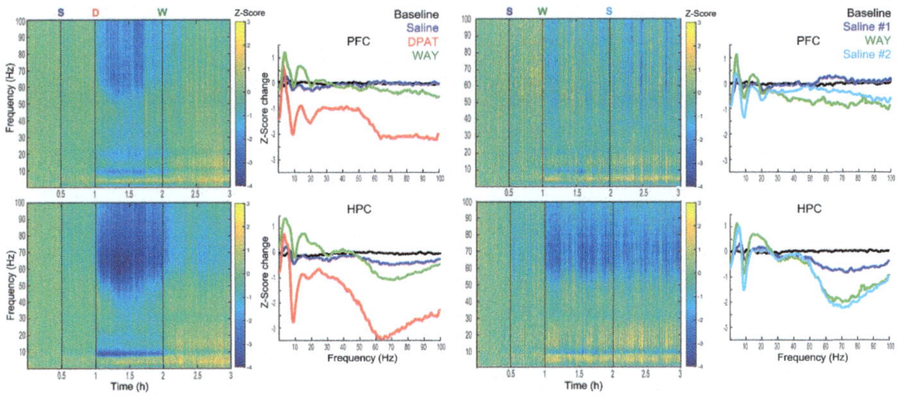

Fig. 20.1 Spectrograms showing the effects of 8-OH-DPAT 1 mg/kg (*n = 7 mice, n = 18 electrodes*) in the PFC (upper left) and HPC (lower left) and WAY-100635 0.5 mg/kg (*n = 4 mice, n = 9 electrodes*) in the PFC (upper right) and HPC (lower right). Spectrograms were normalized by the baseline (*Z*-scored) and averaged across animals. Lateral panels show the *Z*-score change of power during the postdrug period with respect to baseline averaged across 15 min of recording during baseline (black), saline (dark and light blue), 8-OH-DPAT (red), and WAY-100635 (green)

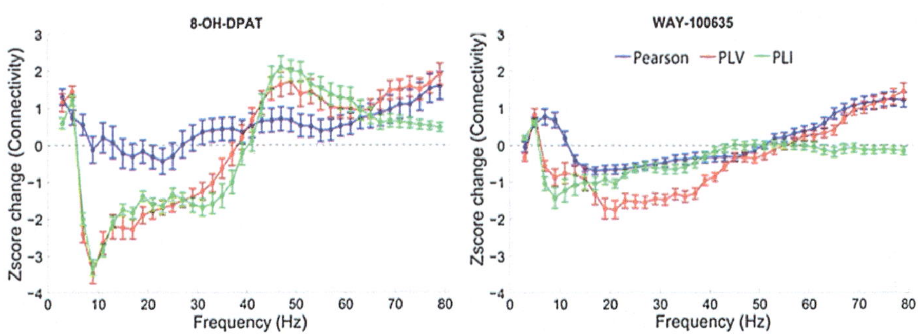

Fig. 20.2 Shown are the effects of 8-OH-DPAT 1 mg/kg (left) and WAY-100635 0.5 mg/kg (right) on prefronto-hippocampal functional connectivity. Pearson correlation, phase-locking value (PLV), and phase-lag index (PLI) were normalized by the baseline (*Z*-scored) and averaged across animals

although this decrease was more pronounced after the former. Interestingly, 8-OH-DPAT increased, whereas WAY-100635 decreased, connectivity at low gamma ($p < 0.001$). Importantly, these alterations occur for connectivity measures that include nonzero-lag effects (PLV and PLI), suggesting that 5-HT$_{1A}$R play a role in the neural communication between the PFC and HPC independent from global effects (Fig. 20.2).

20.4 Discussion

We report that selective pharmacological activation or inhibition of 5-HT_{1A}R exerts strong and complex influences on network dynamics in the PFC and HPC and prefronto-hippocampal functional connectivity. Several oscillatory bands (i.e., delta, theta, beta, and gamma) are disrupted during 5-HT_{1A}R abnormal neurotransmission, suggesting a complex microcircuit effect that likely reflects the sophisticated pattern of expression of 5-HT_{1A}R in both brain structures [12]. Interestingly, both 8-OH-DPAT and WAY-100635 exert greater changes in HPC compared to PFC, revealing that HPC is more sensitive to 5-HT_{1A}R manipulation.

This work also unravels a specific role of 5-HT_{1A}R in shaping PFC-HPC functional connectivity independent from global effects. Overall, either excessive or insufficient 5-HT_{1A}R activation disrupts PFC-HPC network activity both locally and as a circuit. We conclude that psychiatric treatments targeting 5-HT_{1A}R may exert strong influences on PFC-HPC neurodynamics having an impact on cognitive processing shown to depend on this circuit.

Acknowledgments We thank J. Chanovas and P. Nebot for excellent technical assistance. This work was supported by a NARSAD Young Investigator Award and a SAF2016-80726-R (AEI/FEDER, UE) to M.V. Puig.

References

1. Pompeiano, M., Palacios, J.M., Mengod, G.: Distribution and cellular localization of mrna coding for 5-ht1a receptor in the rat brain: correlation with receptor binding. J. Neurosci. **12**, 440–453 (1992)
2. Puig, M.V., Gener, T.: Serotonin modulation of prefronto-hippocampal rhythms in health and disease. ACS Chem. Neurosci. **6**, 1017–1025 (2015)
3. Leiser, S.C., Li, Y., Pehrson, A.L., Dale, E., Smagin, G., Sanchez, C.: Serotonergic regulation of prefrontal cortical circuitries involved in cognitive processing: a review of individual 5-HT receptor mechanisms and concerted effects of 5-HT receptors exemplified by the multimodal antidepressant vortioxetine. ACS Chem. Neurosci. **6**, 970–986 (2015)
4. Robbins, T.W., Arnsten, A.F.T.: The neuropsychopharmacology of fronto-executive function: monoaminergic modulation. Annu. Rev. Neurosci. **32**, 267–287 (2009)
5. Robbins, T.W.: Shifting and stopping: fronto-striatal substrates, neurochemical modulation and clinical implications. Philos. Trans. R. Soc. Lond. Ser. B Biol. Sci. **362**, 917–932 (2007)
6. Meneses, A., Perez-Garcia, G.: 5-HT1A receptors and memory. Neurosci. Biobehav. Rev. **31**, 705–727 (2007)
7. Ögren, S.O., Eriksson, T.M., Elvander-Tottie, E., D'Addario, C., Ekström, J.C., Svenningsson, P., Meister, B., Kehr, J., Stiedl, O.: The role of 5-HT1A receptors in learning and memory. Behav. Brain Res. **195**, 54–77 (2008)
8. Meltzer, H., Massey, B.: The role of serotonin receptors in the action of atypical antipsychotic drugs. Curr. Opin. Pharmacol. **11**, 59–67 (2011)

9. Carr, G.V., Lucki, I.: The role of serotonin receptor subtypes in treating depression: a review of animal studies. Psychopharmacology. **213**, 265–287 (2011)
10. Tauste-Campo, A., Martinez-Garcia, M., Nácher, V., Luna, R., Romo, R., Deco, G.: Task-driven intra- and interarea communications in primate cerebral cortex. Proc. Natl. Acad. Sci. U. S. A. **112**, 4761–4766 (2015)
11. Cohen, J.: A power primer. Psychol. Bull. **112**, 155–159 (1992)
12. Celada, P., Puig, M. V., Artigas, F.: Serotonin modulation of cortical neurons and networks. Front. Neural Circuits. **19**, 7–25 (2013)

Part III
Neuroengineering, Neuroinformation and Brain Computer Interaction

Chapter 21
A New Paradigm Based on Dynamic Visual Stimulation in BCI

Zhaoyang Qiu, Jing Jin, Hanhan Zhang, Yu Zhang, Bei Wang, and Xingyu Wang

Abstract Brain–computer interface (BCI) provided a new communication channel based on the brain activities of the disabled patients. Visual-based P300 BCI is one of the most commonly used BCI systems. Usually, the stimulus used in visual-based P300 BCI was the same character or picture, which could make users feel bored or lose attention. Hence, it would be very helpful in improving the performance of visual-based P300 BCI by concentrating users' attention on the target stimulus. In this study, a new paradigm using dynamic visual stimulation was presented to focus users' attention. Three red dots in a honeycomb-shaped picture would shrink to the center of the honeycomb picture dynamically and were finally merged in the center position as one red dot, which was used as stimulus to evoke event-related potentials (ERPs). Six healthy subjects (three male, aged 24 ± 2.4) participated in this study. To verify the performance of this new paradigm, the face stimulus paradigm was used for comparison. The results showed that the dynamic contraction paradigm obtained 5.0% higher average offline single-trial accuracies and 2.8% higher average online accuracies compared to the face paradigm. According to the reports from subjects, the new paradigm could help to concentrate their attention.

Keywords Event-related potentials (ERP) · P300 · Stimulus paradigm

21.1 Introduction

Brain–computer interface systems (BCIs) can translate brain activities into commands that could be used to control external devices [1–3]. BCIs provide a new communication channel for people, which does not rely on the conventional

---

Z. Qiu · J. Jin · H. Zhang · Y. Zhang · B. Wang · X. Wang (✉)
Key Laboratory of Advanced Control and Optimization for Chemical Processes, Ministry of Education, East China University of Science and Technology, Shanghai, People's Republic of China
e-mail: xywang@ecust.edu.cn

J. M. Delgado-García et al. (eds.), *Advances in Cognitive Neurodynamics (VI)*,
Advances in Cognitive Neurodynamics,
https://doi.org/10.1007/978-981-10-8854-4_21

neuromuscular pathways of peripheral nerves and muscles [4]. Various different neural activities can be used as features in electroencephalogram (EEG)-based BCIs. Event-related (de)synchronization (ERD/ERS) [5, 6], event-related potential (ERP) [7, 8], or visual evoked potentials (VEP) [9, 10] are the most prominent approaches in BCI systems.

ERP-based BCIs commonly rely on the P300 component, which is a positive deflection in voltage occurring in the time of 200–500 ms after stimulation [1, 11]. One well-known P300-based BCI system is the so-called P300 matrix speller that was first described by Farwell and Donchin in 1988. In this system, the user should focus attention to a character in the matrix while each row and column was intensified in a random sequence. The BCI system could identify the character the user wanted by processing EEG signals.

The conventional "flash only" paradigm in which the target reverses color or is briefly masked by a solid box has dominated P300 brain–computer interfaces (BCIs) for decades [8]. Recent research has shown that the face paradigm in which the target is overlapped with a famous face could achieve better performance. However, these paradigms all used static images as stimulus to elicit P300 potentials, such as characters and face pictures. It could make users feel bored or lose attention in the experiment and decrease the performance of P300 BCI. In this study, we proposed a new paradigm using dynamic visual stimulation to focus users' attention. Three red dots in a honeycomb-shaped picture would shrink to the center of the honeycomb picture dynamically and were finally merged in the center position as one red dot. The results showed that the dynamic contraction paradigm could obtain higher average online accuracies compared to the face paradigm.

21.2 Methods

21.2.1 Subjects

Six healthy subjects (three male, aged 24 $\pm$ 2.4) participated in this study. All patients signed a written consent form prior to this experiment. All of them were informed about the purpose of the study and participated of their own free will. The local ethics committee approved the consent form and experimental procedure before any patients participated. All patients were right handed and had normal or corrected-to-normal vision according to self-reports.

21.2.2 Experimental Paradigms

After being prepared for EEG recording, the subjects were seated in a comfortable chair in a shielded room. During data acquisition, subjects were asked to relax and

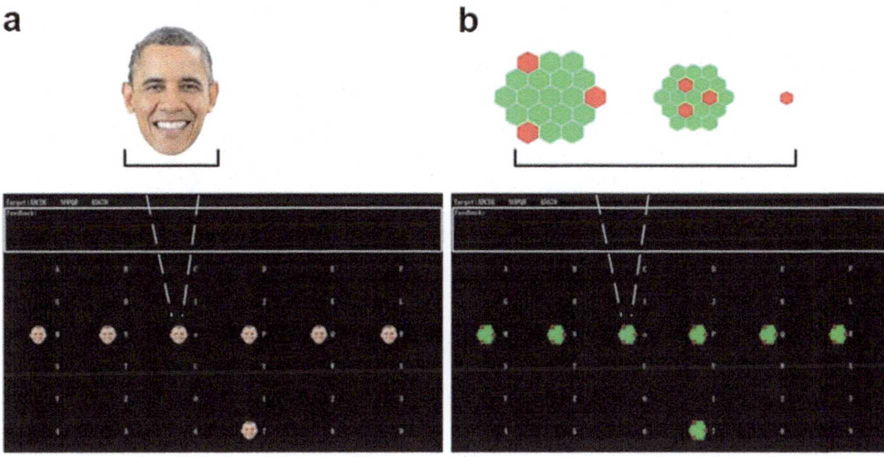

Fig. 21.1 (**a**) The face pattern. (**b**) The dynamic honeycomb-shaped stimuli with red dots pattern

avoid unnecessary movement. The display portrayed a 6 × 6 matrix comprised of gray English letters and symbols against a black background (Fig. 21.1). During a stimulus event, target characters are replaced momentarily with face pictures or honeycomb-shaped pictures; this is referred to as "flashing" in this paper.

Unlike static face pictures, the stimulus in the new paradigm was dynamic. Three red dots in a honeycomb-shaped picture would shrink to the center of the honeycomb picture dynamically and were finally merged in the center position as one red dot, which was used as stimulus to evoke event-related potentials (ERPs).

The experiment contained the offline training session and the online test session. In the offline section, 12 different flashes, within which all the items in the matrix were traversed and a target can be determined, were played once in each trial. A total of three offline experiments were conducted. In the online experiment, the participants were asked to spell all the 36 alphanumeric characters using the classifier obtained in the offline experiment.

21.2.3 Data Acquisition and Analysis

In this study, EEG signals were sampled at 256 Hz through a g.USBamp (Guger Technologies, Graz, Austria). The band pass filter was set to 0.1–30 Hz. Fourteen electrodes (F3, Fz, F4, C3, Cz, C4, CP3, CP4, P3, Pz, P4, P7, Oz, and P8) following the 10–20 international system were used in this study.

Data were referenced to electrode REF located over the right mastoid with a forehead ground (GND), shown in Fig. 21.2. In this paper, the filtered EEG data was down-sampled as features. The method of feature classification was Bayesian linear discriminant analysis (BLDA) [12].

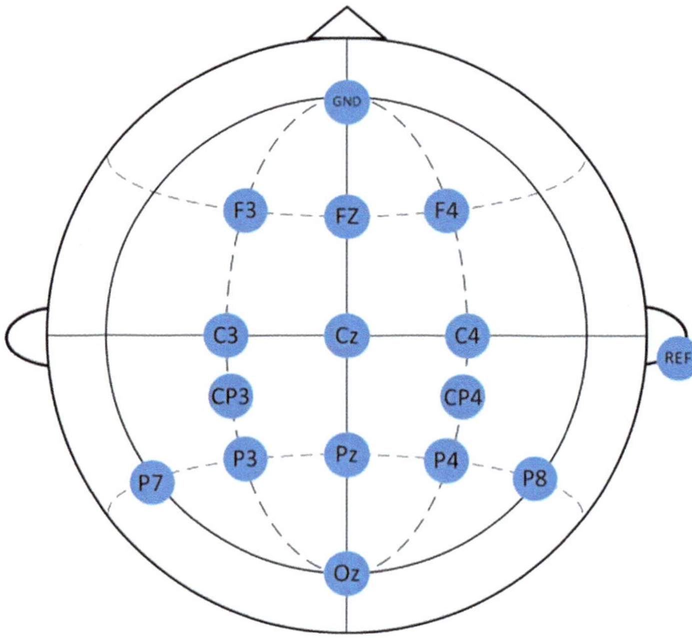

Fig. 21.2 The electrode distribution used in this study

21.3 Results

21.3.1 Brain Patterns

Figure 21.3 displays the grand averages of ERP amplitudes among 6 subjects from 14 electrodes. The red solid lines represent the ERP amplitudes of the target in the dynamic contraction paradigm, and the black dotted lines represent that in the face paradigm. The two patterns had similar vertex positive potential (VPP) components.

However, the dynamic contraction paradigm had relatively higher peak values than those of the face paradigm in channels Cz, C4, CP4, P3, and Pz. The maximum amplitude of the target responses of the face paradigm was higher than the dynamic contraction paradigm in channels F3, Fz, F4, C3, Oz, and P8.

21.3.2 Classification Results

Figure 21.4 shows the classification accuracy of each subject based on single-trial classification (offline data). It may be observed that the classification accuracies of the dynamic contraction paradigm are higher than the face paradigm for subjects S1, S3, S4, S5, and S6. The average offline single-trial accuracies increased by 5% using the dynamic contraction paradigm.

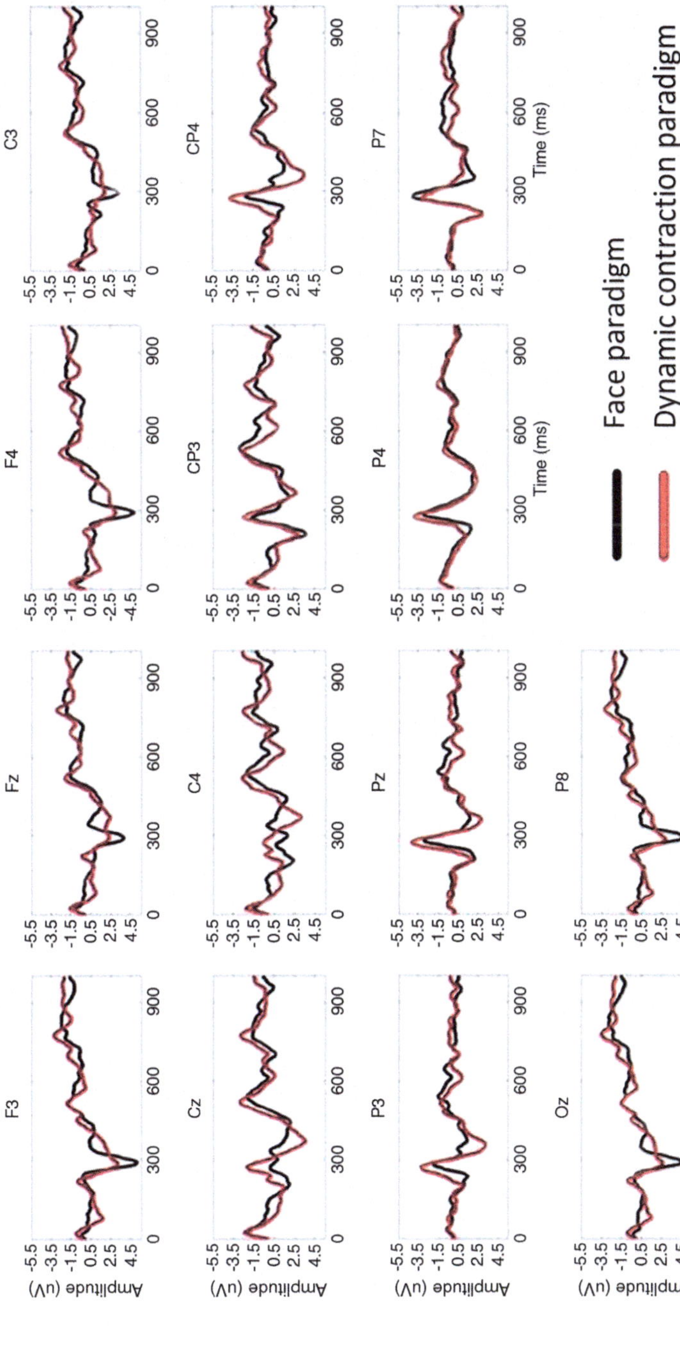

Fig. 21.3 The grand averages of ERP amplitudes of target among 6 subjects over 14 electrodes in the face and dynamic contraction paradigm

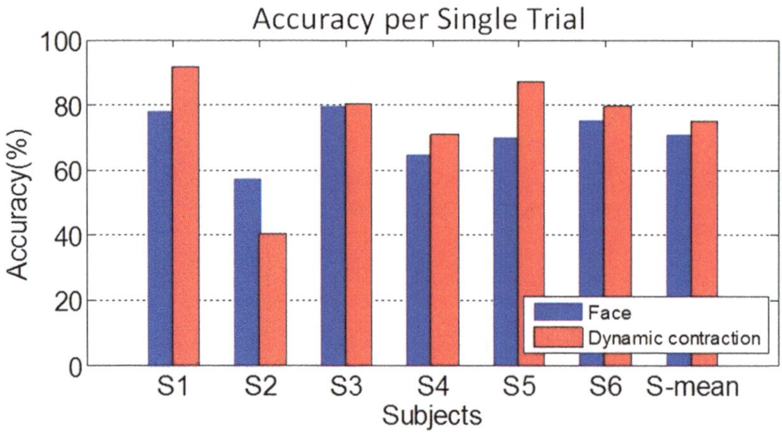

Fig. 21.4 The classification accuracy based on offline single-trial classification

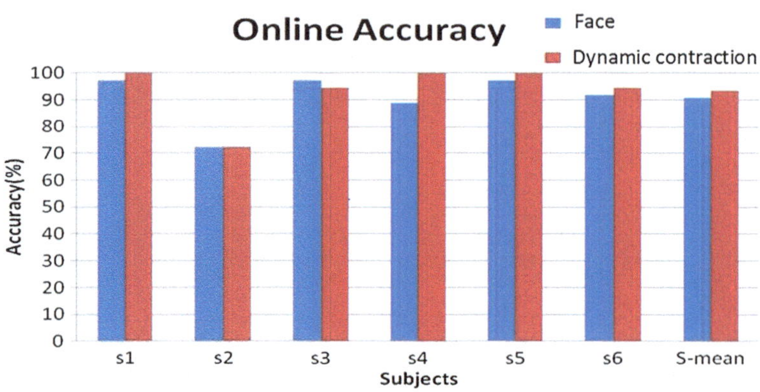

Fig. 21.5 The online classification accuracy of six subjects

Figure 21.5 shows the online classification accuracy for each subject. The online classification accuracies of the dynamic contraction paradigm are higher than the face paradigm for subjects S1, S4, S5, and S6. The average offline single-trial accuracies increased by 2.8% using the dynamic contraction paradigm.

21.4 Conclusion

This study proposed a new paradigm based on dynamic visual stimulation in BCI. Unlike common paradigms which used static images (such as characters and face pictures) as stimulus to elicit P300 potentials, the new paradigm used dynamic visual stimulation to focus users' attention. According to the reports from subjects, the dynamic visual stimulation could help to concentrate their attention effectively. The results also showed that the dynamic contraction paradigm could obtain higher average offline single-trial accuracies and higher average online accuracies compared to the face paradigm. Future work could explore more dynamic visual stimulation to improve the performance of BCIs.

Acknowledgments This work was supported in part by the Grant National Natural Science Foundation of China, under Grant Nos. 91420302 and 61573142. This work was also supported by the Fundamental Research Funds for the Central Universities (WH1516018, 222201717006) and Shanghai Chenguang Program under Grant 14CG31.

References

1. Wolpaw, J.R., Birbaumer, N., McFarland, D.J., Pfurtscheller, G., Vaughan, T.M.: Brain–computer interfaces for communication and control. Clin. Neurophysiol. **113**, 767–791 (2002)
2. Jin, J., Allison, B.Z., Sellers, E.W., Brunner, C., Horki, P., Wang, X.Y., Neuper, C.: An adaptive P300-based control system. J. Neural Eng. **8**, 036006 (2011)
3. Laar, B.V.D., Bos, D.P., Reuderink, B., Poel, M.: How much control is enough? Influence of unreliable input on user experience. IEEE Trans. Cybern. **43**, 1584–1592 (2013)
4. Mak, J.N., Wolpaw, J.R.: Clinical applications of brain–computer interfaces: current state and future prospects. IEEE Rev. Biomed. Eng. **2**, 187–199 (2010)
5. Pfurtscheller, G.: Graphical display and statistical evaluation of event-related desynchronization (ERD). Electroencephalogr. Clin. Neurophysiol. **43**, 757–760 (1977)
6. Pfurtscheller, G.: Event-related synchronization (ERS): an electrophysiological correlate of cortical areas at rest. Electroencephalogr. Clin. Neurophysiol. **83**, 62–69 (1992)
7. Jin, J., Daly, I., Zhang, Y., Wang, X.Y., Cichocki, A.: An optimized ERP brain–computer interface based on facial expression changes. J. Neural Eng. **11**, 1082–1088 (2014)
8. Jin, J., Allison, B.Z., Zhang, Y., Wang, X.Y., Cichocki, A.: An ERP-based BCI using an oddball paradigm with different faces and reduced errors in critical functions. Int. J. Neural Syst. **24**, 1450027 (2014)
9. Cruz, J.N.D., Wan, F., Wong, C.M., Cao, T.: Adaptive time-window length based on online performance measurement in SSVEP-based BCIs. Neurocomputing. **149**, 93–99 (2015)
10. Zhang, Y.S., Guo, D.Q., Cheng, K.W., Yao, D.Z., Xu, P.: The graph theoretical analysis of the SSVEP harmonic response networks. Cogn. Neurodyn. **9**, 305–315 (2015)
11. Woodman, G.F.. A brief introduction to the use of event-related potentials in studies of perception and attention. Atten. Percept. Psychophys. **72**, 2031–2046 (2010)
12. Kolev, V.H., Demiralp, T., Yordanova, J., Ademoglu, A., Isoglu-Alka, U.: Time-frequency analysis reveals multiple functional components during oddball P300. Neuroreport. **8**, 2061–2065 (1997)

Chapter 22
Asynchronous Stimulation Method for N100-P300 Speller

Natsuki Morita and Yoshikazu Washizawa

Abstract Brain-computer interface (BCI) allows people to send commands to computer and communicate with outside world without any physical activities. N100-P300 Speller is a spelling BCI system that utilizes two kinds of event-related potential (ERP), N100 and P300. This system achieved higher accuracy and information transfer rate (ITR) compared with P300 Speller (Sato H, Washizawa Y, N100-P300 speller BCI with detection of user's input intention. In: Proceedings of 6th international brain-computer interface conference 43, 2014). Since N100-P300 Speller requires nine visual stimulations to input one command, it is necessary to boost the classification accuracy or reduce the flashing intervals in order to improve ITR. The present study aims to improve ITR by an asynchronous stimulation method in N100-P300 Speller. The result showed that ITR in the proposed method was improved 0.15 bit/s at an average compared with the conventional N100-P300 Speller. The proposed N100-P300 Speller achieved faster spelling system with lower burden for the user.

Keywords EEG · BCI · P300 · N100

22.1 Introduction

Brain-computer interface (BCI) is an interface to send commands to computers and communicate with outside world using brain waves. Because this interface does not require any physical activities, BCI is expected to become a communication tool for those who have physical disabilities such as amyotrophic lateral sclerosis (ALS) or spinal cord injury [1, 2]. There are several methods to measure brain signals such as magnetoencephalography (MEG), functional magnetic resonance

---

N. Morita (✉) · Y. Washizawa
Department of Computer and Network Engineering, The University of Electro-Communications, Tokyo, Japan
e-mail: natsuki.morita@uec.ac.jp

J. M. Delgado-García et al. (eds.), *Advances in Cognitive Neurodynamics (VI)*,
Advances in Cognitive Neurodynamics,
https://doi.org/10.1007/978-981-10-8854-4_22

imaging (fMRI), and electroencephalography (EEG). EEG measures electrical activities in neural cells as potential change [3]. It is often used in BCI because its equipment is relatively small size and low cost. Brain activity is divided into two kinds, spontaneous activity and evoked one. The former is measured from the brain in natural state, and it is arisen in different frequency band at the different psychological state of the user. The latter is induced by the internal or external stimulation to the user or the user's own reaction.

It needs to utilize some kind of features appearing on the brain signals to build a BCI system. There are some methods to use frequency analysis such as steady-state visual evoked potential (SSVEP) or auditory steady-state response (ASSR). In these methods, the system detects the desired target related to the specific frequency band [4]. There are also ways to utilize the time sequence features; it is called event-related potential (ERP). ERP appears on EEG when the user is stimulated or does some tasks. P300 is a kind of ERP that appears as a result of cognition of the user to the stimulation; it has a latency of about 300 ms and positive peak amplitude. The oddball paradigm is a well-known method to evoke P300. In this procedure, the subject is presented with two kinds of stimulation: one is called standard stimulation that appears at high frequency, and another is the target one that appears at low frequency. The subject is told to count the number of times when the target is displayed. P300 Speller is a popular spelling BCI system utilizing P300. This system uses visual stimulation displaying all commands arranging in a 6×6 matrix. Each row and column is flashed line by line randomly, and the user makes a response when the row or column including the target commands is flashed. The target command is identified by detecting corresponding P300 components [5].

N100 is also a kind of ERP that appears after 100 ms of the onset of the stimulation without the voluntary recognition of the user to the stimulation; it has a negative peak amplitude. N100-P300 Speller is a spelling BCI system utilizing N100 and P300. This system sequentially presents visual stimulation images having six positions like a 2×3 matrix. The user gazes on the position where the target command will be displayed and counts the number of times when the target command is presented. By to detect N100 component, the position where the user gazed on is identified, and by to detect P300 component, the one image including the target command is specified.

Comparing with what P300 Speller requires flashing 12 times to present all 36 commands, N100-P300 Speller needs flashing only nine times. Moreover, it was showed that the classification accuracy of N100-P300 Speller is higher than the one of P300 Speller, it means that N100-P300 Speller has higher accuracy and speed to input compared with P300 Speller [1]. For further improvement of ITR, we need to shorten the inter stimulus interval (ISI) or stimulus onset asynchrony (SOA); however it degenerates accuracy and usability. The purpose of this study is to decrease whole the presentation time with the proposing asynchronous stimulation.

22.2 Methods

22.2.1 Conventional N100-P300 Speller

N100-P300 Speller uses nine visual stimulation images having six positions (2 × 3 matrix). There are four commands and two blanks on each image. This system sequentially presents these images while inserting interval between each image. The user gazes on the position that will display the target command, and they do mental task such as to count the number of times when the target command is displayed. The user's mental task evokes P300, thus by to detect P300, the target command is identified. In this system, different character position has different pattern of flashing commands. For this reason, N100 is evoked in accordance with a specific sequence by the position the user gazes on.

22.2.2 Proposed Method

In order to enhance ITR, it is necessary to decrease whole the presenting time of the stimulation or improve the classification accuracy. To achieve former purpose, because the original N100-P300 Speller uses nine visual stimulation images, it is necessary to shorten ISI or the presentation time of the each stimulation image. In the previous method, the durations of flashing commands off are synchronized with the ones of flashing commands on (Fig. 22.1a). We propose a new stimulation method that desynchronizes these durations, and we set specific sequence of flashing commands in each position (Fig. 22.1b). This method shortens whole the time to present stimulations without to decrease the duration of flashing commands on and ISI in each position, which means ISI not between stimulation images but letters.

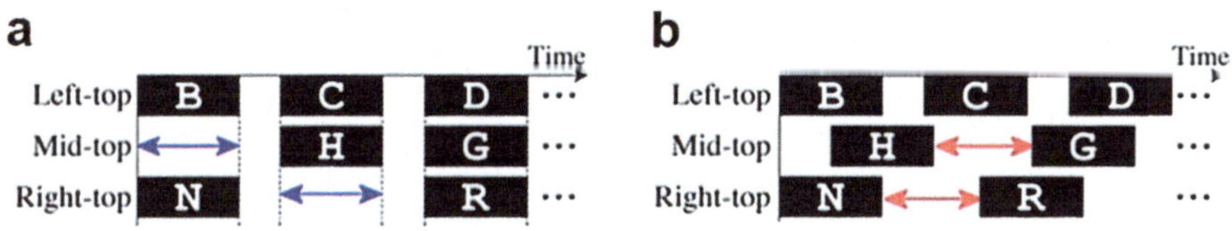

Fig. 22.1 An example of sequence to flash commands in three positions. (**a**) The conventional method. (**b**) The proposed method

22.2.3 Experimental Condition

The experiment was conducted, about one conventional method and three proposed ones that have different presenting time of the stimulation. The one target letter was instructed to the subject in the random order before presenting the stimulation. Then the stimulation of one of four methods was presented to the subject randomly. The subject was asked to intend to input the target character; the task of the subject was to gaze on the position that will be displayed the target and to count the number of times presenting the target object. All commands were displayed two times, and each one letter was displayed for 125 ms, and the ISIs in each position were taken for at least 62.5 ms. These procedures were treated as the one trial, and 20 trials were carried out per each method (total 80 trials per each subject). The conventional method had 3.25 s to stimulate in whole, and the proposed method (1) had about 80.8% time compared to the conventional one, as the same way, proposed method (2) had 75%, and proposed method (3) had about 71.2%. Psychophysics Toolbox extensions were used for presenting the stimulations on MATLAB [6].

The EEG signal was recorded with an active EEG (Guger Technologies OG), amplified with a biosignal amplifier (Digitex Lab. Co., Ltd) and AD transformed with an AD converter (Contec Co., Ltd). Sixteen electrodes were arranged according to the extended international 10–20 system (FCz, FC1, FC2, Cz, C3, C4, C5, C6, CP1, CP2, Pz, P3, P4, TP7, TP8, POz); AFz was used for the ground, and A2 was done for the reference.

22.2.4 Evaluation of the Performance

The classification accuracy of the gaze position and the target commands and the ITR were used as evaluation indexes. The soft margin linear support vector machine (SVM) was used for the classification. ITR was calculated by the following equation, where T [s] is the time for one trial, N is the number of the command candidates, and P is the classification accuracy [7].

$$B = \frac{1}{T}\left\{\log_2 N + P\log_2 P + (1-P)\log_2\frac{1-P}{N-1}\right\} \text{ [bit/s]} \qquad (22.1)$$

22.3 Results

Table 22.1 lists the classification accuracy. The accuracies of the proposed methods are higher than of the conventional one except for the proposed (2) of the subject 1. The accuracy of all the proposed method is higher than of conventional one at an average.

Table 22.1 Classification accuracy of the gaze position and the target [%]

Subject	Conventional		Proposed (1)	
	Gaze position	Target command	Gaze position	Target command
1	85.0 ± 11.1	30.0 ± 16.9	60.0 ± 9.3	30.0 ± 17.6
2	45.0 ± 22.6	15.0 ± 12.7	65.0 ± 17.6	20.0 ± 12.2
3	55.0 ± 21.2	15.0 ± 10.0	45.0 ± 17.3	25.0 ± 12.9
4	55.0 ± 19.6	30.0 ± 4.0	35.0 ± 3.6	10.0 ± 1.3
5	55.0 ± 27.6	25.0 ± 6.3	35.0 ± 22.3	5.0 ± 1.0
Average	59.5 ± 32.0	23.8 ± 16.4	49.3 ± 19.7	19.5 ± 14.7
Subject	Proposed (2)		Proposed (3)	
	Gaze position	Target command	Gaze position	Target command
1	75.0 ± 16.5	25.0 ± 13.2	70.0 ± 15.4	40.0 ± 9.3
2	50.0 ± 12.7	25.0 ± 10.0	70.0 ± 16.2	25.0 ± 7.0
3	75.0 ± 23.8	20.0 ± 8.1	75.0 ± 17.3	35.0 ± 19.1
4	95.0 ± 10.0	45.0 ± 11.6	50.0 ± 6.6	25.0 ± 10.0
5	76.5 ± 23.2	25.0 ± 14.3	63.6 ± 22.8	15.0 ± 9.0
Average	66.0 ± 22.2	31.3 ± 22.5	71.4 ± 18.9	30.2 ± 12.2

Table 22.2 Information transfer rate (ITR) [bit/s]

Subject	Conventional	Proposed (1)	Proposed (2)	Proposed (3)
1	0.240 ± 0.196	0.303 ± 0.240	0.232 ± 0.202	0.493 ± 0.191
2	0.089 ± 0.092	0.150 ± 0.162	0.222 ± 0.152	0.227 ± 0.112
3	0.079 ± 0.053	0.214 ± 0.171	0.150 ± 0.103	0.427 ± 0.369
4	0.258 ± 0.172	0.065 ± 0.075	0.686 ± 0.358	0.232 ± 0.167
5	0.218 ± 0.089	0.032 ± 0.004	0.386 ± 0.215	0.232 ± 0.215
Average	0.176 ± 0.180	0.160 ± 0.178	0.325 ± 0.408	0.326 ± 0.279

Table 22.2 lists ITR. ITR of the proposed method is higher than of the conventional one except for the proposed (2) of the subject 1. It is confirmed that ITR of the proposed (3) is significantly improved compared to the conventional method by t-test at significance level of $\alpha = 0.05$.

22.4 Conclusion

From Table 22.1, the classification accuracies about the gaze position of the proposed (2) and (3) are higher than of the conventional at an average. Because the conventional method is a series of the repetition of presenting four commands on screen simultaneously and then turning off at the same time, the sequence of flashing commands getting into the user's field of vision is constant. On the other hand, the proposed method has various patterns of flashing commands. For these reasons, the accuracy about gaze position of the proposed method, which has different patterns of flashing commands by the position, is higher than of conventional one.

Table 22.3 The presentation time of the stimulation T [%], ITR if the classification accuracies P are the same, and ITR from the result, [%]

	T	ITR (if P is equal)	ITR (from the result)
Conventional	100	100	100
Proposed (1)	80.8	123.7	90.9
Proposed (2)	75.0	133.3	184.6
Proposed (3)	71.2	140.4	185.2

Comparing the conventional method and proposed (1), what has higher accuracy about gaze position is the former, but the latter is higher one about the target command. This means that the classification performance about P300 of the proposed (1) is higher than the conventional one. Because as stated above, the user gets the constant flashing pattern in the conventional method, it is expected that the timing of doing mental tasks is delayed affected by that pattern. In contrast, because the proposed method has a variety of flashing patterns, it is easy for the user to do the mental task.

From the result, the accuracy of the target commands does not have significant difference from the conventional method and the proposed one. This is because that the variance of the intersubject is too large to confirm the difference. However, all the accuracies about the target commands of proposed (3) are higher than the conventional one for all subjects, it is predicted that proposed (3) has the highest accuracy about the target commands.

ITR is calculated using the input time and the number of command candidates in addition to the accuracy; hence it enables to evaluate the comprehensive performance by comparing ITR. From Table 22.2, ITR of the proposed method is higher than conventional one. This result depends on what the presentation time of the proposed methods are shorter the one of the conventional. Moreover, higher classification accuracy of the proposed method affects the improvement of ITR. Table 22.3 lists the whole presentation time of the stimulation, expected ITR if the classification accuracies are the same between four methods, and ITR from the experimental result. Because resulted ITRs are higher than expected, not only reduction of the presentation time but the accuracy improves ITR. It is confirmed that ITR of the proposed (3) significantly improved from the conventional method; therefore it concludes the proposed (3) has the highest performance.

References

1. Sato, H., Washizawa, Y.: N100-P300 speller BCI with detection of user's input intention. In: Proceedings of 6th International Brain-Computer Interface Conference **43** (2014)
2. Mak, J.N., Wolpaw, J.R.: Clinical applications of brain computer interfaces: current state and future prospects. I.E.E.E Rev. Biomed. Eng. **2**, 87–199 (2009)
3. Takeda, T.: Noukougaku [The brain engineering]. Coronasha. (2003)

4. Suo, H., Sun, B., Hu, N., Lu, S., Chen, G.: An auditory steady state response detection method via quadrature matched filter. In: Sixth International Conference on Information Science and Technology (ICIST) (2016) 55–60
5. Farwell, L.A., Donchin, E.: Talking off the top of your head: toward a mental prosthesis utilizing event-related brain potentials. Electroenceph. Clin. Neurophysiol. **70**, 510–523 (1988)
6. Brainard, D.: The psychophysics toolbox. Spat. Vis. **10**, 443–446 (1997)
7. Wolpaw, J.R., Ramoser, H., McFarland, D.J., Pfurtscheller, G.: EEG-based communication: improved accuracy by response verification. I.E.E.E Trans. Rehabil. Eng. **6**, 326–333 (1998)

Chapter 23
Attention Evaluation Based on Single Prefrontal EEG

Jianhai Zhang, Gaomin Liu, Shaokai Zhao, and Wenhao Huang

Abstract Level of concentration is a research hotspot for human being at all aspect of human society and all ages. This paper makes use of the TGAM chip that developed a concentration value collector. Our experiment proves the validity and correctness of this equipment and can extract the attention feature in real time. Compared with the traditional attention evaluation which was also known as a questionnaire, our method had stronger real-time ability and high reliability and was more intuitive and more convenient. In addition, it is possible for our products to be put into practical application because of the high accuracy and real-time.

Keywords EEG · Attention degree · TGAM · Feature extraction · BCI

23.1 Introduction

Concentration or attention plays an important role for human cognitive activities, especially for adolescents. As well known, many learning difficulties for people are not intellectual problems but problems in attention [1]. Psychologists usually use four key characteristics to evaluate the development of a person's concentration: span, stability, distribution, and transfer. To get the four key qualities, the traditional way mainly combines the results of questionnaire and professional observation and judgment. Though some encouraging results have been achieved, these measurements are an indirect mapping of human attention and the results tend to be subjective. Recently, electroencephalogram (EEG)-based attention detection has received increasing attention. Because the EEG signal is collected from the central nervous system, it is expected to provide more objective and comprehensive information for attention detection [2].

J. Zhang (✉) · G. Liu · S. Zhao · W. Huang
College of Computer Science, Hangzhou Dianzi University, Hangzhou, China
e-mail: jhzhang@hdu.edu.cn

J. M. Delgado-García et al. (eds.), *Advances in Cognitive Neurodynamics (VI)*,
Advances in Cognitive Neurodynamics,
https://doi.org/10.1007/978-981-10-8854-4_23

The TGAM chip developed by Neurosky Company has been widely applied in the field of attention detection due to its low cost, ease of use, and credible results [3]. Zhao et al. [4] design a multiparameter embedded biological information measurement system. Experimental results show that the system has effective stability for continuous operation. However, the existing research didn't tell how to evaluate a person's attention according to these detected discrete attention values. In this paper, a method is proposed to solve the above problem by quantifying the concentration values detected by the TGAM chip into three indexes.

23.2 Materials and Methods

23.2.1 Materials

For our research, we made a device to detect attention level based on the TGAM chip. The device is an embedded system with a single biosensor. The sensor is placed on the forehead at the FP1 location, and a reference electrode is connected to the earlobe [5], using Bluetooth to transmit recorded data. The data includes the attention level value in the range of 1–100 with a sample rate of 1 Hz (Fig. 23.1).

Five healthy subjects (three males and two females with an average age of 23.8 $\pm$ 1.7) participated in the experiment. Each participant was asked to do a standard 7 $\times$ 7 Schulte scales [6] test at different times for five times and a Sudoku test for one time. The attention level values for every subject were collected during

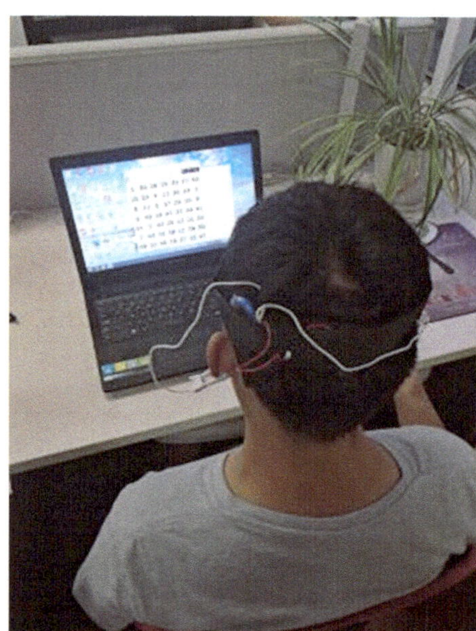

Fig. 23.1 Subject wearing our device and going through Schulte scales experiment

the test. In the above both tests, the first 60 s of each trial is rest time. During the rest time, subjects were asked to close their eyes and try not to think about anything. After the relaxation, the subjects were asked to finish the Schulte table or Sudoku tests as quickly as possible.

23.2.2 Methods

Due to the noise and great fluctuations in recording data, it is hard to directly extract the characteristic of attention. The following processes were implemented for further analysis:

Step 1: Median filtering

First, it is necessary to smooth the recording data. Here the median filtering method is used with window length 3 and step size is 1. See Fig. 23.2b.

Step 2: Trend extraction based on the PauTa criterion

From the smoothing results, in fact it only removed the data "spikes" (accidental error), but many "peaks" and "valleys" still exist. It is still difficult to see what the trends of the data should be. By observation, we found that the peaks and valleys fluctuate around a slowly changing signal. This signal is the trend which we are

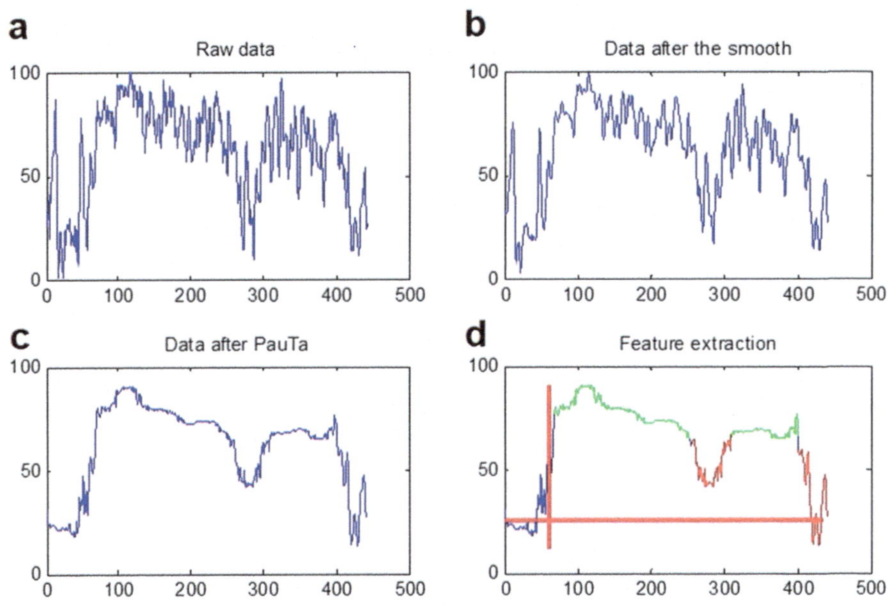

Fig. 23.2 This figure clearly shows the process of how we dealt with the data

interested in. We proposed a new method based on the PauTa criterion to extract the trend. The PauTa criterion assumes that the residual error of the data obeys the normal distribution, which is feasible for our experimental data [7]. The standard PauTa gives the way to remove bad point as follows:

$$\left|X_i - \overline{X}\right| > 3\sigma, \quad \text{then } X_i \text{ as a bad point to be removed} \tag{23.1}$$

$$\left|X_i - \overline{X}\right| \leq 3\sigma, \quad \text{then } X_i \text{ as a normal point to be retained} \tag{23.2}$$

In the above equations, X_i is the current data point and $\overline{X}$ is the average of the nearest 30 points around the X_i point. In the practical is application of the PauTa criterion, we have made an improvement. The retained data point is replaced by the average of the 30 points near that point. See Fig. 23.2c.

Step 3: Determination of attention states

Observing the data variation obtained above (Fig. 23.2), we can find such a trend just as expected. After the start of the task, the degree of concentration will first rise to a high value, then slowly decline to a long steady state, and finally decline sharply, while the task is finished. During the steady state, there exist some distraction stages. In the distraction stage, the attention level declines sharply and lasts for a period and then rises to the steady state again. To define the four states, two types of baseline need to be defined. The first baseline is the average of the rest state, and it was used to determine whether a subject has entered the mission state from a rest state; another is the average of the first 10 s of steady state which can be used to dynamically determine the level of concentration of the current subject. According to the above observations, we give the definition of several attention states in every trial:

1. Rest state: the first 60 s of each trial before the task
2. Mission state: the long steady state excluding all the distraction stages
3. Transfer state: the state from the beginning of a task to the beginning of the steady state
4. Distraction state: during the steady state, the level of concentration below the baseline value of the state for a period ($\geq$10 s)

As shown in Fig. 23.2d, the red horizontal line represents the baseline in the rest state. The vertical line on the left side of the graph is the start of the task. The blue line before the start point is the rest state while the line with same color after the start point is the transfer state. The red curve indicates the distraction state. The green curve indicates the mission state (Fig. 23.3).

Step 4: Indexes extraction

After getting the above states, we can obtain the psychological indicators as follows:

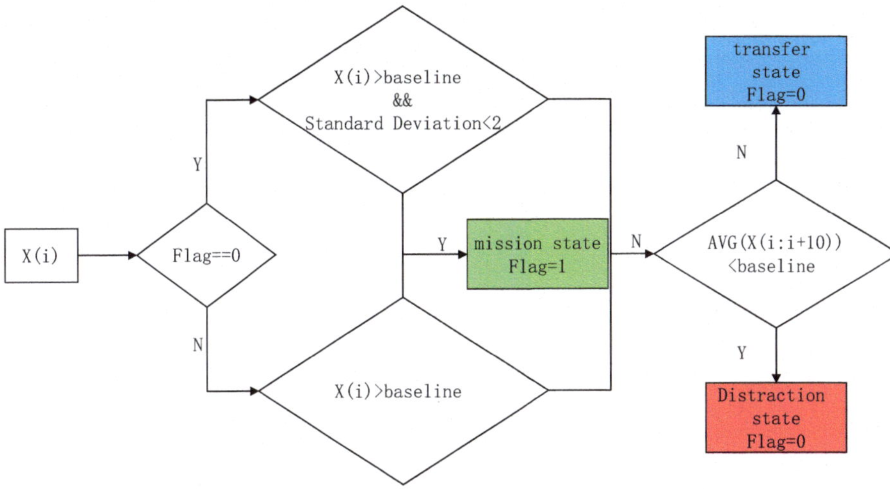

Fig. 23.3 This figure clearly shows the process of how we process the time series and divide it into several attention state

Transfer of concentration T (unit of this feature is second):

$T =$ Transfer time

Stability of concentration *S*:

$S =$ Mission status/(Total duration−relaxed state)

Intensity of concentration I:

$I =$ Average of the sequence of concentration in the mission state/average of the sequence of concentration in the rest state

23.3 Results and Conclusion

According to the above process, here we get three features of each trial for each subject: Transfer *T*, Stability *S*, and Intensity *I*.

The average results of the six subjects are shown in Table 23.1.

By comparing the five subjects listed above, we found that there are increasing trend of attention but a decrease of averaged task endurance from subject 1 to subject 5. It suggests that it easily leads to mental fatigue and distraction when subjects were demanded to focus on one event for a long time. This is consistent with what we can expect.

Since the time spent in the Sudoku experiment is longer that the time spent in the Schulte table, the stability of attention has decreased which is consistent with our analysis of Table 23.2. This result almost exists in all of the subjects except for

Table 23.1 Averaged result of five subjects

Subjects	Transfer	Stability	Intensity	Time endurance
Sub1	15.0 ± 5.571	91.06 ± 6.627	1.50 ± 0.33	305.2 ± 82.12
Sub2	10.0 ± 3.808	89.19 ± 16.061	1.35 ± 0.17	243.2 ± 30.90
Sub3	11.0 ± 3.807	95.77 ± 2.276	1.64 ± 0.63	252.2 ± 62.30
Sub4	5.2 ± 1.92	95.87 ± 5.46	1.23 ± 0.16	214.2 ± 30.79
Sub5	2.6 ± 1.84	97.82 ± 0.89	1.23 ± 0.13	195.3 ± 25.66
AVE	8.76 ± 4.90	93.94 ± 3.64	1.39 ± 0.18	242.0 ± 42.00

Table 23.2 Result of five subjects in Sudoku experiment

Subjects	Transfer	Stability	Intensity
Sub1	8.0233	81.63	1.48
Sub2	8.4337	89.50	1.42
Sub3	6.2849	82.88	0.95
Sub4	5.9606	82.83	0.92
Sub5	3.3134	89.38	1.38
AVE	6.40 ± 2.03	85.24 ± 3.86	1.23 ± 0.27

subject 2. In addition, we also found that the subjects 1, 2, and 3 spent more time than subjects 4 and 5 in the transfer state of Schulte grid experiment, while this phenomenon also appeared in the Sudoku experiment.

In the Introduction section, we introduce the quality of attention in psychology. In this paper, the emphasis of this paper on the conversion and attention to the stability of the attention given to the psychological attention to the quality of the reference, which provides a good platform for the measurement of the concentration of subjects in the future.

References

1. Hengchan, Y.: The research on concentration test and evaluation index for adolescence. China Sport Sci. Technol. **39**, 51–53 (2003)
2. Friedman, D., Shapira, S., Jacobson, L., et al.: A data-driven validation of frontal EEG asymmetry using a consumer device. In: IEEE International Conference on Affective Computing and Intelligent Interaction, 930–937 (2015)
3. Su, Z., Xu, X., Ding, J., et al.: Intelligent wheelchair control system based on BCI and the image display of EEG. In: IEEE Advanced Information Management, Communicates, Electronic and Automation Control Conference (2017)
4. Zhao, J., Zhang, H., Fu, W., et al.: Design of multi-parameter embedded biological information measurement system. In: World Congress on Intelligent Control and Automation, 473–476 (2016)
5. Furlan, P.B., Kimura, A., de Melo, C.L.S., et al.: Portable communication system for disabled speech people controlled by electroencephalograph signals. In: Andescon IEEE (2017)
6. BrainApps.io Brain fitness. https://brainapps.io/game/Clicker
7. Yang, S.: A new method of removing singular points in dynamic testing data: absolute mean value method and its application study. China Meas Urement. **32**, 47–49 (2006)

Chapter 24
Multi-Linc: A New Approach for Exploring Inter-areal Spike Communication

Yoshikazu Isomura

Abstract Each area in the brain sends functional spike outputs to other areas through axonal projections. Until now, however, it has been technically difficult to precisely measure the spike outputs among brain areas in behaving animals. We recently established the *Multi-Linc* analysis method, which uses optogenetically evoked spike collision in multineuronal recording, to elucidate inter-areal spike communication at a single-cell level at millisecond time resolution. Here, I review the concept and demonstration of this method and discuss its advantages and future directions.

Keywords Antidromic spike · Collision test · Optogenetics · Multineuronal recording

24.1 Spike Collision Test

If two spikes collide with each other along an axon, they will disappear. The collision test utilizes this principle to electrophysiologically identify the axonal projection of an individual neuron ([1] for review). Briefly, when we record a single neuron (unit) activity in a brain area and stimulate its target area electrically, we may observe an antidromic spike response evoked several to 10 milliseconds after the stimulation. However, stimulation can instead evoke monosynaptic or polysynaptic spike responses via axonal collaterals of nearby companion neurons, axonal projections from neurons in the target area, and/or bifurcate axonal branches of neurons in a third area. The collision test allows us to determine whether the spike response is antidromic or synaptic in single-unit recording. In this test, we electrically stimulate the target area immediately after a spike occurs spontaneously at the soma. If it is

Y. Isomura (✉)
Brain Science Institute, Tamagawa University, Machida, Tokyo, Japan
e-mail: isomura@lab.tamagawa.ac.jp

J. M. Delgado-García et al. (eds.), *Advances in Cognitive Neurodynamics (VI)*,
Advances in Cognitive Neurodynamics,
https://doi.org/10.1007/978-981-10-8854-4_24

antidromic, we will no longer detect the antidromic spike response at the soma, owing to collision of the spontaneously occurring (orthodromic) spike and the antidromically evoked spike along the axon.

Using the collision test, Evarts [2] elegantly demonstrated the functional spike activity of antidromically identified corticospinal neurons in behaving monkeys. To date, this is the only method capable of discriminating axonal projections electrophysiologically (not histologically) in vivo. However, because of its limited availability and low efficiency, only skillful experts have mastered this decisive test in behaving animals. The test has several technical problems regarding both stimulation and recording. First, electrical stimulation irregularly localizes a current near the tip of the stimulating electrode, which sometimes has low effectiveness and is prone to make an electrolytic lesion. Second, the stimulation always creates a huge electrical artifact, which often obstructs the collision test. Third, single-unit recording can detect only one neuron at a time, and it is laborious to search with a recording electrode, advancing little by little.

24.2 *Multi-Linc* Analysis

To overcome these problems and improve the availability and efficiency of this approach, we recently established the *Multi-Linc* (*multi*-areal and *multi*neuronal *l*ight-*in*duced *c*ollision) analysis [3], in which we perform multiple collision tests for different neurons/projections in parallel by substituting multi-areal optogenetic stimulations for single-site electrical stimulation, and multineuronal recordings in several areas for single-unit recording, in a collision test (Fig. 24.1a).

In fact, optogenetics was recently shown to be suitable for spike collision in cortical or subcortical pathways [4, 5]. Notably, we adopted transgenic (Tg) rats

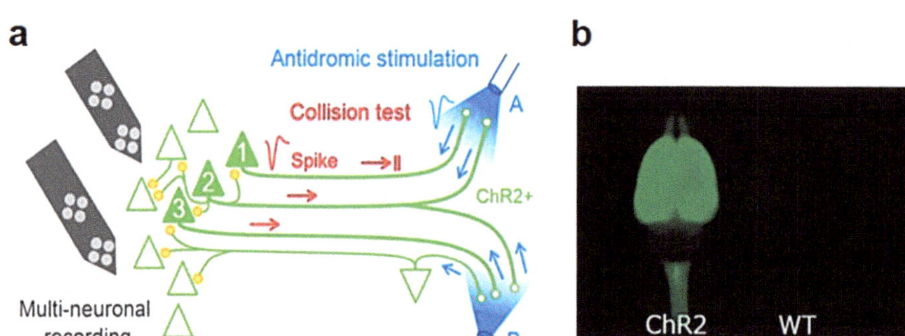

Fig. 24.1 *Multi-Linc* analysis. (**a**) The concept of *Multi-Linc* analysis. An antidromically evoked spike (blue) collides with a spontaneously occurring spike (red) and disappears (spike collision). (**b**) A transgenic rat expressing ChR2 [6] in the whole brain (left). WT, wild type (right)

expressing channelrhodopsin-2 (ChR2)-Venus proteins under the control of Thy1.2 promoter in many of PNS and CNS neurons (W-TChR2V4 line) [6] to achieve efficient multi-areal optogenetic stimulation. The adult ChR2 Tg rats express ChR2 spontaneously and abundantly in the whole brain (Fig. 24.1b), requiring neither viral infection, in utero electroporation, nor crossbreeding.

In situ hybridization confirmed that a large number of neurons expressed ChR2 mRNA at high levels in major brain structures such as the cerebral cortex, hippocampus, thalamus, and cerebellum. We combined this optogenetic stimulation with our multineuronal (multiple isolated single-unit) recording via 32-channel tetrode-like silicon probes [7] in order to greatly increase the number of neurons/projections that would be subject to the collision test. In this way, the *Multi-Linc* analysis has the potential to reveal spike outputs from identified projection neurons in multiple brain areas simultaneously.

To demonstrate the usefulness of *Multi-Linc* analysis, we applied this approach to identifying the axonal projections of layer 5 pyramidal cells in the primary and secondary motor cortices of behaving ChR2 Tg rats. Anatomically, layer 5 pyramidal cells are classified as either intratelencephalic (IT)-type or pyramidal tract (PT)-type neurons [8, 9]. IT-type neurons project into the telencephalon (cerebral cortex and striatum) bilaterally, but never out of these regions, and the PT-type neurons project out of the telencephalon (e.g., to the thalamus and spinal cord) but never into the contralateral telencephalon. Thus, IT-type neurons should display spike collisions specifically after antidromic stimulation of the contralateral cerebral cortex or striatum, whereas PT-type neurons should do so upon stimulation of the ipsilateral thalamus (Fig. 24.2a).

Using *Multi-Linc* analysis, we successfully identified a hundred IT- and PT-type neurons, some of which exhibited functional spike activity in association with

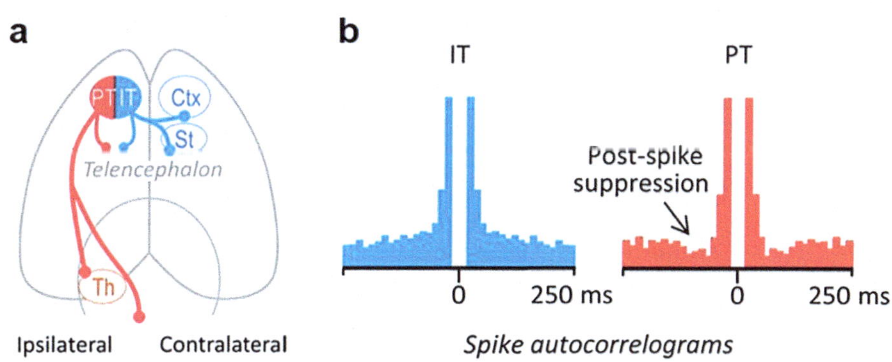

Fig. 24.2 Two types of cortical projection neurons identified by *Multi-Linc* analysis. (**a**) Axonal projections of intratelencephalic (IT)-type neurons (blue) and pyramidal tract (PT)-type neurons (red). Ctx, cortex; St, striatum; Th, Thalamus. (**b**) Schematic illustrations of spike autocorrelograms in IT-type and PT-type neurons. Post-spike suppression was observed specifically in the PT-type neurons. See [3] for details

performance of a specific behavior, i.e., lever manipulation with one forelimb [3]. Furthermore, we observed post-spike suppression as a characteristic of PT-type neurons in their spike autocorrelograms in both task-performing and resting conditions (Fig. 24.2b) [3]. These results demonstrated that *Multi-Linc* analysis is a reliable and useful tool for elucidating distinct spike dynamics of cortical projection neurons.

24.3 Advantages and Future Directions

Several cutting-edge technologies are capable of revealing functional connectivity among brain areas. However, none of these approaches offers simultaneous measurement of spike outputs from individual neurons in one area to other areas with high time resolution. For example, two-photon laser-scanning calcium imaging, in combination with pathway-specific expression of a calcium indicator, allows us to measure functional changes in spike rate simultaneously in hundreds of neurons projecting to a focused area. However, due to its low time resolution (10 to 100 ms), even this powerful technology cannot precisely discriminate the occurrence of each spike. Consequently, we cannot determine functional spike properties such as spike synchrony among neurons or spike phase locking during fast oscillations (e.g., gamma bands; cf. [10, 11]). In addition, the pathway-specific expression of calcium indicator is usually limited to only one or a few pathways, based on the number of available fluorescent proteins of different colors, which must be introduced by viral infection. Thus, these methods are unsuitable for precisely analyzing spike interactions among many brain areas.

By contrast, *Multi-Linc* analysis has several advantages for exploring such inter-areal spike interactions. First, spike outputs via multiple pathways from several areas can be efficiently assayed in parallel, using precious animals that have been trained over long periods of time to perform specific behavioral tasks. Second, it provides reliable information about spike synchrony among identified projection neurons, as well as their spike phase locking during the fast oscillations. Third, the ChR2 Tg animals make it possible optogenetically identify axonal projections without time-consuming preparations such as viral infection, in utero electroporation, or crossbreeding. Fourth, unlike conventional electrical stimulation, optogenetic stimulation results in neither electrical artifacts nor electrolytic lesions. Finally, unlike juxtacellular recording methods [7, 12], *Multi-Linc* does not require any morphological visualization or reconstruction of recorded neurons.

The current *Multi-Linc* approach still has room for its optimization. For example, axonal terminal-specific expression of ChR2 molecules would greatly improve the spatial specificity of optogenetic stimulation. In addition, if the optogenetic stimulation of different sites could be selected automatically in an optimal manner, the efficiency of multineuronal collision tests would be enhanced dramatically. In the future, such technical improvements will pave the way for a “high-throughput” *Multi-Linc* method that will reveal a global view of inter-areal spike communications in behaving animals.

Acknowledgments This work was supported by Brain/MINDS from AMED and Grant-in-Aid for Scientific Research on Innovative Areas (JP26112005) from MEXT. I am grateful to Drs. A. Saiki, Y. Sakai, R. Fukabori, S. Soma, J. Yoshida, M. Kawabata, H. Yawo, K. Kobayashi, and M. Kimura for the development of the *Multi-Linc* system.

References

1. Lipski, J.: Antidromic activation of neurones as an analytic tool in the study of the central nervous system. J. Neurosci. Methods. **4**, 1–32 (1981)
2. Evarts, E.V.: Relation of pyramidal tract activity to force exerted during voluntary movement. J. Neurophysiol. **31**, 14–27 (1968)
3. Saiki, A., Sakai, Y., Fukabori, R., Soma, S., Yoshida, J., Kawabata, M., Yawo, H., Kobayashi, K., Kimura, M., Isomura, Y.: *In vivo* spiking dynamics of intra- and extratelencephalic projection neurons in rat motor cortex. Cereb. Cortex. (2017). https://doi.org/10.1093/cercor/bhx012
4. Jennings, J.H., Sparta, D.R., Stamatakis, A.M., Ung, R.L., Pleil, K.E., Kash, T.L., Stuber, G.D.: Distinct extended amygdala circuits for divergent motivational states. Nature. **496**, 224–228 (2013)
5. Li, N., Chen, T.W., Guo, Z.V., Gerfen, C.R., Svoboda, K.: A motor cortex circuit for motor planning and movement. Nature. **519**, 51–56 (2015)
6. Tomita, H., Sugano, E., Fukazawa, Y., Isago, H., Sugiyama, Y., Hiroi, T., Ishizuka, T., Mushiake, H., Kato, M., Hirabayashi, M., Shigemoto, R., Yawo, II., Tamai, M.: Visual properties of transgenic rats harboring the channelrhodopsin-2 gene regulated by the thy-1.2 promoter. PLoS One. **4**, e7679 (2009)
7. Isomura, Y., Harukuni, R., Takekawa, T., Aizawa, H., Fukai, T.: Microcircuitry coordination of cortical motor information in self-initiation of voluntary movements. Nat. Neurosci. **12**, 1586–1593 (2009)
8. Morishima, M., Kawaguchi, Y.: Recurrent connection patterns of corticostriatal pyramidal cells in frontal cortex. J. Neurosci. **26**, 4394–4405 (2006)
9. Shepherd, G.M.: Corticostriatal connectivity and its role in disease. Nat. Rev. Neurosci. **14**, 278–291 (2013)
10. Igarashi, J., Isomura, Y., Arai, K., Harukuni, R., Fukai, T.A.: θ-γ oscillation code for neuronal coordination during motor behavior. J. Neurosci. **33**, 18515–18530 (2013)
11. Kimura, R., Saiki, A., Fujiwara-Tsukamoto, Y., Sakai, Y., Isomura, Y.: Large-scale analysis reveals populational contributions of cortical spike rate and synchrony to behavioural functions. J. Physiol. **595**, 385–413 (2017)
12. Isomura, Y., Takekawa, T., Harukuni, R., Handa, T., Aizawa, H., Takada, M., Fukai, T.: Reward-modulated motor information in identified striatum neurons. J. Neurosci. **33**, 10209–10220 (2013)

Chapter 25
Intra-body Communication as an Emerging Approach to Neuromodulation

Javier Reina-Tosina, M. Amparo Callejón, Laura Fernández, and Laura M. Roa

Abstract The application of galvanic intra-body communication (IBC) to brain tissues opens new alternatives for the coupling of electromagnetic energy to target areas for brain stimulation and neuromodulation. The lack of knowledge about the electric field distribution under neural IBC stimulation can be alleviated with computational electromagnetic models. In this work, we perform a parametric study of relevant design parameters, such as frequency range or electrode configuration, which is supported by a simplified spherical model emulating human brain tissues. The objective is to obtain electric field and current density distributions as a function of frequency for different electrode configurations, allowing their comparison with other neuromodulation techniques such as transcranial direct-current stimulation.

Keywords Intra-body communication · Neural stimulation · Neuromodulation · Computational electromagnetics

25.1 Introduction

Neuromodulation techniques are based on the stimulation of the nervous system through the application of electric, electromagnetic, or optical energy sources. Their objective is the activation, inhibition, and/or regulation of the neural activity depending on the target disorder. Focusing on electric/electromagnetic sources, several options have been applied to a wide range of neural pathologies, with deep brain stimulation (DBS), transcranial magnetic stimulation (tMS), and transcranial direct-current stimulation (tDCS) being the most popular [1]. Despite the advances in brain computational modelling [2–4], the bioelectric mechanisms underlying the

J. Reina-Tosina (✉) · M. A. Callejón · L. Fernández · L. M. Roa (✉)
Biomedical Engineering Group, Universidad de Sevilla, Escuela, Técnica Superior de Ingeniería, Seville, Spain
e-mail: jreina@us.es; mcallejon@us.es; lroa@us.es

J. M. Delgado-García et al. (eds.), *Advances in Cognitive Neurodynamics (VI)*, Advances in Cognitive Neurodynamics,
https://doi.org/10.1007/978-981-10-8854-4_25

neuromodulation process remain unknown [5, 6] and there is a lack of methodology to establish the stimulation parameters, which are commonly set following safety recommendations and clinical experience [7].

Intra-body communication (IBC) uses the biological tissues as transmission media for electromagnetic signals and have been proposed for the interconnection of on-body personal health devices and medical implants [8]. This paper shows a preliminary study about the feasibility of galvanic coupled IBC for brain stimulation. Our hypothesis is that neuromodulation in other frequency bands [4] with IBC techniques can represent a promising approach to activate/inhibit target brain areas with a higher flexibility in terms of accuracy and customization to the anthropometrical and dielectric properties of the tissues.

To shed light on the use of specific frequency ranges, signal levels, or current pathways during the neuromodulation process based on IBC techniques, we have developed a 3D electromagnetic computational model following a similar procedure than the computational finite element (FE) approach proposed in [9], now applied to specific brain tissues. The objective is to use the proposed FE model to analyze key variables in the neuromodulation process, including their dependence with frequency and electrode configuration, and a comparative study between IBC and tDCS.

25.2 Computational FE Model of the Human Head

A multilayered sphere composed of different tissues—scalp, skull, cerebrospinal fluid (CSF), and gray and white matter—has been used to resemble the human head. The use of this simplified geometry is justified due to its ease of computational effort and because it allows the derivation of qualitative knowledge that is not hindered by a complex geometry. Figure 25.1a presents a transverse section of the different tissue layers considered in the spherical head model, with thicknesses based on true anatomical proportions. The electrodes were located on the head surface composing a galvanic coupled pair with a thickness of 1 mm and an area of 5 cm × 7 cm [10].

The frequency response of tissue dielectric properties was emulated as four-dispersion Cole-Cole models based on [11] for scalp, fat, skull, and white and gray matter. Specifically, a weighted average of the dielectric properties of the skin and of the subcutaneous fat was considered for the scalp following the approach in [10]. In the case of CSF, a constant value of 1.79 for electric conductivity and 110 for relative permittivity are widely accepted [12]. The electrode's properties were those of perfect conductors, and an injected current I_0 of 1 mA was selected.

The proposed model has been implemented with the aid of the electric currents interface within the AC/DC module of COMSOL Multiphysics 4.3a, which solves a quasi-static approximation of Maxwell's equations using the FE method. Specifically, the formulation includes the charge continuity equation and the electric form of Gauss' law in the frequency domain.

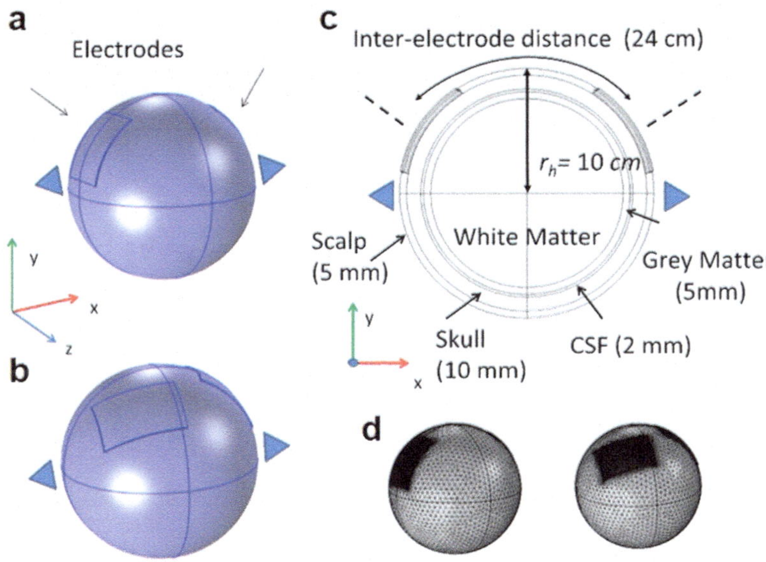

Fig. 25.1 (**a** and **b**) Spherical head model with the placement of two electrode configurations. (**c**) Transverse section showing the different brain tissue layers. (**d**) Resultant FE meshed model

The spherical geometry has been meshed using a tetrahedral element within the finer default meshing option provided by COMSOL Multiphysics for the selected physics. The resultant mesh can be seen in Fig. 25.1d. The quasi-static formulation is only valid when the wavelength is considerably greater than the dimensions of the proposed geometry, thus allowing both the inductive and wave propagation effects to be neglected. For this reason, the frequency limit of this work is fixed to 100 kHz, a common frequency range used in IBC galvanic coupling [8, 9].

25.3 Results

Figure 25.2 shows the frequency response (between 1 Hz and 100 kHz) of the electric field, in each brain tissue corresponding to the configuration of Fig. 25.1b, which is a common location in tDCS clinical protocols for the primary motor cortex [10].

These results are within the same magnitude range than previous works in the literature [6, 10] and suggest that the electric field is mainly confined in the scalp at low frequencies, remaining approximately constant in inner tissues such as gray and white matter. Since the signal excitation consists of a constant current amplitude waveform, the electric field is modulated by tissue resistivity, which decreases with frequency.

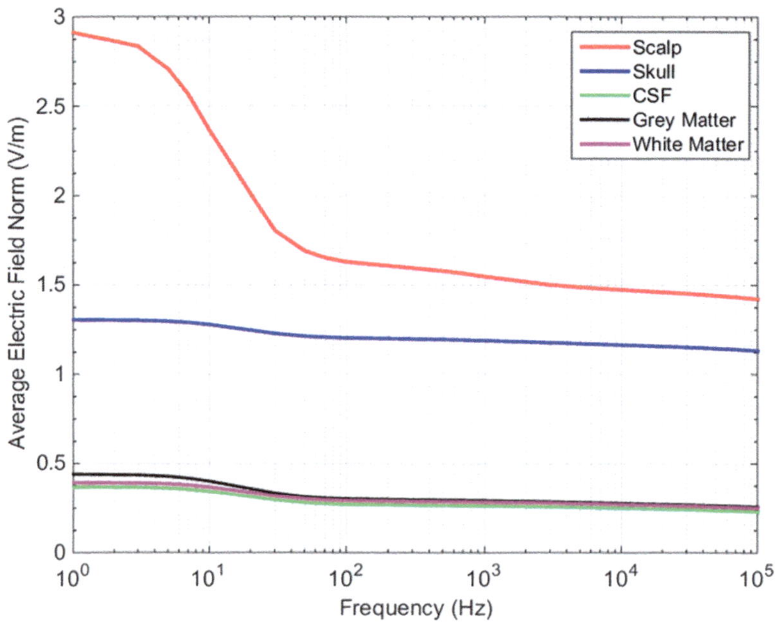

Fig. 25.2 Average volume value of computed electric field norm (V/m) as a function of frequency for each brain tissue

Regarding the electric current density, their levels for the different brain tissues are shown in Fig. 25.3. Interestingly, current is mainly confined in the CSF due to its high conductivity, with values (not depicted) over ten times higher than those corresponding to the other tissues. This trend starts to change at higher frequencies where electric current increases in the white matter. This evidences that the overall value of electric current flowing through each brain tissue is not only a function of frequency and tissue conductivity but also of tissue area.

With the aim of studying the effect of different electrode positions, the frequency was swept over three points (1 Hz, 1 kHz, and 100 kHz), and the electrode placements of Fig. 25.1a and Fig. 25.1b were simulated to compare the differences between both configurations.

Figure 25.4 represents the level of current density in brain tissues for the cases under study. In addition to the dependence of current penetration with frequency, the configuration of Fig. 25.1a shows a symmetrical current pattern, while that of Fig. 25.1b accomplishes a more selective stimulation. Although not shown, simulation results let us conclude that as the interelectrode increases, the percentage of current diminishes in the outer layers and increases in the inner brain tissues, possibly caused by the fact that the current lines are distributed widely through the complete section of the head, allowing them to penetrate toward deeper brain tissue areas.

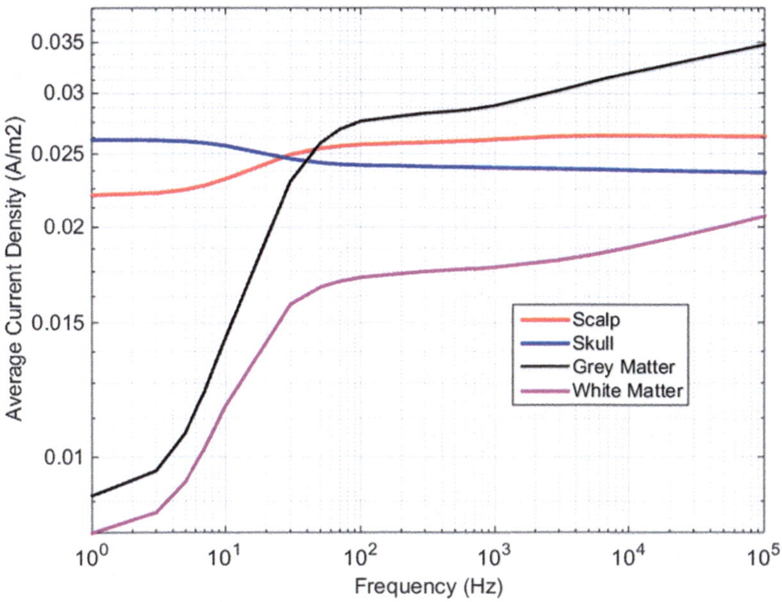

Fig. 25.3 Average volume value of computed current density norm (A/m^2) as a function of frequency for each brain tissue

Table 25.1 Current density levels (A/m^2) for brain tissues

Tissue	DC	100 Hz	1 kHz
Scalp	0.250	0.013	0.013
Fat	0.013	0.015	0.015
Skull	0.012	0.017	0.016
CSF	0.263	0.535	0.520
GM	0.038	0.038	0.042
WM	0.027	0.028	0.029

Electrode separation: 24 cm

Finally, a comparison of the current levels obtained with IBC and tDCS is summarized in Table 25.1.

Again, the differences between both methods are mainly due to the frequency-dependent characteristic of the tissue dielectric properties. Considering that scalp conductivity is four magnitude orders lower in AC than DC for frequencies up to 100 kHz, under IBC stimulation, much lower tangential current flows between the electrodes. Moreover, surface charge accumulation in the interfaces between the inner tissues is minimized, favoring the radial penetration.

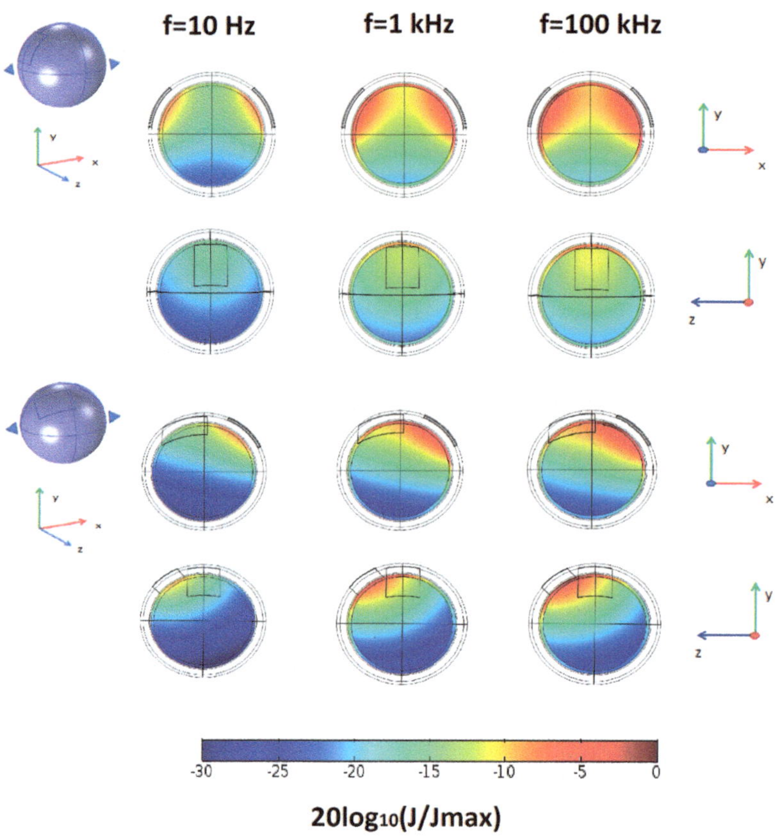

Fig. 25.4 Current density distribution in terms of frequency and electrode configuration

25.4 Conclusions

We have shown that IBC provides a flexible method for neuromodulation as the electrical pathways established through the different brain tissues strongly depend on parameters such as frequency, tissue conductivity, and area. The computed results show that most of the electric current flows through the CSF and the white matter. Another key variable able to modulate the electric current pattern through brain tissues is the electrode configuration, which shapes the current lines to be directed through the targeted brain areas. Thus, the use of computational models becomes a practical tool due to their ability to provide practical hints about the placement of electrodes to focus a target brain area.

Acknowledgments Supported by grants PI15/00306 and DTS15/00195, by Instituto de Salud Carlos III, and INT-2-CARE, NeuroIBC, and ALBUMARK, by CIBER-BBN.

References

1. Lewis, P.M., Thomson, R.H., Rosenfeld, J.V., Fitzgerald, P.B.: Brain neuromodulation techniques: a review. Neuroscientist. **22**, 406–421 (2016)
2. Song, B., Wen, P., Ahfock, T., Li, Y.: Numeric investigation of brain tumor influence on the current distributions during transcranial direct current stimulation. I.E.E.E. Trans. Biomed. Eng. **63**, 176–187 (2016)
3. Kwon, O.I., et al.: Current density imaging during transcranial direct current stimulation using DT-MRI and MREIT: Algorithm development and numerical simulations. I.E.E.E. Trans. Biomed. Eng. **63**, 168–175 (2016)
4. Wagner, T., et al.: Impact of brain tissue filtering on neurostimulation fields: a modeling study. NeuroImages. **85**, 1048–1057 (2014)
5. Ye, H., Steiger, A.: Neuron matters: electric activation of neuronal tissue is dependent on the interaction between the neuron and the electric field. J. Neuroeng. Rehabil. **12**, 65 (2015)
6. Yi, G.-S., et al.: Exploring how extracellular electric field modulates neuron activity through dynamical analysis of a two-compartment neuron model. J. Comput. Neurosci. **36**, 383–399 (2014)
7. Rossini, P., et al.: Non-invasive electrical and magnetic stimulation of the brain, spinal cord, roots and peripheral nerves: basic principles and procedures for routine clinical and research application. An updated report from an I.F.C.N. Committee. Clin. Neurophysiol. **126**, 1071–1107 (2015)
8. Seyedi, M., Kibret, B., Lai, D.T.H., Faulkner, M.A.: Survey on intrabody communications for body area network applications. I.E.E.E. Trans. Biomed. Eng. **60**, 2067–2079 (2013)
9. Callejón, M.A., Reina-Tosina, J., Naranjo-Hernández, D., Roa, L.M.: Galvanic coupling transmission in intrabody communication: a finite element approach. I.E.E.E. Trans. Biomed. Eng. **61**, 775–783 (2014)
10. Parazzini, M., et al.: Transcranial direct current stimulation: estimation of the electric field and of the current density in an anatomical human head model. I.E.E.E. Trans. Biomed. Eng. **58**, 1773–1780 (2011)
11. Gabriel, S., Lau, R.W., Gabriel, C.: The dielectric properties of biological tissues: III. Parametric models for the dielectric spectrum of tissues. Phys. Med. Biol. **41**, 2271–2293 (1996)
12. Wenger, C., et al.: The electric field distribution in the brain during TTFields therapy and its dependence on tissue dielectric properties and anatomy: a computational study. Phys. Med. Biol. **60**, 7339–7757 (2015)

Chapter 26
Electrophysiology Techniques in Visual Prosthesis

Alejandro Barriga-Rivera and Gregg Jorgen Suaning

Abstract Research in the field of visual prosthesis is advancing quickly with several groups around the world joining efforts to produce more effective and safe implants. In particular, new stimulation strategies are being investigated to elicit more meaningful percepts of light and to safely increase the visual acuity achieved by these devices. The synergy between in vitro, in silico and in vivo electrophysiology techniques can be exploited to accelerate research outcomes and to make them quickly available to the recipients of these implants.

Keywords Visual prosthesis · Electrophysiology · Electrical stimulation

26.1 Introduction

Electrical stimulation of the retina is promising a functional solution for those blinded by conditions of the eye such as retinitis pigmentosa. Current research is pushing the boundaries of bionic vision further with new stimulation paradigms that can improve the performance of the current technology available in the market [1–3]. One of the challenges of retinal electrostimulation is to develop the ability to elicit neural messages in the same way as healthy vision while reducing risks derived from injecting charge. In this context the synergies that exist in the combination of in silico, in vitro and in vivo studies can help accelerate the outcomes of visual prosthesis research. As shown in Fig. 26.1., results obtained from in vitro

A. Barriga-Rivera (✉)
Faculty of Engineering and Information Technology, The University of Sydney, Sydney, NSW, Australia

Division of Neuroscience, University Pablo de Olavide, Seville, Spain
e-mail: alejandro.barriga-rivera@sydney.edu.au

G. J. Suaning
Faculty of Engineering and Information Technology, The University of Sydney, Sydney, NSW, Australia

J. M. Delgado-García et al. (eds.), *Advances in Cognitive Neurodynamics (VI)*,
Advances in Cognitive Neurodynamics,
https://doi.org/10.1007/978-981-10-8854-4_26

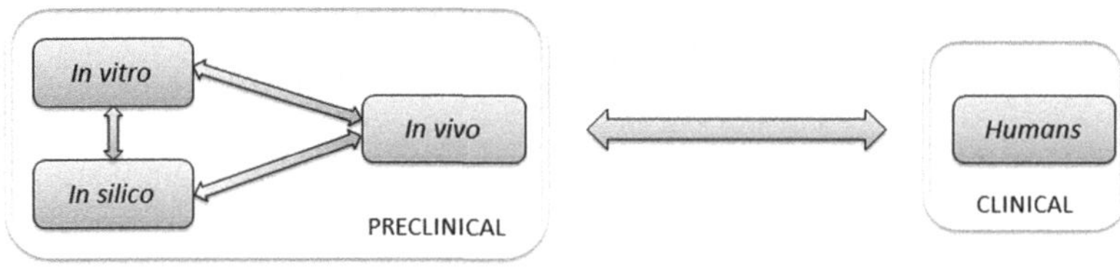

Fig. 26.1 Diagram with interaction between the three preclinical scenarios and the access to the patient. In vitro research assists with the validation of computational models. Their outcomes can be used to design experiments using animal models thus providing a quick and effective pathway to clinical trials

retinal preparations can assist with validating computational models, and these can be used to predict the response of the neural tissue to a variety of stimulation paradigms [1]. The efficacy of these stimulation strategies can be tested using in vivo electrophysiology experiments.

26.2 In Vitro Techniques

26.2.1 Patch Clamp

In visual prosthesis research, whole-cell recordings from retinal ganglion cells (RGCs) are commonly used to investigate the response of different functional RGC types to visual and synthetic stimuli [4]. Common animal models include the rabbit, the rat and the mouse. The animal is euthanised and the eyes immediately enucleated and placed in Ames medium. Next the eyes are hemisected and the retinae isolated and cut into several sections. These sections are placed inside a chamber and perfused with Ames medium bubbled with 5% CO_2 and 95% O_2 to achieve a pH equal to 7.4. The medium is heated at 33–35 °C [5]. Then, a micropipette electrode pulled from borosilicate glass, filled with intracellular solution [4], is advanced into the cell and interfaced to a patch-clamp amplifier. This procedure is carried out under a microscope using a micromanipulator, which, to reduce vibrations, can be electronically actuated. A second electrode placed in the extracellular space is typically used to deliver the electrical stimulus.

26.2.2 Calcium Imaging: Laser Scanning Confocal Microscopy

Intracellular calcium levels can be imaged by using fluorescent markers which respond to the binding of Ca^{++} ions. Thus, calcium-dependent electrical activity of neurons and glial cells can be monitored using this technique. Although the target tissue is usually stained with the appropriate dye, calcium indicators can also be genetically encoded allowing for monitoring of a variety of neuronal events [6]. Two microscopy techniques are used, confocal microscopy and two-photon microscopy. The second approach will be described in the in vivo section.

Laser scanning confocal microscopy allows for the reconstruction of three-dimensional images with high optical resolution. It relies on the use of point illumination and a pinhole to eliminate out-of-focus light beams. Only those photons emitted by the fluorescent molecules in close proximity to the focal plane can be detected. This produces an increased resolution with lower intensity when compared to wide-field fluorescence microscopy. In visual prosthesis, this technique has been used to assess biocompatibility of new electrode designs [7] or to study amacrine cells among others. Retinal preparations are similar to those requited to conduct patch-clamp recordings. In fact, both calcium activity and intracellular potentials can be simultaneously recorded [5].

26.2.3 In Vitro Electrophysiological Recordings

Retinal local field potentials following electrical stimulation of the retina can be studied using multi-electrode arrays (MEAs). The excised retina, prepared in the same way as that for the patch-clamp technique, is placed and fixed over an array of electrodes [8]. As technology advances, more densely packed electrode arrays are being developed allowing for smaller-scale recordings [9].

26.3 In Vivo Techniques

26.3.1 In Vivo Electrophysiological Recordings

While in vitro techniques allow for analysing the activity of the isolated retinal network, in vivo electrophysiology techniques are required to further the understanding on how this information is being transmitted to higher visual centres. A repertoire of MEAs have been used in the study of the effects of retinal neurostimulation including supradural electrodes [10], penetrating cortical electrode arrays [2] and subcortical electrode arrays [11]. These are typically implanted in the superior colliculus (SC), the lateral geniculate nucleus (LGN), or the visual cortex (VC). Briefly, the animal is anaesthetised using a combination of anaesthetic agents delivered simultaneously through a venous line and a tracheostomy. Gaseous anaesthesia (e.g. isoflurane) combined with infused anaesthetic agents (e.g. ketamine/alphaxalone or ketamine/xylazine) [2, 12] allow for rapid adjustment of the depth of anaesthesia while maintaining good neural responsiveness. Then, an electrode array is surgically inserted to stimulate the retina through an incision in the sclera, typically in the suprachoroidal space overlaying the area centralis [2] or in the retrobulbar space

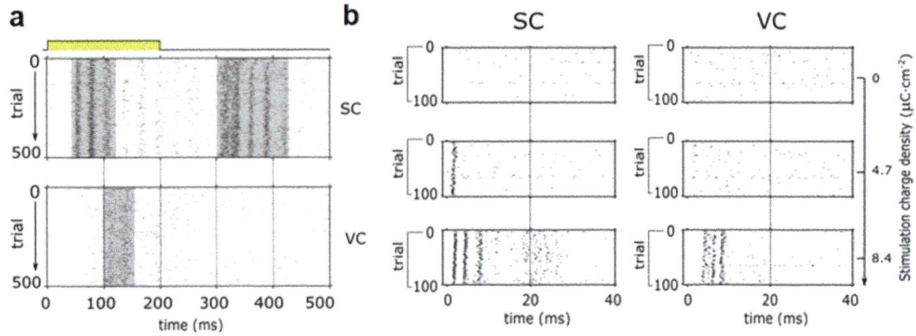

Fig. 26.2 Example of spike rasters obtained simultaneously from superior colliculus (SC) and visual cortex (VC) of a rat which were detected using the methodology described in [12]. Responses in panel (**a**) were obtained using a pulse of light generated with a white diode placed 1 cm away from the surface of the eye. Responses in panel (**b**) were elicited from a retrobulbar disc electrode, 2 mm in diameter, using constant-current biphasic pulses with charge balanced and 400 μs in duration

[12]. Location of the stimulating electrode array is verified using fundus imagery techniques or optical coherence tomography [1]. With the animal placed within a stereotaxic frame, a craniotomy is performed to expose the primary VC following coordinates from Tusa et al. (cats) [13]. If working with rats, coordinates of the visual centres were determined by Paxinos and Watson [14]. Figure 26.2 shows an example of recordings in a rat from both the SC and the VC.

Next, the dura is carefully reflected under a surgical microscope using a needle with the tip bent. The cortical representation of the retinal area under the influence of retinal stimulation is mapped using a surface electrode. Then, a penetrating MEA is inserted and interfaced to a signal acquisition system [15]. Access to subcortical areas can be guided by assessing visually evoked responses.

26.3.2 Two-Photon Microscopy

In traditional fluorescence microscopy, the exciting photon has higher energy compared to the emitted photon. In two-photon microscopy, two photons of lower energy, typically in the near-infrared spectrum, are used to excite a fluorescent dye which emits a higher-energy photon. This technique provides deeper penetration and can be used both in vitro using retinal preparations and in vivo for cortical electrophysiology [16]. This emerging technique is showing an excellent capacity to answer important questions in neuroscience in general and in visual neuroscience in particular. Although the number of publications that have considered the use of this technique in retinal prosthesis is scarce [17, 18], this approach can help in overcoming some of the significant challenges to do with stimulation artefacts. The in vitro methodology is similar to that used in confocal microscopy. However, when used in vivo, animal preparations require special attention. Briefly, with the animal surgically anaesthetised, a craniotomy is performed to expose the VC. A well is created using silicone tubing or dental cement around the opening and the cortical area perfused with synthetic cerebrospinal fluid (sCSF) at 35 °C. The exposed VC will be stained with calcium sensitive dyes (OGB-1, Fluo-4 AM or Cal-520) and loading acetoxymethyl (AM) esters [19] using a micropipette inserted coaxially into the cortex. The visual cortex will be then imaged using the two-photon microscope following electrical stimuli delivered at the retina.

26.3.3 Voltage-Sensitive Dyes

Voltage-sensitive dyes (VSDs) represent an alternative to avoid mechanical damage produced by the in vivo insertion of conventional electrodes. Two common VSDs have been used to study cortical activity, blue RH-1691 and red di-4-ANEPPS. Blue dyes have been reported in the study of cortical stimulation [20] and are preferred over red dyes as the absorption spectrum of haemoglobin is closer to the latter [21]. As in the case of two-photon microscopy, the cortical area under study is encircled using silicon tubing or dental cement. The space is filled with sCSF enriched with a voltage-sensitive dye (e.g. RH-1691) and heparin for staining for 90 min [20]. Afterwards the space will be washed with sCSF for at least 30 min. The sCSF is then changed every 60 min. The exposed cortical region will be illuminated with an excitation light source (640 nm) and imaged under epifluorescence microscopy to identify electrical activation. Images are acquired at least at 1 kHz for its analysis.

26.4 In Silico Techniques

Since the Hodgkin-Huxley mathematical description of the cell membrane of excitable cells was presented in the 1950s, this model has been extensively used in neuroscience research. The model is formulated using a series of differential equations that describe the ionic conductances of the neuronal membrane. This model has been particularised for the case of RGCs and refined using realistic cell morphologies based on the anatomic reconstruction of the rabbit retina [22]. Although the use of animals is still unavoidable, these models can assist researchers with complying with two of the 3-Rs principle: *replacing* some animals and *reducing* the number required by limiting the research space. An example of the potential of computational research is the recent publication by Guo et al. [23] which suggests that sequential stimulation, first at the soma followed by stimulation at the axon of the RGCs using high frequency, can elicit more physiologically natural neural messages.

26.5 Conclusion

Progress in bionic vision has been enabled in the described techniques, which, combined, allow for devising new ways of improving integration between the device and the biology underlying visual perception.

Acknowledgements This work has been supported by the National Health and Medical Research Council (RG1063046).

References

1. Barriga-Rivera, A., Guo, T., Yang, C.-Y., Al-Abed, A., Lovell, N.H., Morley, J.W., et al.: High-amplitude electrical stimulation can reduce elicited neuronal activity in visual prosthesis. Sci. Rep. **7**, 42682 (2017)
2. Matteucci, P.B., Barriga-Rivera, A., Eiber, C.D., Lovell, N.H., Morley, J.W., Suaning, G.J.: The Effect of electric cross-talk in retinal neurostimulation. Invest. Ophthalmol. Vis. Sci. **57**, 1031–1037 (2016)
3. Barriga-Rivera, A., Morley, J.W., Lovell, N.H., Suaning, G.J.: Cortical responses following simultaneous and sequential retinal neurostimulation with different return configurations. In: Proceedings of the 38th Annual International. Conference of the Engineering in Medicine and Biology Society (EMBC-IEEE), 5435–5438 (2016)
4. Tsai, D., Morley, J.W., Suaning, G.J., Lovell, N.H.: Direct activation and temporal response properties of rabbit retinal ganglion cells following subretinal stimulation. J. Neurophysiol. **102**, 2982–2993 (2009)
5. Toychiev, A.H., Sagdullaev, B., Yee, C.W., Ivanova, E., Sagdullaev, B.T.: A time and cost efficient approach to functional and structural assessment of living neuronal tissue. J. Neurosci. Methods. **214**, 105–112 (2013)
6. Lin, M.Z., Schnitzer, M.J.: Genetically encoded indicators of neuronal activity. Nat. Neurosci. **19**, 1142–1153 (2016)
7. Bendali, A., Rousseau, L., Lissorgues, G., Scorsone, E., Djilas, M., Dégardin, J., et al.: Synthetic 3D diamond-based electrodes for flexible retinal neuroprostheses. Model, production and in vivo biocompatibility. Biomaterials. **67**, 73–83 (2015)
8. Weitz, A.C., Nanduri, D., Behrend, M.R., Gonzalez-Calle, A., Greenberg, R.J., Humayun, M.S., et al.: Improving the spatial resolution of epiretinal implants by increasing stimulus pulse duration. Sci. Transl. Med. **7**, 318ra203–318ra203 (2015)
9. Tsai, D., John, E., Chari, T., Yuste, R., Shepard, K.: High-channel-count, high-density microelectrode array for closed-loop investigation of neuronal networks. In: Proceedings of the 37th Annual International Conference of the Engineering in Medicine and Biology Society (EMBC-IEEE), 7510–7513 (2015)
10. Barriga-Rivera, A., Eiber C.D., Dodds C.W., Adrian, F., Veronica, T., Lovell N.H., et al.: Electrically evoked potentials in an ovine model for the evaluation of visual prosthesis efficacy. In: Proceedings of the 37th Annual International Conference of the Engineering in Medicine and Biology Society (EMBC-IEEE) (2015)
11. Barriga-Rivera, A., Guo, T., Morley, J.W., Lovell, N.H., Suaning, G.J.: Retinal electrostimulation in rats: activation thresholds from superior colliculus and visual cortex recordings. In: Proceedings of the 39th Annual International Conference of the Engineering in Medicine and Biology Society (EMBC-IEEE), 1166–1169 (2017)

12. Barriga-Rivera, A., Tatarinoff, V., Lovell, N.H., Morley, J.W., Suaning, G.J.: Long-term anesthetic protocol in rats: feasibility in electrophysiology studies in visual prosthesis. Vet. Ophthalmol. (In press). https://doi.org/10.1111/vop.12507
13. Tusa, R., Palmer, L., Rosenquist, A.: The retinotopic organization of area 17 (striate cortex) in the cat. J. Comp. Neurol. **177**, 213–235 (1978)
14. Paxinos, G., Watson, C.: The Rat Brain in Stereotaxic Coordinates. Academic, San Diego (1998)
15. Barriga-Rivera, A., Eiber, C.D., Matteucci, P.B., Chen, S.C., Morley, J.W., Lovell, N.H., et al.: A 4+ 1 architecture for in vivo electrophysiology visual prosthesis. J. Accessibility Des. All. **6**, 81–101 (2016)
16. Svoboda, K., Yasuda, R.: Principles of two-photon excitation microscopy and its applications to neuroscience. Neuron. **50**, 823–839 (2006)
17. Lagali, P.S., Balya, D., Awatramani, G.B., Münch, T.A., Kim, D.S., Busskamp, V., et al.: Light-activated channels targeted to ON bipolar cells restore visual function in retinal degeneration. Nat. Neurosci. **11**, 667–675 (2008)
18. Yang, C.-Y., Woolley, A.J., Tsai, D., Suaning, G.J., Morley, J.W., Lovell, N.H.: Calcium imaging of retinal ganglion cell dendritic responses to extracellular microelectrode stimulation. In: Proceedings of the 7th International IEEE/EMBS Conference on Neural Engineering, 438–441 (2015)
19. Stosiek, C., Garaschuk, O., Holthoff, K., Konnerth, A.: In vivo two-photon calcium imaging of neuronal networks. Proc. Natl. Acad. Sci. U. S. A. **100**, 7319–7324 (2003)
20. Hayashida, Y., Takeuchi, K., Ishikawa, N., Okazaki, Y., Tamas, F., Tanaka, H., et al.: Voltage-sensitive dye imaging of the visual cortices responding to electrical pulses at different intervals in mice in vivo. In: Proceedings of the 36th Annual International Conference of the Engineering in Medicine and Biology Society (EMBC-IEEE), 402–405 (2014)
21. Grandy, T.H., Greenfield, S.A., Devonshire, I.M.: An evaluation of in vivo voltage-sensitive dyes: pharmacological side effects and signal-to-noise ratios after effective removal of brain-pulsation artifacts. J. Neurophysiol. **108**, 2931–2945 (2012)
22. Guo, T., Tsai, D., Morley, J.W., Suaning, G.J., Kameneva, T., Lovell, N.H., et al.: Electrical activity of ON and OFF retinal ganglion cells: a modelling study. J. Neural Eng. **13**, 025005 (2016)
23. Guo, T., Barriga-Rivera, A., Suaning, G.J., Tsai, D., Dokos, S., Morley, J.W., et al.: Mimicking natural neural encoding through retinal electrostimulation. In: Neural Engineering (NER) 8th International IEEE/EMBS Conference, 284–287 (2017)

Chapter 27
Application of Video-Oculography for the Analysis of the Vestibulo-Ocular Reflex in Acute Hypoxic Mice

Juan Carlos López-Ramos, Ana Belén García Cebrián, and José M. Delgado-García

Abstract The mouse is the animal most commonly used in biomedical research. In particular, recent advances in molecular and genetic techniques require new experimental approaches for the study of pathologies involving the vestibulo-ocular reflex (VOR) and other visual, otologic, or cerebellar-related impairments in this laboratory animal. For this, we have developed a video-oculographic recording system that enables analyzing several parameters of the VOR. For quantification and comparative purposes, we have studied the VOR in hypoxic and control mice submitted to 5000 m and 7000 m of simulated altitude in a hypobaric chamber. VOR gain and phase, and the cephalic impulse, were checked in experimental mice. Results indicate an increased VOR gain, from 0.1 Hz to 0.6 Hz, and an impaired VOR performance, in hypoxic mice. In conclusion, the setup reported here for vestibular stimulation provides a video-oculographic recording system suitable for the study of the VOR in mice and delivers enough quantitative data to determine specific impairments, as in the case of acute hypobaric hypoxia.

Keywords Vestibulo-ocular reflex · Video-oculography · Hypobaric hypoxia

27.1 Introduction

The vestibulo-ocular reflex (VOR) generates compensatory eye movements in response to head movements. Head motion is detected by the semicircular canals and translated into adequate motor commands via a well-defined neural network located in the brainstem [1–3]. This reflex is the result of the activation, by the semicircular canals, of a reflex arc between primary and secondary vestibular afferent neurons and the motoneurons that innervate the extraocular muscles. The

J. C. López-Ramos (✉) · A. B. G. Cebrián · J. M. Delgado-García
Division of Neuroscience, Pablo de Olavide University, Seville, Spain
e-mail: jclopez@upo.es

J. M. Delgado-García et al. (eds.), *Advances in Cognitive Neurodynamics (VI)*,
Advances in Cognitive Neurodynamics,
https://doi.org/10.1007/978-981-10-8854-4_27

actual VOR occurs under conditions in which the individual is not subjected to additional visual stimulations. If visual stimuli are present, the VOR is enhanced [4]. It has been reported that hypoxia may produce various vestibular symptoms including swaying, spontaneous nystagmus, and positional nystagmus [5, 6], and healthy subjects have been observed to experience an altered vestibular function at high altitudes, which reverses with acclimatization. These deficits have been linked with central, rather than peripheral, vestibular disturbance [7]. In this regard, an abnormal spontaneous firing rate in medial vestibular nuclei neurons due to a decreased calcium-activated potassium channels caused by hypoxia has been described [8].

Since the mouse is the animal most commonly used in biomedical research, recent advances in molecular and genetic techniques require the development of new experimental approaches for the study of pathologies involving the VOR and other visual, otologic, or cerebellar-related impairments susceptible of being evoked in this laboratory animal. In the present study, we describe an experimental setup for the vestibular stimulation of alert-behaving mice, provided with a video-oculographic recording system. In particular, the specific aim was to study the effects of acute hypobaric hypoxia on the VOR of wild-type mice.

27.2 Methods

27.2.1 Experimental Subjects

Experiments were performed on C57Bl/6 adult male mice (4–6 months old; 25–33 g) obtained from an official supplier (University of Granada Animal House, Granada, Spain). Before surgery, animals were housed in common cages ($n = 10$ per cage), but after surgery they were housed individually. Mice were kept on a 12:12 h light-dark cycle with constant ambient temperature (21 ± 1 °C) and humidity ($50 \pm 7\%$). Food and water were available ad libitum. Experiments were carried out in accordance with the guidelines of the European Union Council (2010/276:33–79/EU) and Spanish regulations (BOE 34:11370–421, 2013) for the use of laboratory animals in chronic experiments. Experiments were also approved by the local Ethics Committee of Pablo de Olavide University (Seville, Spain).

27.2.2 Surgery

Animals were anesthetized with a mixture of ketamine (35 mg/kg) and xylazine (2 mg/kg) i.p. In a single surgical step, a holding system, consisting of a head plate matched to a stereotaxic apparatus, was fixed to the skull with the help of two small screws and dental cement. The aim was the proper stabilization of the head during animal rotation and eye movement recordings.

27.2.3 *Video-Oculographic Analytic System*

For our experiments, we have developed a video-oculographic analytic system, which synchronizes a MATLAB-based tracking of pupil movements with the movements of the animal's head that evoked them. The recordings (eye and table movements) were digitized with an analog/digital converter (CED 1401 Plus, Cambridge, England). The turntable was rotated by hand, following a sinusoidal wave displayed on an oscilloscope screen. Rotation of the table was recorded with a potentiometer attached to the rotating axis. The eye of the mouse was illuminated indirectly with a red cold light attached to the head-holding system. Eye positions during head rotation were recorded with a fast CCD camera (Pike F-032, Allied Technologies, Stadtroda, Germany) at a rate of 60, or 100, pictures/s. Tracking of the pupil image mass center was carried out with the help of a MATLAB-based homemade program. Finally, recordings of head rotations and eye positions were synchronized and analyzed offline with the CED software (Fig. 27.1).

Statistical analyses were carried out using the SigmaPlot 11 software (Systat Software, Inc., Germany) for a statistical significance level of $p < 0.05$. Mean values are followed by their standard error of the mean (SEM). Collected data were analyzed using a one-way ANOVA, coupled with the suggested pairwise comparison tests.

27.2.4 *Vestibular Stimulation and Recording of the VOR*

For vestibular stimulation, a single animal was placed on the turntable system (Fig. 27.1). Its head was immobilized, using the implanted head plate, at an inclination of 30° from the horizontal plane, while the animal was able to pace freely on a treadwheel. The turntable was rotated at different angular velocities at three selected frequencies (0.1, 0.3, and 0.6 Hz), with the light on, for ten cycles, with intervals of 2 min between frequencies. As already described [9] and illustrated in Fig. 27.2a, eye positions for each frequency and animal were recorded for offline analysis of gain and phase. Additionally, cephalic impulse was tested. The cephalic impulse was evoked with fast random ±10° rotations of the head, with the light off, using the same recording and stimulation system. For quantification and comparative purposes, we studied the VOR in control mice, which remained throughout at 35 m of altitude, and in acute hypoxic mice, submitted to 30 min exposure to 5000 m and 7000 m of simulated altitude in a homemade hypobaric chamber [10], just before the VOR tests.

27.3 Results

Compared with controls, acute hypoxic mice presented a significantly impaired performance for the visually enhanced VOR. According to the results illustrated in Fig. 27.2b–c, hypoxic mice ($n = 13$) presented evident abnormalities in gain

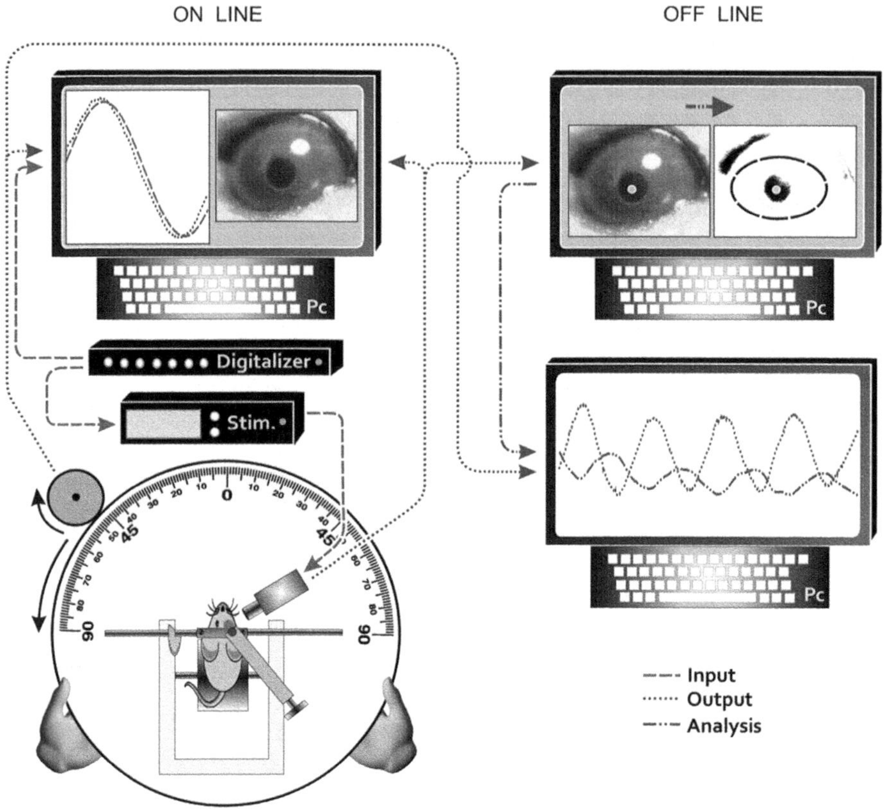

Fig. 27.1 Schematic representation of the video-oculographic system developed to study the VOR in mice. The turntable is rotated manually, and its movement, following a waveform model previously defined and shown on the scope (dashed line), is digitized online (dotted line) and recorded. Simultaneously, a fast CCD camera is triggered by a stimulator *(stim)*, which is itself triggered at the same time by the CED software *(Digitizer)*. Collected photographs are saved (dotted line) for offline analysis, when both cephalic and eye movements are synchronized (dotted and dotted-dashed line) for the analysis of gain, phase, and gaze stabilization after the cephalic impulse

Fig. 27.2 (continued) movements during VOR tests at the three selected frequencies (0.1, 0.3, and 0.6 Hz) in control (35 m) and acute hypoxic (5000 m and 7000 m) mice. Gray bars indicate the gain and phase analysis. (**b** and **c**) Gain and phase analysis of VOR in control and hypoxic animals. Values are mean ± SEM of the data acquired for the three selected frequencies. (**d**) Analysis of gaze fixation after cephalic impulse in control and hypoxic mice. Lines represent superimposition of ±15 trials of cephalic and ocular movements, showing ocular destabilization in acute hypoxic animals

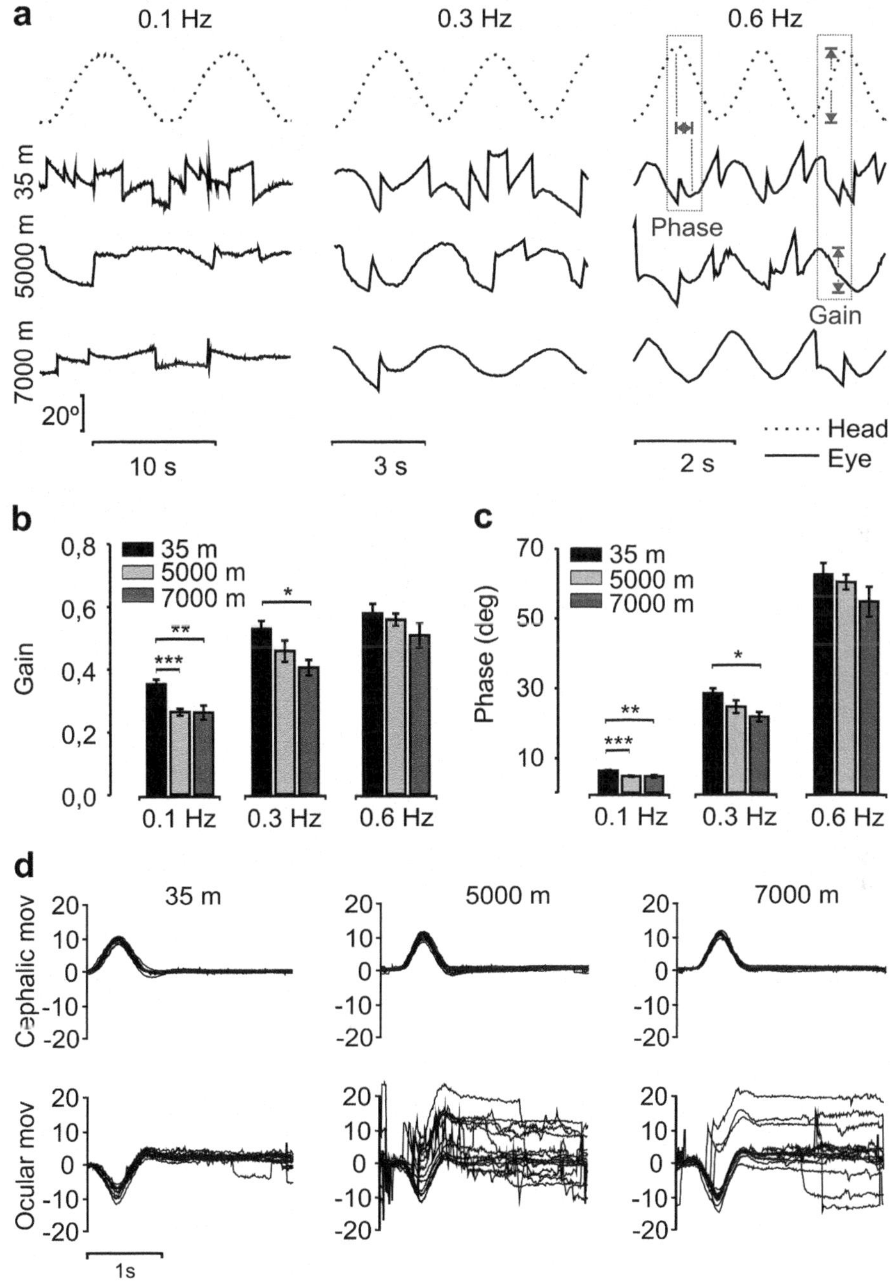

Fig. 27.2 Vestibulo-ocular reflex and cephalic impulse tests showed impaired performance in hypoxic mice. (**a**) Examples of head (black dashed lines) and eye (black continuous lines)

and phase in comparison with values reached by the control group ($n = 13$). Thus, hypoxic mice failed to reach control levels of VOR gain for head-rotation frequencies of 0.1 Hz ($F_{[2, 33]} = 9.44$; $p < 0.001$) and 0.3 Hz ($F_{[2, 33]} = 4.72$; $p < 0.05$) and phase for 0.1 Hz ($F_{[2, 33]} = 9.45$; $p < 0.001$) and 0.3 Hz ($F_{[2, 33]} = 4.72$; $p < 0.05$). Moreover, gain and phase abnormalities were more evident, at 0.3 Hz, with increased simulated altitude, reaching significant differences only following 7000 m of hypoxic exposure. Both groups of hypoxic mice suffered destabilization of gaze following the cephalic impulse test (Fig. 27.2d).

27.4 Discussion

In this work, we have described a new video-oculographic system suitable for the study of eye movements in partially restrained mice. In particular, we have carried out a study of the VOR, evoked by a sinusoidal movement of the head and by cephalic impulses, in acute hypoxic mice. As far as we know, there are no previous reports about recording cephalic impulse in mice, and/or of evoking VOR in hypoxic animals.

Interestingly enough, hypoxia has been reported to cause abnormalities in the function of vestibular neurons, consisting of a decrease of calcium-activated potassium channels, which causes an abnormal increase of firing frequency and depolarization of their resting membrane potentials [8].

In this sense, the origin of the abnormalities may be vestibular, rather than cerebellar, since the abnormalities are stronger with slower frequencies (0.1–0.3 Hz), whereas cerebellar impairment affects VOR more at higher ones (0.3–0.6 Hz) ([11], and our own data). Thus, the impairment at high altitude may be related to a selective change in sensitivity to the stimulus, the stimulus (rotations) being weaker at slower frequencies and stronger at higher ones. Finally, the impairment observed in the cephalic impulse test may be related with the increase of positional nystagmus observed in humans at high altitude [5, 6].

References

1. Ito, M., Nisimaru, N., Yamamoto, M.: Specific patterns of neuronal connections involved in the control of the rabbit's vestibulo-ocular reflexes by the cerebellar flocculus. J. Physiol. Lond. **265**, 833–854 (1977)
2. Lorente de Nó, R.: Vestibulo-ocular reflex arc. Arch. Neurol. Psychiatr. **30**, 245–291 (1933)
3. Szentágothai, J.: The elementary vestibulo-ocular reflex arc. J. Neurophysiol. **13**, 395–407 (1950)
4. Leigh, R.J., Brandt, T.: A reevaluation of the vestibulo-ocular reflex: new ideas of its purpose, properties, neural substrate, and disorders. Neurology. **43**, 1288–1295 (1993)
5. Cingi, C., Erkan, A.N., Rettinger, G.: Ear, nose, and throat effects of high altitude. Eur. Arch. Otorhinolaryngol. **267**, 467–471 (2010)
6. Singh, D., Kochhar, R.C., Kacker, S.K.: Effects of high altitude on inner ear functions. J. Laryngol. Otol. **90**, 1113–1120 (1976)
7. Mees, K., Suckfüll, M.: Cochlear and vestibular risk at high altitude. Laryngorhinootologie. **81**, 465–468 (2002)
8. Xie, H., Zhang, Y., Pan, X., Wu, S., Chen, X., Wang, J., Liu, H., Qian, X., Liu, Z., Liu, L.: Decreased calcium-activated potassium channels by hypoxia causes abnormal firing in the spontaneous firing medial vestibular nuclei neurons. Eur. Arch. Otorhinolaryngol. **272**, 2703–2711 (2015)
9. de Jeu, M., De Zeeuw, C.I.: Video-oculography in mice. J. Vis. Exp. **65**, e3971 (2012)
10. López-Ramos, J.C., Martínez-Romero, R., Molina, F., Cañuelo, A., Martínez-Lara, E., Siles, E., Peinado, M.A.: Evidence of a decrease in nitric oxide-storage molecules following acute hypoxia and/or hypobaria, by means of chemiluminescence analysis. Nitric Oxide. **13**, 62–67 (2005)
11. Van Alphen, A.M., Schepers, T., Luo, C., de Zeeuw, C.I.: Motor performance and motor learning in Lurcher mice. Ann. N. Y. Acad. Sci. **978**, 413–424 (2002)

Chapter 28
RatButton: A User-Friendly Touchscreen Presentation Software

Celia Andreu-Sánchez, Miguel Ángel Martín-Pascual, Agnès Gruart, and José María Delgado-García

Abstract Psychological and neuroscientist research requires precise presentation of stimuli and their connections to cognitive responses. Here, we introduce RatButton, a software for touchscreen research written for iOS devices with a triple role: design, presentation, and monitoring of protocols. It can be used in human and animal experimentation. The main benefits from RatButton are that it has been designed for capacitive touchscreens, that no programming language is needed, and that no PC connection is required. RatButton is also available for easy and flexible VirtuBox environment.

Keywords Touchscreen · Control software · Stimulus presentation · Psychometric · Behavioral control · Instrumentation · Neurophysiology · Cognitive science

28.1 Introduction

For the past decades, psychology and cognitive neuroscience have been using several analogic systems to present and evaluate protocols. During the last years, there has been a clear flow to digital systems. Those have expanded possibilities for researchers. Digital systems, however, tend to require learning programming language. So, behavioral researchers are asked not only to design, analyze, and study their protocols but to learn some specific software language to also write them. Here, we are presenting a software for touchscreen research in psychological and neuroscientist environment, with no need of programming language learning.

C. Andreu-Sánchez (✉) · M. Á. Martín-Pascual
Neuro-Com Research Group, Universitat Autònoma de Barcelona, Barcelona, Spain
e-mail: Celia.Andreu@uab.cat

A. Gruart · J. M. Delgado-García
Division of Neuroscience, Pablo de Olavide University, Seville, Spain

J. M. Delgado-García et al. (eds.), *Advances in Cognitive Neurodynamics (VI)*,
Advances in Cognitive Neurodynamics,
https://doi.org/10.1007/978-981-10-8854-4_28

In the past years, some systems have been developed for presenting visual stimuli in touch screens. Many of those systems have been created for infrared interactions, but this kind of interactivity has already been improved with strong and stable touchscreen's systems.

Here, we are introducing RatButton, a software for presenting interactive stimuli in touchscreens, designed for psychological and cognitive neuroscience experimentation, with human and animal models. RatButton is available for real touchscreen devices with real touch interactivity, it does not need programming, and it does not require a connection to an external PC.

28.2 Materials and Methods

28.2.1 Hardware and Software Requirements

RatButton software has been developed in C, C++, and Objective-C environments for touchscreen devices with iOS. It requires an iOS 9.0 version or later. This makes the following current devices suitable for RatButton: iPad® 4th generation, iPad® Air, iPad® Mini, and iPad® Pro. All are devices from Apple Inc. and have capacitive screens and different sizes.

The reasons for displaying RatButton in this type of devices are the major stability of the system, the higher responsibility of the screens, and the more homogeneity found in the versions of the operative system, among others, compared to their current market competitors. RatButton does not require a PC connection, as researcher can design the protocol for the experiment, display the experiment, look up the data, and send it via cloud, all in the iOS device. If desired, a connection to a PC can be made. This could be of interest for, for instance, registering the results with a different software, or for creating interactions to different devices.

28.2.2 Visual Display

The visual presentation of stimuli is made in capacitive touchscreens. The visual characteristics of the stimuli depend on the visual characteristics of the used device. We have analyzed iPad® 4th generation chromatic luminance properties (Table 28.1) with a colorimeter (Minolta Chroma Meter xy-1; Minolta Camera Co., Ltd., Osaka, Japan), a standard Munsell CIE 1931 (x,y)–Chromaticity Diagram xy-AC (Minolta Co., Ltd., Osaka, Japan), and a photometer (Sekonic DualMaster L-558/L; Sekonic Corporation, Nerima-Ku, Tokyo, Japan) to learn the real chromatic performance of the display. This is of special interest when creating protocols for specific vision requirements of subjects, e.g., rats.

Table 28.1 Chromatic luminance measures in iPad® (model 4th generation)

	White	Red	Green	Blue
Kelvin	5700	1600	2840	4000
Cd/m^2	320	70	150	84
x	0.32	0.65	0.301	0.144
y	0.35	0.32	0.603	0.045

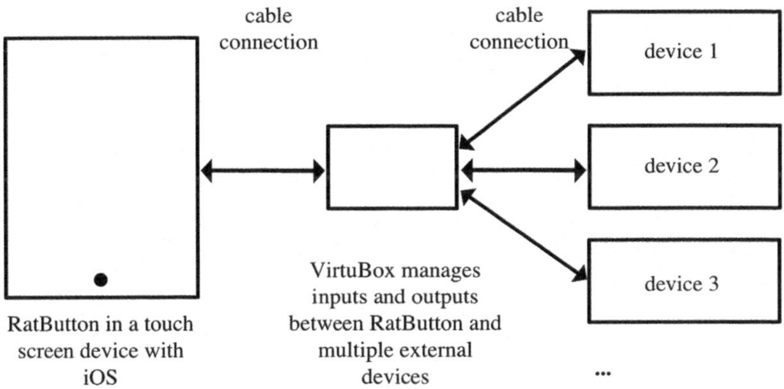

Fig. 28.1 VirtuBox environment. RatButton is displayed in an iOS device, which is connected through cable to VirtuBox device. It manages, also through a cable, the inputs and outputs connections between RatButton and different external devices

28.2.3 Connections with VirtuBox

As mentioned, RatButton does not require a connection to an external PC, though it is possible to manage. In the case of wanting an external connection, the software is carefully developed to be used with a system that we designed with our colleagues of Cibertec Company (Madrid, Spain), and we have called VirtuBox. VirtuBox is a stimuli presentation system using RatButton, able to control different hardware connections in a scientific environment [1].

VirtuBox manages inputs and outputs between RatButton and external devices (Fig. 28.1), such as a feeder, a lighting system, or a sound system. The possibilities for this are very wide. VirtuBox also creates an easy environment for researcher so there is no need of coding for controlling the connections.

28.2.4 User Interface

As mentioned above, researcher does not need to learn any programming language to present protocols. The software is very easy to use and no coding is needed. We have created a user-friendly interface for researchers. So, they can focus in developing the conceptual needs of their experiments and do not have to learn any programming language to create their protocols.

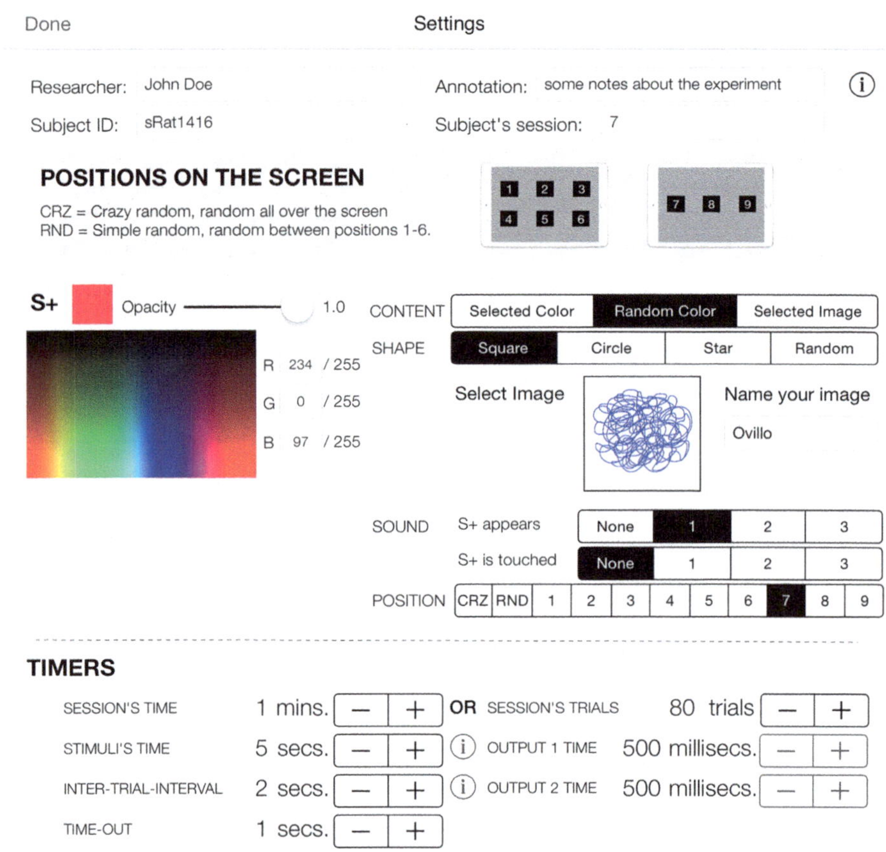

Fig. 28.2 A part of settings availability for researchers. Settings page offers users the possibility of making changes to S+ and S− stimuli (visual and sound characteristics), to the background, to the timers, to the connections, and to the session

RatButton offers different standard protocols; however researchers can make several changes in the protocols through the Settings page (Fig. 28.2). It is possible to change timers for, for example, inter-trial interval, stimuli duration, session duration, time-outs; visual characteristics of stimuli such as color, shape, or picture; or the sound in different situations, among others. The software team keeps on working to improve those customizable characteristics.

28.2.5 *Protocols*

RatButton is designed for animal and human research. It performs different designed protocols. At the moment, the most common protocols for psychological experiments are available (those are expected to be increase): visual discrimination

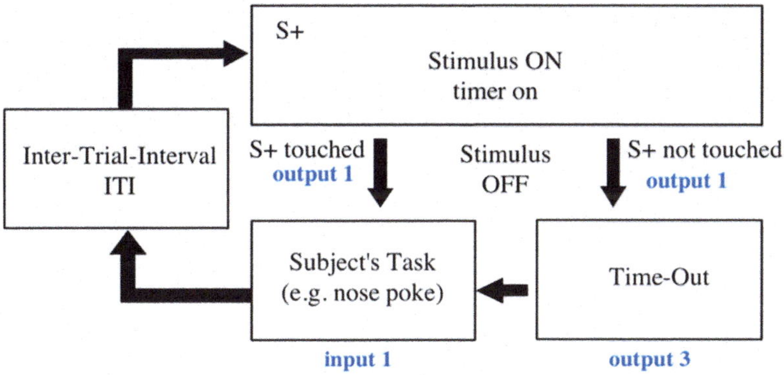

Fig. 28.3 Map for a sample protocol. In this case, the rewarded stimulus (S+) appears during a selected time and after the time expires and the stimulus disappears. In the case of having touched the stimulus, an optional output is created to inform a possible external device about this to reward the subject; in this example there is, also, a task to be done by the subject in an external device, for example, a nose-poke task. This creates an input to the protocol and activates the inter-trial-interval (ITI) timer, with a duration previously selected in Settings page; after that time, a new trial is started. In the case of not touching the stimulus, it is possible to (optionally) inform an external device about this fact to maybe punish this behavior and start a time-out situation; this can happen inside RatButton or with the collaboration of an external device (e.g., a house lighting), and for the later optional situation, an output is created. The inputs and outputs connections are totally optional and are properly managed by VirtuBox system. All mentioned parameters (stimulus' characteristics, timers, and connections) are super-easy customizable in Settings area of RatButton

learning (and reversal) (VDL), trial-unique nonmatching to location (TUNL), 5-choice serial reaction time task (5-CSRTT), location discrimination (LD), and pattern discrimination (PD).

Those protocols are presented with a visual map, in which researchers can view and analyze the flow of the protocol (Fig. 28.3 shows an example). They can easily detect whether the map fits their interests and, if so, how to change and adapt their specific preferences through Settings page.

28.2.6 Connections

RatButton can use different connections: cable and Wi-Fi; however, those are not needed for its use. The cable connection can be used to plug in information of the experiment to an external gadget. An example of this connection would be the case of VirtuBox system commercialized by Cibertec S.A. and explained above. In that case, the mentioned system will provide user with the appropriate cable for their needs. The Wi-Fi connection can be of interest to upload the results of the experiments developed by researchers to an external system via Internet connection. This could be made after a register session or at the end of the day. RatButton is not developed to necessarily work with Internet connection.

28.2.7 Data Management

RatButton manages results itself. However, those can also be externally managed through cable connection. As some researchers have already installed in their labs data acquisition systems, a system of outputs for different situations, such as when a stimulus is presented or when it is touched, has been created. Those outputs can be managed by an external device.

28.3 Software Validation

Different neuroscientist projects are validating RatButton software. It has been used by Division of Neuroscience at University Pablo de Olavide (Seville, Spain), by IDIBAPS and University Autònoma de Barcelona (Barcelona, Spain), and by Neuro-Com Research Group (Barcelona, Spain). Some of the publications to date are [2–4]. RatButton is distributed and offers support via online. It is available at www.ratbutton.com.

28.4 Discussion

We have presented here a software for stimuli presentation in psychological and neuroscientist experimental environments for animal and human research. The most important advantages of this software are the real interactivity made in a capacitive touchscreen presentation, the lack of need of code programming by researcher, and the lack of need a PC connection for designing, creating, presenting protocols, and collecting the data. RatButton is available online for iOS, 9.0 and later, devices. RatButton has also been used in brain-machine interface (BMI) environments.

Acknowledgments Many thanks to José Antonio Santos Naharro and to Cibertec Company team.

References

1. Martín-Pascual, M.A., Andreu-Sánchez, C., Santos Naharro, J.A., Gruart, A., Delgado-García, J.M.: Device for animal experimentation in neuroscience research. European Patent Office 1–11. PCT ES2012 070544. Priority: 19.07.2011 – ES 201131227 P. (2011)
2. Hernández-González, S., Andreu-Sánchez, C., Martín-Pascual, M.A., Gruart, A., Delgado-García, J.M.: A cognition-related neural oscillation pattern, generated in the prelimbic cortex, can control operant learning in rats. J. Neurosci. **37**, 5923–5935 (2017)

3. Martín-Pascual, M.A., Andreu-Sánchez, C., Hernández-González, S., Gruart, A., Delgado-García, J.M.: Device for animal experiments in neuroscience research. In: Cognitive Info Communications 3rd IEEE international conference on Cognitive Info Communication 725–726. (2012)
4. Hernández-González, S., Andreu-Sánchez, C., Martín-Pascual, M.A., Gruart, A., Delgado-García, J.M.: Experimental design for brain-machine interaction with alert behaving rats. Dynamic Brain Forum (2012)

Chapter 29
ERFo: An Algorithm for Extracting a Range of Optimal Frequencies for Filtering Electrophysiological Recordings

C. Rocío Caro-Martín, Agnès Gruart, José M. Delgado-García, and Alessandro E. P. Villa

Abstract In the analysis of raw electrophysiological recordings, signal filtering is a crucial step to eliminate frequency components associated with noise and recording artifacts. Two problems have to be addressed: determining the optimal frequency range of the signal and which frequency values (extreme values, maximum and minimum) that are characteristic of the raw signal power spectrum. We developed an algorithm called ERFo (extractor of range for filtering optimization) that determines the frequency range boundaries expected to be optimal for the observation of the spectrum with the largest power in the range of interest. The regular differentiations (first and second derivatives) of the raw signal are used to detect the maximum and minimum amplitudes, which are reported in the algorithm to determine the frequency range of the raw signal filtration.

Keywords Software · Research tool · Signal · Filter · Neurophysiology · Electrophysiology · Extracellular recordings

29.1 Introduction

An extracellular electrophysiological recording covers a broad frequency range which depends on the neuropil of the cerebral recording site, on the cell type that is generating the discharge, on the geometry and electrical properties of the electrode, and on the distance of the electrode tip from the recorded cell [1]. For this reason, a filtering step is essential in signal preprocessing. It is a complicated step because the resulting signal may change depending on the applied filter type or the impact of this

C. Rocío Caro-Martín (✉) · A. Gruart · J. M. Delgado-García
Division of Neuroscience, Pablo de Olavide University, Seville, Spain
e-mail: crcarmar@upo.es

A. E. P. Villa
Neuroheuristic Research Group, University of Lausanne, Lausanne, Switzerland

J. M. Delgado-García et al. (eds.), *Advances in Cognitive Neurodynamics (VI)*,
Advances in Cognitive Neurodynamics,
https://doi.org/10.1007/978-981-10-8854-4_29

filter on the phase of each frequency component of the recording, the commercial and academic hardware and software used for filtering, and—last but not least—the marked frequency range which is selected by the user [2].

This range is important to eliminate frequency components associated with noise and recording artifacts. In this report we present a new application for extracting the extreme values of the frequency range best suited for filtering an extracellular electrophysiological recording.

The algorithm can be used for off-line analysis of extracellular recordings to get the optimal filtering parameters as a preprocessing stage for waveform spike sorting of single units and to perform an exhaustive high-pass or band-pass filtration for isolation of fast neuronal spiking activity.

29.2 Methods

29.2.1 Algorithm for the Extraction of an Optimal Frequency Range

The analysis of regular differentiations was used for the extraction of an optimal frequency range of a raw signal. The computation of the regular differentiations was performed following the first-order and second-order derivatives of a kernel function [1, 3, 4] generated by a Gaussian curve with the same mean and standard deviation as the raw electrophysiological recording (Fig. 29.1a).

The first-order $D_\alpha^1 f(t)$ and second-order $D_\alpha^2 f(t)$ derivatives of the recording were calculated through a convolution between the derivatives of the kernel function $G_\alpha^j(t)$ and the recording $f(t)$.

$$D_\alpha^j f(t) = \left(f * G_\alpha^j\right)(t) = \sum\nolimits_\alpha f(\alpha)\, G(t-\alpha) \qquad (11.1)$$

In Eq. (11.1), α was a dummy variable of the convolution that reflected the displacement of $G_\alpha^j(t)$ across $f(t)$ and j was the order of the derivative (1, first-order; and 2, second-order). The study of these regular differentiations (based on the convolution) provided much more exhaustive results during the computation of the frequency ranges. Selection of both the kernel function $G_\alpha^j(t)$ and the value of the parameter α depends on the order of the derivative to be calculated, the level of additive noise, and the required smoothness of the recording. Once the regular differentiations were calculated (Fig. 29.1b–d), two adaptive thresholds were applied which were set automatically to $\pm 3\sigma$ (σ is the standard deviation of the first-order and second-order derivatives of the recording, respectively). This strategy is meant to ensure that the optimal power spectrum is included in the interest range (extracted range). After applying the thresholds, the extreme values of the frequency range were obtained by determining the cycle properties of the

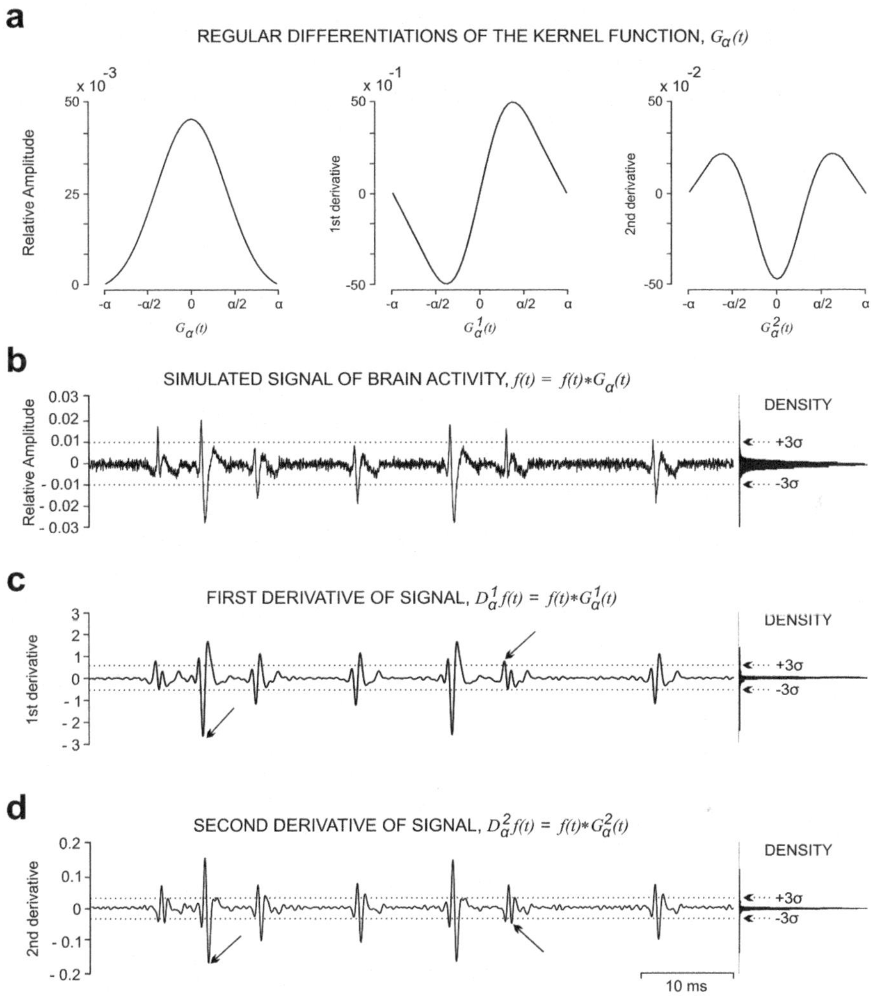

Fig. 29.1 An example illustrating the process of extraction of the frequency range in a simulated recording. (**a**) Kernel function, which is used to estimate the first-order and second-order derivatives of the electrophysiological neural recording. Note that α is a dummy variable of time (in ms) that indicates the displacement of the kernel function across the recording. (**b**) Simulated recording with low noise (SNR 3.55 dB; [2]) is shown during 60 ms at a sampling rate of 44 kHz. (**c**) First-order derivative of the simulated recording shown in (**a**) for convolution, $D^1_{\propto} f(t)$. (**d**) Second-order derivative of the simulated recording shown in (**a**) for convolution, $D^2_{\propto} f(t)$. In each panel (**b**, **c**, and **d**), the horizontal dotted lines indicate two adaptive amplitude thresholds ($\pm\, 3\sigma$) for direct spike-event detection, σ is the standard deviation of the signal, and on the right the probability density function (PDF) is shown for the signal amplitudes. The black arrows indicate the biggest spike event and the smallest spike event

detected spike events. Note that the maximum frequency of the range was in correspondence with the minimum period and the minimum frequency of the range with the maximum period.

29.2.2 Algorithm Implementation

The proposed algorithm was designed and developed in MATLAB version 7.12.0/R2011a on a Windows platform. To streamline the developed software, all ERFo (.mat) functions were packed into an executable (.exe) MATLAB file. The correct performance of this software requires having the release version R2011 of MATLAB Runtime installed on the computer. During the installation of the executable ERFo file, MATLAB Runtime will be automatically installed as a mandatory package, making an updated version of Windows and Internet access necessary. Consequently, MATLAB toolboxes do not need to be installed on the PC.

29.3 Results

29.3.1 Graphical User Interface (GUI)

The extractor of range for filtering optimization (ERFo) allows knowing the optimal frequency range of a signal and which frequency values (extreme values, maximum and minimum) are characteristic of the raw signal power spectrum. ERFo recognizes and supports raw signal files of limitless duration in text (.txt) and wave (.wav) formats in the off-line operation mode. In addition, upon completion of data loading, the waveform of the raw signal and the general information about the file (sampling rate in kHz, sampling time in ms, and duration in s) are displayed in the main window of this software (Fig. 29.2).

In order to obtain the optimal frequency range of the raw signal corresponding to the user's experimental file, ERFo internally calculates the first- and second-order derivatives by convolution between the first- and second-order derivatives of the kernel function and the raw signal, respectively (Fig. 29.3b).

If the signal duration is longer than 60 s, ERFo breaks up the signal in traces of 60 s and extracts the optimum frequency range across the median of all optimum frequency values collected from each trace (Fig. 29.3a).

The optimal frequency range, the first-order and second-order derivatives of the raw signal, and the two adaptive thresholds for detecting the waveforms which surpass them in the first-order and second-order derivatives of the raw signal (horizontal red lines) are displayed in the results window.

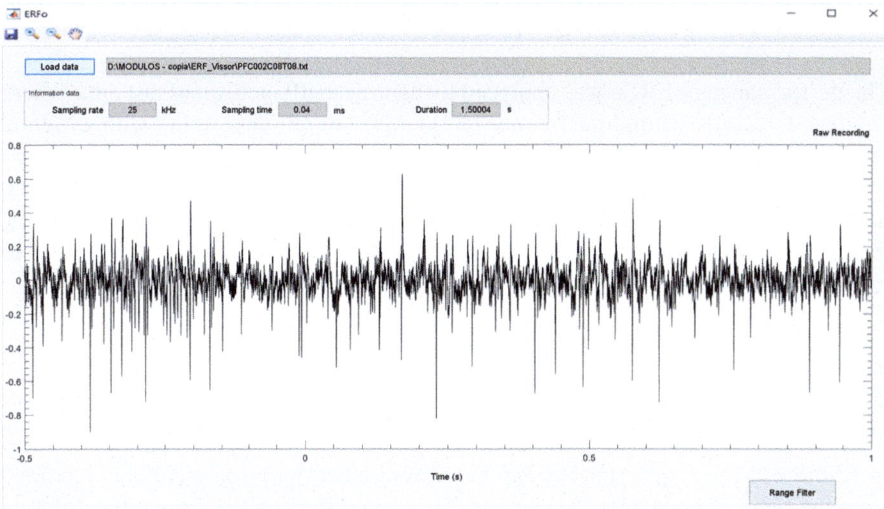

Fig. 29.2 Main window of the graphical user interface of the software. This window shows the waveform of the raw signal and the general information about the file. There is a button to calculate the extreme values of the optimal frequency range

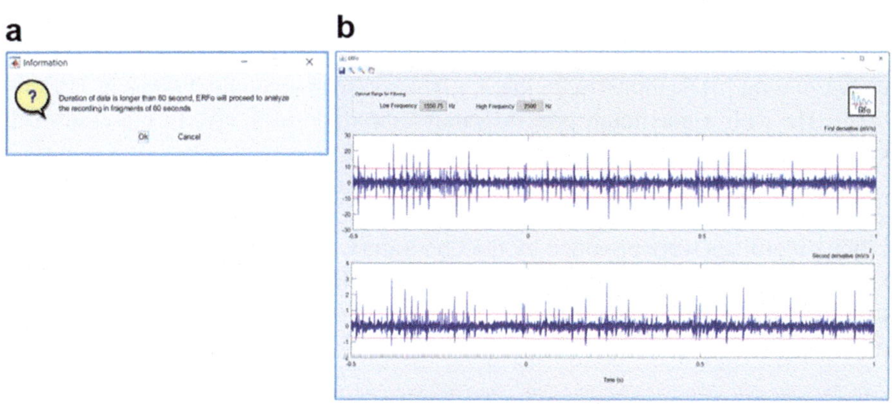

Fig. 29.3 Results window of the graphical user inter face of the software. (**a**) Warning information window regarding the file duration. (**b**) First-order and second-order derivatives of the raw signal and the extreme values of the optimal frequency range are shown in the results window

Note that the kernel function is a Gaussian curve generated with the standard deviation and mean of the raw signal. In general, ERFo can be applied to any signal or one that is intended to be filtered.

29.3.2 Units Test

The performance of ERFo was analyzed using artificially generated test data. These simulated records (sampling frequency, 44 kHz; duration, 3 min) with different levels of noise [1, 3, 4] were provided by the Neuroheuristic Research Group (University of Lausanne, Switzerland). Specifically, one simulated record without noise and two simulated records with high noise (SNR = 2.51 dB) and low noise (SNR = 3.55 dB) were tested. The optimal frequency ranges of the simulated records were {1898, 2580 $\pm$ 15.5} Hz.

29.3.3 Example of Electrophysiological Recordings

In addition, ERFo software was applied to real electrophysiological data recorded from the brain of mice, rats, and rabbits. These recordings were sampled at different frequencies and had different maximum peak-to-valley amplitude.

In the case of an electrophysiological recording from the selected sensory cortical area of a mouse [4, 5] (sampled at 10 kHz), ERFo returned an optimal frequency range of {2250, 2917.5} Hz with a maximum peak-to-valley amplitude of 0.3 mV. This recording was provided by the Neuroheuristic Research Group (Department of Anatomy, Histology and Neuroscience, Autonomous University of Madrid, Spain).

In the case of an electrophysiological recording extracted from the prelimbic cortex of a rat [6] (sampled at 25 kHz), ERFo returned an optimal range of {1736.5, 2639} Hz with a maximum peak-to-valley amplitude of 1 mV. In the case of an electrophysiological recording from rostro-medial prefrontal cortex of a rabbit [7] (sampled at 25 kHz), ERFo returned an optimal frequency range of {1550.75, 2500} Hz with a maximum peak-to-valley amplitude of 1.5 mV. Both rat and rabbit recordings were provided by the Laboratory of the Division of Neuroscience (University of Pablo de Olavide, Seville, Spain). Note that the duration of the three experimental samples was 1500 ms.

29.4 Discussion

The optimal frequency range extracted with ERFo provides relevant information about the raw signal. The optimal maximum frequency is the most important value because it indicates the boundary between the noise and the maximum frequencies of interest embedded in the signal. As a rule of thumb, it is recommended to add 50 Hz to the value of the optimal maximum frequency into a range as upper value in the setting of the band-pass filter. The optimal minimum frequency into a range is determined with the maximum amplitude of the waveform (cycle) found in the raw signal. Thus, there are no lower frequencies in the signal; the setting of the minimum frequency for the filter must be less than this optimal minimum frequency and not very close to it. The minimum frequency recommended would be to subtract a value of between 500 Hz and 1000 Hz to the optimal minimum frequency.

Acknowledgments The authors wish to thank Dr. Natali Barros-Zulaica and Dr. Samuel Hernández-González for providing additional electrophysiological recordings. ERFo was developed by Carmen Rocío Caro Martín (PhD student at Pablo de Olavide University, Spain) under the supervision of Dr. Alessandro E.P. Villa (professor at UNIL, the University of Lausanne, Switzerland) within the research project awarded by BFU2011-29286, Junta de Andalucía (BIO122, CVI 2487, and P07-CVI-02686) to Agnès Gruart i Massó and José María Delgado García, and the short-term fellowship "Researcher in Training" grant EEBB-I-16-10562 awarded to Carmen Rocío Caro Martín by the Spanish Ministry of Economy and Competitiveness.

References

1. Aksenova, T.I., Chibirova, O., Dryga, A.O., Tetko, I.V., Benabid, A.L., Villa, A.E.P.: An unsupervised automatic method for sorting neuronal spike waveforms in awake and freely moving animal. Methods. **30**, 178–187 (2003)
2. Widmann, A., Schroeger, E., Maess, B.: Digital filter design for electrophysiological data – a practical approach. J. Neurosci. Methods. **250**, 34–46 (2015)
3. Asai, Y., Aksenova, T.I., Villa, A.E.P.: On-line real-time oriented application for neuronal spike sorting with unsupervised learning. In: Duch, W., Kacprzyk, J., Oja, E., Zadrożny, S. (eds.) ICANN 2005, LNCS, vol. 3696, pp. 109–114. Springer, Berlin (2005)
4. Chibirova, O.K., Aksenova, T.I., Benabid, A.L., Chabardes, S., Larouche, S., Rouat, J., Villa, A.E.P.: Unsupervised spike sorting of extracellular electrophysiological recording in subthalamic nucleus of Parkinsonian patients. Bio Systems. **79**, 159–171 (2005)
5. Chaves-Coira, I., Barros-Zulaica, N., Rodrigo-Angulo, M., Núñez, A.: Modulation of specific sensory cortical areas by segregated basal forebrain cholinergic neurons demonstrated by neuronal tracing and optogenetic stimulation in mice. Front. Neural Circuits. **10**, 28 (2016)
6. Hernández-González, S., Andreu-Sánchez, C., Martín-Pascual, M.A., Gruart, A., Delgado-García, J.M.: A cognition-related neural oscillation pattern, generated in the prelimbic cortex, can control operant learning in rats. J. Neurosci. **37**, 5923–5935 (2017)
7. Caro-Martín, C.R., Leal-Campanario, R., Sánchez-Campusano, R., Delgado-García, J.M., Gruart, A.: A variable oscillator underlies the measurement of time intervals in the rostral medial prefrontal cortex during classical eyeblink conditioning in rabbits. J. Neurosci. **35**, 14809–14821 (2015)

Chapter 30
VISSOR: An Algorithm for the Detection, Identification, and Classification of the Action Potentials Distributed Across Electrophysiological Recordings

C. Rocío Caro-Martín, José M. Delgado-García, Agnès Gruart, and Raudel Sánchez-Campusano

Abstract Pattern recognition of neuronal discharges is the electrophysiological basis of the functional characterization of brain processes, so the implementation of a spike-sorting algorithm is an essential step for the analysis of neural codes and neural interactions in a network or brain circuit. We developed an unsupervised automatic computational algorithm for the detection, identification, and classification of the neural action potentials distributed across electrophysiological recordings and for the clustering of these potentials based on the shape, phase, and distribution features, which are extracted from the first-order derivative of the potentials under study. This algorithm was implemented in a customized spike-sorting software called *VISSOR* (*Viability of Integrated Spike Sorting of Real Recordings*). The validity and effectiveness of this software were tested by the classification of the action potentials detected in extracellular recordings of the rostro-medial prefrontal cortex (rmPFC) of rabbits during the classical eyelid conditioning.

Keywords Data mining · Pattern recognition · Spike-sorting software · Action potentials · Neurophysiology

---

C. Rocío Caro-Martín · J. M. Delgado-García · A. Gruart · R. Sánchez-Campusano (✉)
Division of Neuroscience, Pablo de Olavide University, Seville, Spain
e-mail: rsancam@upo.es

J. M. Delgado-García et al. (eds.), *Advances in Cognitive Neurodynamics (VI)*,
Advances in Cognitive Neurodynamics,
https://doi.org/10.1007/978-981-10-8854-4_30

30.1 Introduction

The main steps of a spike-sorting algorithm are (1) spike detection and alignment in which it is needed to apply an amplitude threshold, (2) the extraction of features of the detected spike waveforms, and (3) the clustering of these spike waveforms, which is based on the similarity of their profiles, given that each neuron fires spikes of a particular shape.

The spike-sorting approaches are mainly focused on extracting various features of the spike waveform (waveform-based features) to be able to interpret physiologically the obtained results. Most of the features used by different authors have been separated as (1) shape-based features (directly extracted from the detected action potential; see [1]), (2) phase-based features (extracted from the trajectory of the action potential in the phase space; see [2]), (3) geometric-based features (extracted as the area under the spike waveform; see [3]), and (4) transformation-based features (coefficients, factors, or components extracted from different mathematical transformations; see [4]).

In this work, we present a new application/software for the detection, identification, and classification of the action potentials distributed across the real electrophysiological recordings and for the clustering of these action potentials based on shape (features from spike waveform first derivative in time domain), phase (features from spike trajectory in the phase space: first derivative vs. second derivative), and distribution (features from spike amplitude distribution function for both the first and second derivatives) features. For this purpose, an unsupervised off-line spike-sorting algorithm (*VISSOR*) was implemented.

30.2 Methods

30.2.1 Algorithm Implementation

The proposed algorithm was designed and developed in MATLAB version 7.12.0/R2011a on a Windows platform. To streamline the developed software, all *VISSOR* (.mat) functions were packed into an executable (.exe) file with the aid of MATLAB. The correct performance of this software requires having the R2011a version of MATLAB Runtime installed on the PC. During the installation of the executable *VISSOR* file, MATLAB Runtime will be installed automatically as a mandatory package, making updated versions of Window and Internet access necessary. Consequently, MATLAB toolboxes are not needed to be installed on the PC.

30.2.2 Unsupervised Automatic Algorithm

The spike-sorting algorithm of *VISSOR* software was implemented in two blocks: (1) data preprocessing block, which includes the calculation of the regular differentiations and interpolation of the filtered recordings, the application of the adaptive threshold of amplitude, and the spike detection and alignment steps, and (2) spike classification block, which contains the features extraction method/algorithm (in accord with shape, phase, and distribution features) and the spike events clustering of the neural recording.

During the preprocessing, if the sampling frequency was <20 kHz, the recording was interpolated applying a cubic splines function to improve the calculation of its derivatives. An adaptive threshold was applied to the first-order derivative of the filtered recording. This threshold was fitted to the data points of this recording once the possible artifacts were removed and the segments contaminated by them rejected. The adaptive threshold (*Thr*) was set automatically to three times the median absolute deviation of the first-order derivative $\dot{V}(t)$ of the band-pass filtered record.

$$Thr = \pm 3\sigma_n; \qquad \sigma_n = \text{median}\left\{\frac{\dot{V}(t)}{0.6745}\right\} \tag{30.1}$$

During the classification, a vector of 24 physiological parameters was determined for each spike event. This 24-dimensional vector (24D vector) included shape (features from spike waveform first derivative in time domain; Fig. 30.1a), phase (features from spike trajectory in the phase space: first derivative vs. second derivative; Fig. 30.1b), and distribution (features from spike amplitude distribution function for both the first and second derivatives; Fig. 30.1c) spike features.

These feature vectors were the input to an unsupervised *K*-means clustering method with three internal validation measures: Silhouette, Davies-Bouldin, and Dunn indices. This unsupervised clustering method was used to determine the number of classified spike events distributed across time, and their neuronal identity. In this way, spike waveforms originated from different neurons were differentiated. An efficient index of cohesion-dispersion among and within clusters (*CD* index) during the neural events classification was also applied. Furthermore, a modified index of clustering error (*CE* index) taken from [4] upon completion of the classification process was included.

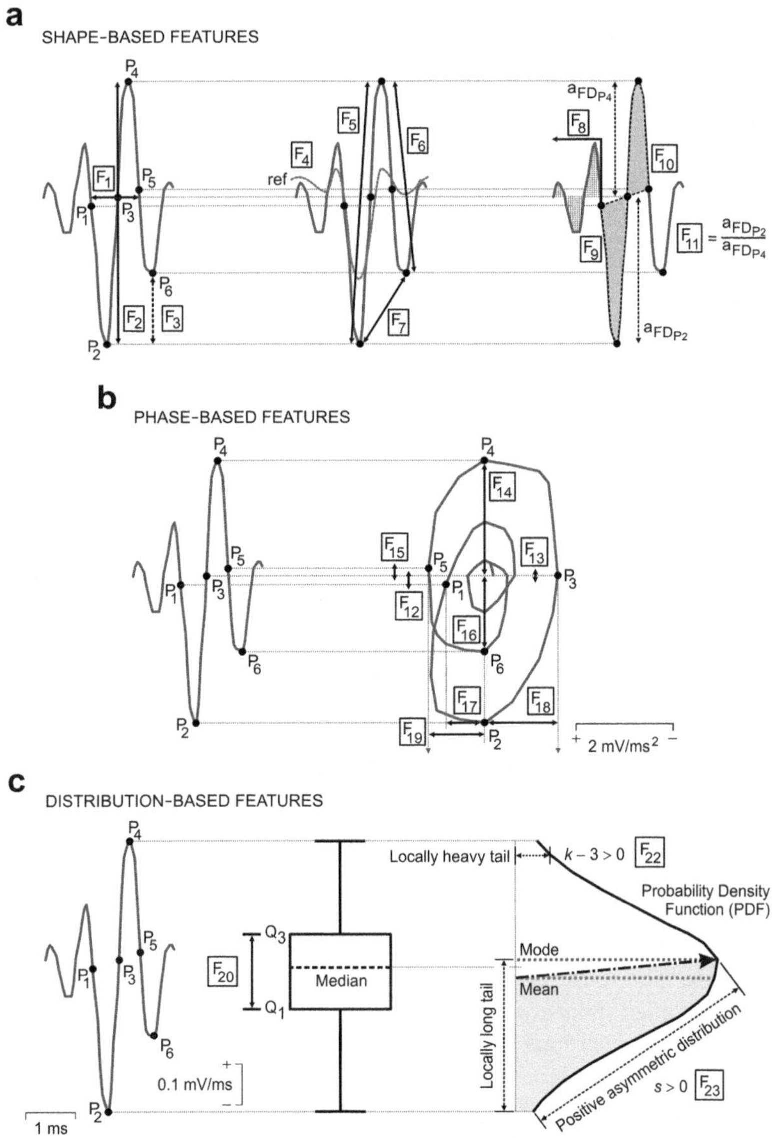

Fig. 30.1 Schematic representation of the extracted features. (**a**) Six fundamental points (P_1–P_6) and 11 shape-based features (F_1–F_{11}) from each spike in the time domain of the spike first derivative (FD, in mV/ms). (**b**) Eight phase-based features (F_{12}–F_{19}) from each spike trajectory in the phase space [second derivative (SD, in mV/ms^2) vs. FD]. (**c**) Five distribution-based features (F_{20}–F_{24}) for the statistical amplitude distribution of the FD (F_{20}, F_{22}, and F_{23}) and SD (F_{21} and F_{24}) of the spike

30.3 Results

30.3.1 Graphical User Interface

VISSOR is a spike-sorting algorithm that recognizes specific patterns of neuronal discharges (spikes) present in electrophysiological recordings. It supports up to 25 minutes of signal duration in text (.txt) and wave (.wav) formats in the off-line operation mode.

Upon finalization of signal loading, the raw signal and the first-order derivative of this signal are displayed in the main window of the *VISSOR* software (Fig. 30.2). All the useful information extracted from the neural signal is also included in the graphical report.

The next step is the selection of the amplitude threshold. The amplitude threshold can be selected manually (by the user) or generated automatically according to Eq. (30.1). Other key parameters, such as the relative refractory period (rrp, in ms) and the percentage of artifacts, can be introduced at this stage of the process. Following the input of this information, a complete list of 24 independent physiological features (see Fig. 30.1 for details) is displayed for each selected spike. Each 24D vector of features can be used for the subsequent spike classification.

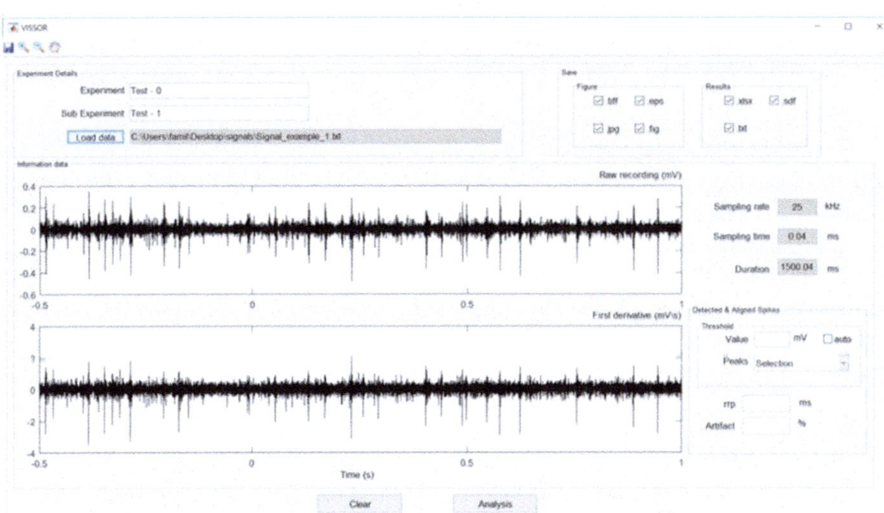

Fig. 30.2 Main window of the graphical user interface of *VISSOR*. This window shows the filtered recording (in mV), the first-order derivative of the filtered recording (in mV/s), and other useful information extracted from them (sampling rate, in Hz; sampling time, in ms; and the total duration, in ms). At the top left can be found the "Load data" button for uploading the neural recording. At the top right is the "Save" section where different file extensions can be selected for saving both "Figures" and "Results." At the bottom are two key buttons: the "Clear" button, for removing all the values introduced in the different fields and checkboxes, and the "Analysis" button, for continuing with the spike-sorting analysis

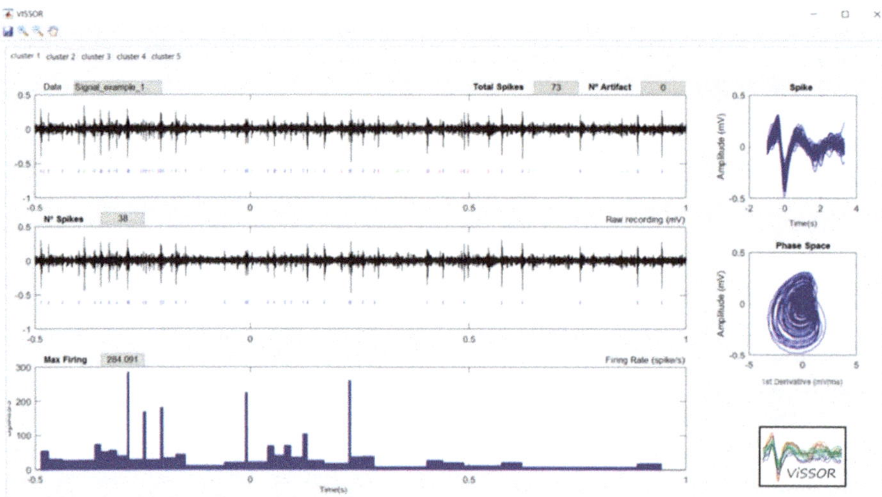

Fig. 30.3 Results window of the graphical user interface of *VISSOR*. An example showing the classification results: 73 spikes (vertical lines) classified into five different clusters (cluster 1, 38 spikes, blue lines; cluster 2, 5 spikes, red lines; cluster 3, 21 spikes, green lines; cluster 4, 8 spikes, magenta lines; cluster 5, 1 spike, grid line). Note that the file name of the neural data, the total number of detected spikes, and the number of artifacts are also included. In particular, from spikes detected in cluster 1, the firing rate (in spikes/s) is calculated and represented and its maximum value (Max Firing) is indicated. On the right are displayed the spike profiles (for cluster 1) and the phase-space portraits of the identified action potentials

The *VISSOR* software provides a results window (Fig. 30.3) that displays all the spikes, classified into different clusters, with relevant information about the number of spikes per cluster (or neuron) and their corresponding neuronal firing rate. Each spike is indicated with a short vertical line and each cluster is shown in a different color. Spikes of the same color belong to the same cluster. At the top right of the results window can be seen all the spikes of a selected cluster and at the bottom right the corresponding phase-space portraits.

30.3.2 Units Test

The performance of the *VISSOR* software was tested on both simulated records also used in [1, 5] and real extracellular recordings [2]. The simulated records (sampling frequency, 44 kHz; duration, 3 min) without noise and with a low level of added noise (SNR = 3.55 dB) were provided by the Neuroheuristic Research Group (University of Lausanne, Switzerland).

The results obtained after performing the two validation tests on simulated data (with and without noise) showed three activity patterns (cluster 1, 2728 spikes; cluster 2, 2690 spikes; and cluster 3, 2733 spikes) in full accordance with those obtained by other authors (see [1, 5] for details). When the *VISSOR* software was tested on real extracellular recordings from neurons in the rostro-medial prefrontal cortex (rmPFC) acquired in the electrophysiological laboratory of the Division of Neuroscience (Pablo de Olavide University, Spain), the results also showed three firing activities of principal neurons during classical eyeblink conditioning using delay paradigms with different interstimulus intervals (see [2] for details).

30.4 Discussion

The *VISSOR* software is a robust and nonredundant spike-sorting algorithm based on the exhaustive extraction of features with a clear physiological description of the spike. This physiological information is highly appreciated in the qualitative/quantitative characterization of neuronal activity, including the neural modulating properties, and has practical uses in neurophysiology beyond the mere computation of the number of spikes, classes, or neurons.

In contrast to other methods also based on feature extraction, the method employed in *VISSOR* is based on shape, phase, and distribution features of each spike. Removal of the multicollinearity among the extracted features means that the resulting vector (R^{24}) does not hold redundant information, thereby eliminating the need to reduce the dimensionality. Furthermore, these independent features can be quickly and easily calculated and represent no explicit threat to the computational cost and complexity of the software.

In accordance with the results obtained by applying the VISSOR method/algorithm for the classification of spikes from rmPFC neurons, the rmPFC could be involved in the determination of the interstimulus time interval during associative learning tasks, but these neurons seem not to encode the oscillatory properties characterizing the kinematics of conditioned eyelid responses [2].

Acknowledgments The authors wish to thank Dr. Alessandro E.P. Villa for provided simulated records. *VISSOR* was developed by Carmen Rocío Caro Martín (PhD student) under the supervision of Dr. Raudel Sánchez Campusano and Dr. Agnès Gruart i Massó. This study was supported by grants from the Spanish MINECO (BFU2011-29286/BES-2012-052748) and Junta de Andalucía (BIO122, CVI 2487, and P07-CVI-02686) to Agnès Gruart i Massó and José María Delgado García.

References

1. Asai, Y., Aksenova, T.I., Villa, A.E.P.: On-line real-time oriented application for neuronal spike sorting with unsupervised learning. In: Duch, W., Kacprzyk, J., Oja, E., Zadrożny, S. (eds.) ICANN; LNCS 3696, pp. 109–114. Springer, Berlin (2005)
2. Caro-Martín, C.R., Leal-Campanario, R., Sánchez-Campusano, R., Delgado-García, J.M., Gruart, A.: A variable oscillator underlies the measurement of time intervals in the rostral medial prefrontal cortex during classical eyeblink conditioning in rabbits. J. Neurosci. **35**, 14809–14821 (2015)

3. Zviagintsev, A., Perelman, Y., Ginosar, R.: Low–power architectures for spike sorting. In: Proceedings of 2nd International IEEE EMBS Conference on Neural Engineering, 162–165 (2005)
4. Letelier, J.C., Weber, P.P.: Spike sorting based on discrete wavelet transform coefficients. J. Neurosci. Method. **101**, 93–106 (2000)
5. Chibirova, O.K., Aksenova, T.I., Benabid, A.L., Chabardes, S., Larouche, S., Rouat, J., Villa, A.E.P.: Unsupervised spike sorting of extracellular electrophysiological recording in subthalamic nucleus of Parkinsonian patients. Bio Systems. **79**, 159–171 (2005)

Part IV
Modelling Higher-Order Functions and Dysfunctions

Chapter 31
Influence of β-Amyloid Plaques on the Local Network Activity in the APP/PS1 Mouse Model of Alzheimer's Disease

Patricia Castano-Prat, Guillermo Aparicio-Torres, Alberto Muñoz, and Maria V. Sanchez-Vives

Abstract One of the neuropathological hallmarks of Alzheimer's disease is the abnormal accumulation of amyloid-β (Aβ) plaques. To better understand the local effect of Aβ plaques on network activity, we characterized spontaneous slow oscillatory activity (interspersed Up or active states with Down or silent states) near Aβ plaques in APP/PS1 mice. To this end, we used arrays of 16 electrodes with 100 μm separation to record from the cerebral cortex, and the electrodes whose tip was adjacent to an Aβ plaque were identified anatomically. Although the frequency of the oscillations was unaffected, we found lower firing rate during Up and Down states close to Aβ plaques than far from them in the APP/PS1 mouse model, alongside lower activity synchronization in the beta-gamma frequency range, suggesting a local reduction in cortical excitability around Aβ plaques. Reduced synaptic connectivity around Aβ plaques could underlie the decreased excitability, altering the local dynamics of the cortical circuits.

Keywords Slow oscillations · Excitability · Alzheimer's disease · Amyloid

P. Castano-Prat
Institut d'Investigacions Biomèdiques August Pi i Sunyer (IDIBAPS), Barcelona, Spain

G. Aparicio-Torres · A. Muñoz
Laboratorio Cajal de Circuitos Corticales, Centro de Tecnología Biomédica, Universidad Politécnica de Madrid, Madrid, Spain

Instituto Cajal (CSIC) and CIBERNED, Madrid, Spain

M. V. Sanchez-Vives (✉)
Institut d'Investigacions Biomèdiques August Pi i Sunyer (IDIBAPS), Barcelona, Spain

ICREA, Barcelona, Spain
e-mail: msanche3@clinic.ub.es

J. M. Delgado-García et al. (eds.), *Advances in Cognitive Neurodynamics (VI)*,
Advances in Cognitive Neurodynamics,
https://doi.org/10.1007/978-981-10-8854-4_31

31.1 Introduction

One of the principal neuropathological hallmarks of Alzheimer's disease (AD) is the abnormal accumulation of the amyloid-β (Aβ) peptide, which progresses from a soluble to an insoluble aggregate state, known as Aβ plaques [1]. The Aβ peptide exerts a deleterious effect on neuronal communication [2, 3], inhibiting synaptic currents [4, 5], preventing potentiation and increasing synaptic depression [6–8], or causing the endocytosis of glutamatergic receptors [6, 9]. Furthermore, postmortem morphological studies in humans have shown that neurites that traverse the extracellular deposits of Aβ peptide have an abnormal geometry, which could be modifying the conduction velocity of the nerve impulses, thereby preventing the synchronous activation of the convergent inputs onto neuronal ensembles [7, 8]. All these alterations are consistent with the progressive reduction of neuronal activity and spontaneous synchronization at gamma frequencies that have been described in animal models of AD [10, 11] and in patients with the disease [10–12]. However, these patients also have a marked tendency to suffer epileptic seizures [13]. In fact, calcium imaging studies have reported hyperactive neurons in the vicinity of the Aβ plaques in the APP/PS1 mouse model of amyloidosis [14], together with a disruption of slow wave activity propagation which can be reversed by increasing fast GABAergic inhibition [15]. All this points to an amyloid-dependent increase in cortical excitability. To better understand the local effect that Aβ plaques exert on network activity, we characterized spontaneous slow oscillatory activity in the close vicinity of Aβ plaques (within 18 μm of the electrode tip) in the cortex of the APP/PS1 mouse model, and we contrasted it with the activity taking place far from the plaques.

31.2 Methods

31.2.1 APP/PS1

We studied 8–9.5-month-old APP-PS1 (transgenes APP_{Swe} and PS1dE9) male mice ($n = 5$) obtained from an established colony at the Faculty of Medicine at the University of Barcelona (Spain) and provided by Dr. Anna Colell.

31.2.2 In Vivo Extracellular Recordings

Anesthesia was induced by ketamine (100 mg/kg, i.p.) and medetomidine (1.3 mg/kg, i.p.). The surgical methodology has been previously described [16]. Arrays of 16 aligned electrodes separated by 100 μm (NeuroNexus) were used to record slow oscillations. The array was placed at 0.9–1.2 mm from the cortical surface (deep layers) and perpendicular to the midline in prelimbic, primary motor and primary somatosensory cortices for each animal and hemisphere. The signal was amplified (Multi Channel Systems), digitized at 20 kHz, and acquired unfiltered using a CED acquisition board and software from Cambridge Electronic Design.

31.2.3 Identification of the Electrode Tract and of the β-Amyloid Plaques in the APP/PS1 Model

For the identification of the recording electrode tract, the electrode was submerged in the lipophilic fluorescent dye DiI (Invitrogen) before carrying out the extracellular recordings. In each of the five registered animals, several electrode penetrations were performed throughout the anteroposterior and lateral axes. At the end of the recording, the brains were perfused, removed, and fixated during 4 h at 4 °C and then cryoprotected by immersion in 30% saccharose until they sank. Next, brains were frozen in dry ice and then cut along the coronal plane with a freezing sliding microtome. To reveal the presence of Aβ plaques, serial sections (50 μm) were dyed during 10 min in 1% thioflavin-S solution, rinsed with distilled water and ethanol and finally rinsed with phosphate buffer. Sections were then dyed with 4′,6-diamidino-2-phenylindole (DAPI) to reveal the limits between cortical layers and the cytoarchitectonic areas. Then, sections were mounted with *ProLong Gold Antifade* (Invitrogen) and examined in a fluorescent microscope (Zeiss 710). The fluorescence for DiI, thioflavin-S, and DAPI of all serial sections was recorded through separate channels with a 10x lens. Recorded images through the different channels were combined with ZEN 2012 (Zeiss) in order to study the spatial relationship between the electrode tracts and the Aβ plaques (Fig. 31.1), thus selecting the recordings in which the electrode tip was co-localized with an Aβ plaque ($n = 35$) and those in which it was not ($n = 98$).

31.2.4 Detection of Up and Down States and Calculation of Slow Oscillation Parameters

The detection of Up and Down states was carried out based on the estimation of the multiunit activity (MUA) signal from the frequency power between 200 and 1500 Hz, in 5-ms sliding windows [17–19]. The parameters were calculated for each of the bilateral recordings in the four recorded areas. Up and Down states were detected by establishing a threshold in this data sequence. The slow oscillation frequency was calculated as the inverse of the average duration of the Up-Down cycle. The firing rate of the local network was performed from the logMUA signal. The firing rate values during Up and Down states were computed as the average logMUA values during these periods. To analyze the fast components of slow

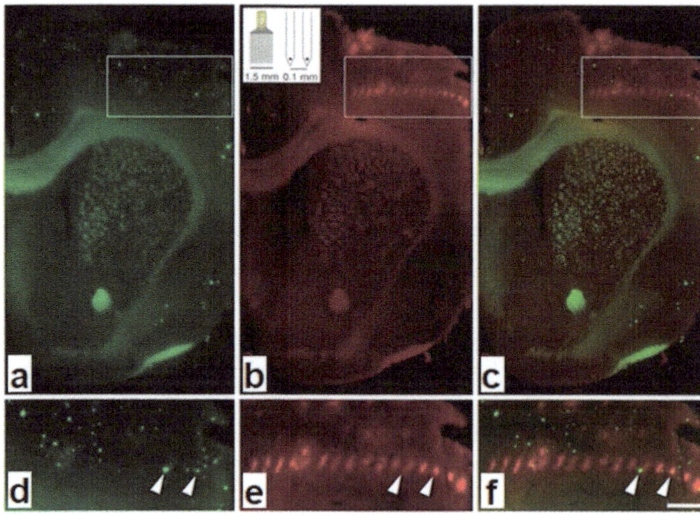

Fig. 31.1 (**a**–**c**) Images taken from a section of an APP/PS1 mouse showing the relationship between Aβ plaques stained with thioflavin-S (**a** and **d**) and the electrode tracks (**b** and **e**). (**c** and **f**) Merged images. Inset in (**b**) shows a 16-channel array. Squared zones in (**a**–**c**) are, respectively, shown at higher magnification in (**d**–**f**). Arrowheads indicate contacts of certain electrodes with Aβ plaques. Scale bar in **f**: 505 μm in (**a**–**c**); 270 μm in (**d**–**f**)

oscillations, we analyzed the spectral density separately for each Up and Down state and then obtained the average spectral density for Up and Down states. For this, we employed Welch's method with 50% overlapped windows of 5000 samples and calculated a relative spectral density with respect to the exponential decay defined by 1/f, which we will refer to as excess power. To compare the slow oscillation parameters in the APP/PS1 model between the recordings in which the electrode tip was co-localized with an Aβ plaque with those in which it was not, we used Student's *t*-test for paired samples correcting for multiple comparisons with the Bonferroni method (corrected $\alpha = 0.0125$).

31.3 Results

The accumulation of Aβ peptide is one of the main histopathological indicators of AD [1]. To study the slow wave activity that takes place close to Aβ plaques (Fig. 31.1), we characterized this activity pattern in the cortex of five mice of the APP/PS1 amyloidosis model (Fig. 31.2a). Our findings show that slow oscillation frequency (Fig. 31.2b) and Down state duration (Fig. 31.2c) remained constant, while there was a tendency for Up states to be shorter around Aβ plaques (Fig. 31.2d). The variability of the oscillatory cycle was also unaltered (Fig. 31.2e).

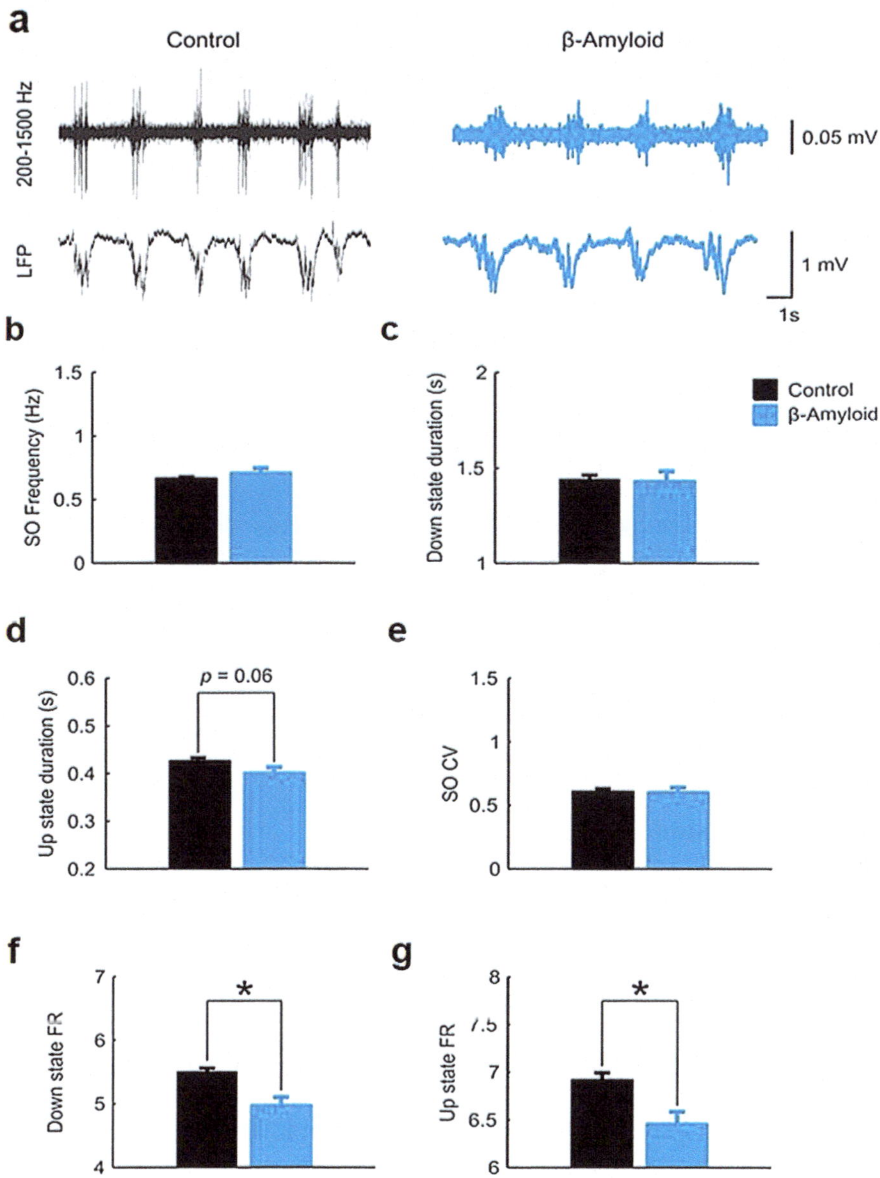

Fig. 31.2 Slow oscillation parameters in the vicinity of Aβ plaques in the APP/PS1 mouse model. (**a**) Raw local field potential (LFP) recordings (bottom) and filtered between 200 and 1500 Hz (top) as an estimation of the population firing rate. (**b–g**) Slow oscillation (SO) frequency, Down state duration, Up state duration, coefficient of variation of the SO, Down state firing rate, Up state firing rate, respectively, comparing the parameters obtained near an Aβ plaque and far away from them. Bars depict the mean, and error bars are SE. *$p < 0.0125$

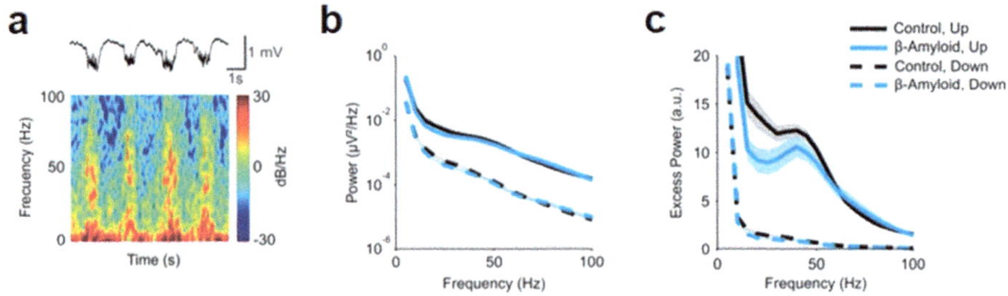

Fig. 31.3 High-frequency content of SO in the vicinity of Aβ plaques in the APP/PS1 mouse model. (**a**) Top: raw LFP recording in the prelimbic cortex of a control mouse under ketamine anesthesia. Bottom: spectrogram showing the presence of high-frequency rhythms, mainly during the Up states. (**b**) Mean raw power spectrum during Up states (solid lines) and the Down states (dashed lines) recorded in the vicinity of Aβ plaques (gray) and far away from them (black). (**c**) Excess power during the Up states and the Down states. Shadow is the SE

The firing rate during Up and Down states was significantly lower close to the Aβ plaques than far from them (Fig. 31.2g and f, respectively), indicating reduced excitability in the vicinity of the Aβ plaques.

Similarly, the high-frequency content of slow oscillations (Fig. 31.3a) recorded close to Aβ plaques was lower than far from them. This reduction occurred between 20 and 55 Hz and was significant during Up states (Fig. 31.3b), which can be clearly appreciated in the excess power (Fig. 31.3c).

31.4 Discussion

Slow oscillatory activity occurring close to Aβ plaques was studied in 8–10-month-old APP/PS1 mice, which corresponds to an age in which these animals present abundant amyloid deposition. We identified which recordings had been obtained within 18 μm of an Aβ plaque, which allowed the comparison between recordings obtained close to Aβ plaques with those obtained far from them. Our results show that firing rate is reduced close to Aβ plaques. This reduction happened during Up and during Down states and was accompanied by a lower synchronization of activity in the beta-gamma frequency range, especially during Up states. The Aβ peptide has been reported to exert a series of noxious effects [2–6, 9, 20] that are coherent with the decrease in neuronal activity and could contribute to the reduction in firing rate detected close to Aβ plaques in APP/PS1 animals. Similarly, *postmortem* morphological studies in humans have demonstrated that the neurites that go through the extracellular Aβ deposits present an abnormal geometry—more intricate—that could be modifying the speed of propagation of nerve impulses, consequently preventing synchronous activity of neuronal groups in which these afferents converge [7]. These predictions have been confirmed in vivo in a murine model of AD [8] and could be related to the lower activity synchronization in the beta-gamma frequency range and to the decrease in firing rate that we observed close to the Aβ plaques in the APP/PS1 model.

The progressive reduction of neuronal activity and the decrease in spontaneous synchronization at gamma frequencies, also described in other AD animal models [10, 11], are well-established alterations in AD patients [12] and could be related to the characteristic cognitive deficits that are also present in the APP/PS1 model [21]. Aβ plaques modify synaptic circuits in a spatially selective manner, given that the cellular membrane of pyramidal neurons' cell bodies or the initial segments of their axon that are in contact with Aβ plaques lack GABAergic perisomatic innervation arising from basket and chandelier interneurons, respectively [22, 23]. As these populations of fast spiking parvalbumin-positive interneurons have been proposed to be involved in the synchronization of the firing patterns of large populations of pyramidal cells in different states of consciousness [24, 25], we hypothesize that the reduction of the synchronization in the beta-gamma frequencies found here could be related to the alterations of GABAergic microcircuits in close proximity of Aβ plaques.

The local alteration of neurons that sit close to Aβ plaques has also been described in other in vivo studies in which imaging techniques were employed to visualize calcium influx in the cytoplasm and used as an indirect measure of neuronal firing. These studies have reported that the APP/PS1 model presents hyperactive neurons exclusively near Aβ plaques [14], as well as a disruption in the propagation of slow oscillations that depends on the Aβ levels and that can be reverted by increasing fast GABAergic inhibition [15]. This points to an increase in cortical excitability associated with amyloid deposition in this model, which is in contrast with the findings obtained here, in which neuronal activity seems to be reduced around Aβ plaques.

Other studies have shown that higher levels of Aβ cause a depression in excitatory synaptic transmission [26], which has led to the hypothesis that in non-pathological conditions, Aβ could be part of a negative feedback system that maintains neuronal excitability within a physiologically healthy range [7]. Thus amyloid deposition could be associated with the rise in cortical excitability but also to its reduction, which could in turn be related to the apparently contradictory findings between studies, as well as among studies with AD patients that have described cortical hyperexcitability and a higher predisposition to suffer epileptic seizures [13], but also with a decrease in neuronal activity [12] and in cortical metabolism [27].

The complexity of the cortical network response and the presence of Aβ plaques highlight the need to dissect the mechanisms generating both the hyperexcitability and the hypoexcitability that have been linked to AD, together with the need to identify which alterations are primary manifestations of the disorder and which ones represent compensatory processes.

Acknowledgments Funded by FLAGERA-PCIN-2015-162-C02-01, EU H2020 Research and Innovation Programme, Grant 720270 (HBP SGA1), BFU2014-52467-R (MINECO), and CERCA (Generalitat de Catalunya). We thank Dr. Anna Colell for providing the animal model and Cristina Gonzalez-Liencres for editorial work.

References

1. Querfurth, H.W., LaFerla, F.M.: Alzheimer's disease. N. Engl. J. Med. **362**, 329–344 (2010)
2. Selkoe, D.J.: Alzheimer's disease is a synaptic failure. Science (80). **298**, 789–791 (2002)
3. Coleman, P.D., Yao, P.J.: Synaptic slaughter in Alzheimer's disease. Neurobiol. Aging. **24**, 1023–1027 (2003)
4. Hsia, A.Y., Masliah, E., McConlogue, L., Yu, G.Q., Tatsuno, G., Hu, K., Kholodenko, D., Malenka, R.C., Nicoll, R.A., Mucke, L.: Plaque-independent disruption of neural circuits in Alzheimer's disease mouse models. Proc. Natl. Acad. Sci. U. S. A. **96**, 3228–3233 (1999)
5. Chang, E.H., Savage, M.J., Flood, D.G., Thomas, J.M., Levy, R.B., Mahadomrongkul, V., Shirao, T., Aoki, C., Huerta, P.T.: AMPA receptor downscaling at the onset of Alzheimer's disease pathology in double knockin mice. Proc. Natl. Acad. Sci. **103**, 3410–3415 (2006)
6. Hsieh, H., Boehm, J., Sato, C., Iwatsubo, T., Tomita, T., Sisodia, S., Malinow, R.: AMPAR removal underlies Abeta-induced synaptic depression and dendritic spine loss. Neuron. **52**, 831–843 (2006)
7. Knowles, R.B., Wyart, C., Buldyrev, S.V., Cruz, L., Urbanc, B., Hasselmo, M.E., Stanley, H.E., Hyman, B.T.: Plaque-induced neurite abnormalities: implications for disruption of neural networks in Alzheimer's disease. Proc. Natl. Acad. Sci. U. S. A. **96**(9), 5274 (1999)
8. Stern, E.A., Bacskai, B.J., Hickey, G.A., Attenello, F.J., Lombardo, J.A., Hyman, B.T.: Cortical synaptic integration in vivo is disrupted by amyloid-beta plaques. J. Neurosci. **24**, 4535–4540 (2004)
9. Snyder, E.M., Nong, Y., Almeida, C.G., Paul, S., Moran, T., Choi, E.Y., Nairn, A.C., Salter, M.W., Lombroso, P.J., Gouras, G.K., Greengard, P.: Regulation of NMDA receptor trafficking by amyloid-beta. Nat. Neurosci. **8**, 1051–1058 (2005)
10. Palop, J.J., Chin, J., Roberson, E.D., Wang, J., Thwin, M.T., Bien-Ly, N., Yoo, J., Ho, K.O., Yu, G.-Q., Kreitzer, A., Finkbeiner, S., Noebels, J.L., Mucke, L.: Aberrant excitatory neuronal activity and compensatory remodeling of inhibitory hippocampal circuits in mouse models of Alzheimer's disease. Neuron. **55**, 697–711 (2007)
11. Verret, L., Mann, E.O., Hang, G.B., Barth, A.M.I., Cobos, I., Ho, K., Devidze, N., Masliah, E., Kreitzer, A.C., Mody, I., Mucke, L., Palop, J.J.: Inhibitory interneuron deficit links altered network activity and cognitive dysfunction in Alzheimer model. Cell. **149**, 708–721 (2012)
12. Stam, C.J., van Cappellen van Walsum, A.M., Pijnenburg, Y.A.L., Berendse, H.W., de Munck, J.C., Scheltens, P., van Dijk, B.W.: Generalized synchronization of MEG recordings in Alzheimer's disease: evidence for involvement of the gamma band. J. Clin. Neurophysiol. **19**, 562–574 (2002)
13. Palop, J.J., Mucke, L.: Epilepsy and cognitive impairments in Alzheimer disease. Arch. Neurol. **66**, 435 (2009)
14. Busche, M.A., Eichhoff, G., Adelsberger, H., Abramowski, D., Wiederhold, K.-H., Haass, C., Staufenbiel, M., Konnerth, A., Garaschuk, O.: Clusters of hyperactive neurons near amyloid plaques in a mouse model of Alzheimer's disease. Science. **321**, 1686–1689 (2008)
15. Busche, M.A., Kekuš, M., Adelsberger, H., Noda, T., Förstl, H., Nelken, I., Konnerth, A.: Rescue of long-range circuit dysfunction in Alzheimer's disease models. Nat. Neurosci. **18**, 1623–1630 (2015)
16. Castano-Prat, P., Perez-Zabalza, M., Perez-Mendez, L., Escorihuela, R.M., Sanchez-Vives, M.V.: Slow and fast neocortical oscillations in the senescence-accelerated mouse model SAMP8. Front. Aging Neurosci. **9**, 141 (2017)

17. Ruiz-Mejias, M., Ciria-Suarez, L., Mattia, M., Sanchez-Vives, M.V.: Slow and fast rhythms generated in the cerebral cortex of the anesthetized mouse. J. Neurophysiol. **106**, 2910–2921 (2011)
18. Reig, R., Mattia, M., Compte, A., Belmonte, C., Sanchez-Vives, M.V.: Temperature modulation of slow and fast cortical rhythms. J. Neurophysiol. **103**, 1253–1261 (2010)
19. Sanchez-Vives, M.V., Mattia, M., Compte, A., Perez-Zabalza, M., Winograd, M., Descalzo, V.F., Reig, R.: Inhibitory modulation of cortical up states. J. Neurophysiol. **104**, 1314–1324 (2010)
20. Kamenetz, F., Tomita, T., Hsieh, H., Seabrook, G., Borchelt, D., Iwatsubo, T., Sisodia, S., Malinow, R.: APP processing and synaptic function. Neuron. **37**, 925–937 (2003)
21. Kilgore, M., Miller, C.A., Fass, D.M., Hennig, K.M., Haggarty, S.J., Sweatt, J.D., Rumbaugh, G.: Inhibitors of class 1 histone deacetylases reverse contextual memory deficits in a mouse model of Alzheimer's disease. Neuropsychopharmacology. **35**, 870–880 (2010)
22. Garcia-Marin, V., Blazquez-Llorca, L., Rodriguez, J.-R., Boluda, S., Muntane, G., Ferrer, I., Defelipe, J.: Diminished perisomatic GABAergic terminals on cortical neurons adjacent to amyloid plaques. Front. Neuroanat. **3**, 28 (2009)
23. León-Espinosa, G., DeFelipe, J., Muñoz, A.: Effects of amyloid-β plaque proximity on the axon initial segment of pyramidal cells. J. Alzheimers Dis. **29**, 841–852 (2012)
24. Cobb, S.R., Buhl, E.H., Halasy, K., Paulsen, O., Somogyi, P.: Synchronization of neuronal activity in hippocampus by individual GABAergic interneurons. Nature. **378**, 75–78 (1995)
25. Klausberger, T., Márton, L.F., Baude, A., Roberts, J.D.B., Magill, P.J., Somogyi, P.: Spike timing of dendrite-targeting bistratified cells during hippocampal network oscillations in vivo. Nat. Neurosci. **7**, 41–47 (2004)
26. Tampellini, D., Capetillo-Zarate, E., Dumont, M., Huang, Z., Yu, F., Lin, M.T., Gouras, G.K.: Effects of synaptic modulation on beta-amyloid, synaptophysin, and memory performance in Alzheimer's disease transgenic mice. J. Neurosci. **30**, 14299–14304 (2010)
27. Johnson, K.A., Fox, N.C., Sperling, R.A., Klunk, W.E.: Brain imaging in Alzheimer disease. Cold Spring Harb. Perspect. Med. **2**, a006213–a006213 (2012)

Chapter 32
Altered Functional Connectivity in a Mouse Model of Fragile X Syndrome

Miguel Dasilva, Alvaro Navarro-Guzman, Luca Maiolo, Andres Ozaita, and Maria V. Sanchez-Vives

Abstract We evaluated the integrity of the cortical network in the *Fmr1* knockout mouse model of fragile X syndrome (FXS) by recording micro-electrocorticogram (ECoG) activity and measuring functional connectivity, alongside the degree of clustering (modularity) and integration capacity (path length). We found that functional connectivity is increased at both the short- and long-range levels, especially in prefrontal areas in the FXS model. This correlates with a decrement in the degree of cortical network clustering, together with an anomalously high information exchange capacity at high-frequency oscillatory bands. We conclude that the cognitive deficits manifested in FXS could be associated with the state of over-synchronization of the cortical network, which is more evident at high-frequency oscillatory bands and at frontal areas of the brain.

Keywords Cortex · Fragile X syndrome · Functional connectivity · Integration · Segregation

M. Dasilva · A. Navarro-Guzman
Institut d'Investigacions Biomèdiques August Pi i Sunyer (IDIBAPS), Barcelona, Spain

L. Maiolo
Istituto per la Microelettronica e Microsistemi (CNR), Rome, Italy

A. Ozaita
Universitat Pompeu Fabra (UPF), Barcelona, Spain

M. V. Sanchez-Vives (✉)
Institut d'Investigacions Biomèdiques August Pi i Sunyer (IDIBAPS), Barcelona, Spain

ICREA, Barcelona, Spain
e-mail: msanche3@clinic.ub.es

J. M. Delgado-García et al. (eds.), *Advances in Cognitive Neurodynamics (VI)*, Advances in Cognitive Neurodynamics,
https://doi.org/10.1007/978-981-10-8854-4_32

32.1 Introduction

Fragile X syndrome (FXS) is the most common form of inherited intellectual disability and the leading cause of autism in humans. It is caused by the silencing of the *Fmr1* gene, resulting in the loss of expression of the fragile X mental retardation protein (FMRP) [1]. FXS patients show a variety of neuropsychiatric symptoms, many of which have also been described in the *Fmr1* knockout mouse model of FXS [2]. Specifically, at the cortical level, studies have demonstrated the presence of an imbalance between excitation and inhibition, ultimately leading to network hyperexcitability [3]. However, it is still far from well understood how this bias in network excitability relates to the neurocognitive alterations present in FXS.

The fine balance between excitatory and inhibitory circuits represents a key element in the gating and maintenance of neuronal synchrony and oscillations at the level of cortical networks [4]. This is especially relevant at high-frequency components, such as beta and gamma bands, which are related to cognitive function [5]. On the other hand, the adequate functionality of cortical systems requires an equilibrium between functional integration and segregation, so there is a compensation between the level of clustering (modularity) and the capacity of clusters to interconnect and exchange information (path length) [6]. Therefore, investigating the oscillatory activity of cortical neuronal networks, alongside functional connectivity and the integrity of neuronal communication, might shed some light on the neurophysiological mechanisms underlying the neurocognitive deficits manifested in mental disability conditions like FXS.

32.2 Methods

32.2.1 Animals

All procedures were approved by the Ethics Committee at the Hospital Clinic of Barcelona and were carried out to the standards laid down in Spanish regulatory laws (BOE-A-2013-6271) and European Communities Directive (2010/63/EU). *Fmr1* knockout mice in FVB background were obtained as previously described [7].

32.2.2 Surgical Procedures and Recording

Animals were anesthetized by inhalation of 4% isoflurane in 100% oxygen for induction and 1% for maintenance. Atropine (0.3 mg/kg), methylprednisolone (30 mg/kg), and mannitol (0.5 g/kg) were administered subcutaneously to avoid respiratory secretions and edema. Body temperature was constantly monitored and kept at 37 °C by means of a thermal blanket (RWD Life Science). Animals were

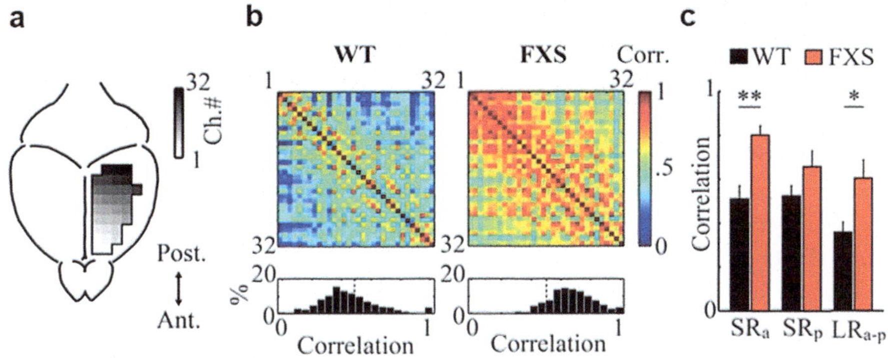

Fig. 32.1 Altered functional connectivity in *Fmr1* knockout mice. (**a**) Disposition of the recording multielectrode array on the left hemisphere of the mouse brain. Electrode number represented in gray scale. (**b**) Wideband functional connectivity matrices (top) and distribution of correlation values (bottom) for WT and FXS mice. (**c**) Group differences in short-range functional connectivity assessed within the anterior (*SRa*, 13 most anterior electrodes) and posterior (SR_p, 13 most posterior electrodes) clusters of the array and long-range functional connectivity between clusters ($LR_{a\text{-}p}$). *$p < 0.05$, **$p < 0.01$, Wilcoxon rank sum test. Bar plots represent data expressed as mean ± SEM

placed on a stereotaxic frame, and a craniotomy and durotomy were performed over the left hemisphere from −3.0 mm to +3.0 mm relative to bregma and +3.0 mm relative to midline. Extracellular micro-electrocorticogram (ECoG) activity was recorded by means of 32-channel multielectrode arrays (550 μm spacing, Fig. 32.1a) covering the entire exposed area of the cortex. The signal was amplified (Multi Channel Systems, GmbH), digitized at 5 kHz, and fed into a computer via a digitizer interface (CED 1401 and Spike2 software, Cambridge Electronic Design, UK).

32.2.3 Data Analysis and Statistics

We defined a 32 × 32 connectivity matrix as the pairwise Pearson's correlation coefficient between micro-ECoG recordings. To evaluate specific frequency coupling, pairwise correlations were calculated from the envelope of the Hilbert transformation of the recorded signals at each frequency band [8]. The first eigenvalue can be conceived as a measure of the size of the primary cluster in the connectivity matrix. Therefore, the degree of clustering can be estimated by mapping the first eigenvalue into a 0–1 range, so that global dependency (low modularity) is represented by 0 and total independency (high modularity) by 1. In addition, path length was obtained by computing the average shortest path between pairs of channels in a distance matrix [9]. Statistical significance was assessed by the Wilcoxon rank sum test.

Kolmogorov-Smirnov test was used to compare distributions. Statistical and data analyses were performed using Matlab (MathWorks).

32.3 Results

We recorded subdural micro-ECoG activity by means of 32-channel multielectrode arrays (550 μm interelectrode spacing, Fig. 32.1a) covering almost the entire cortical surface on the left hemisphere of six *Fmr1* knockout (FXS) mice and ten wild-type (WT) littermates under isoflurane anesthesia. Under these conditions, the emergent activity was organized in slow oscillations as previously described [10]. Analysis of functional connectivity based on the coherence of the raw micro-ECoG signal between all pairs of electrodes revealed an overall tendency for an increase in the magnitude of the pairwise correlation in FXS mice as compared to WT (Fig. 32.1b, top). This was more evident when plotting the distribution of all correlation values of both genotypes (Fig. 32.1b, bottom), illustrating a significant bias to higher correlations in FXS mice ($p = 2.2 \times 10^{-78}$, Kolmogorov-Smirnov test). The relative magnitude of the correlation between elements of cortical networks has been described as a measure of cortical functional connectivity. Thus, these results suggest the presence of higher functional connectivity in the cortical network of FXS animals than of WTs.

Further inspection of the correlation matrices in Fig. 32.1b suggested that beyond the overall increase in functional connectivity in FXS mice, there was a nonhomogeneous distribution of the magnitude of pairwise interactions along the anteroposterior axis of the cortex. To investigate this in more detail, we divided the electrode array into two different clusters, one anterior (13 most anterior electrodes) and one posterior (13 most posterior electrodes), and evaluated the magnitude of the correlation within and between these two clusters. As shown in Fig. 32.1c, the level of intrinsic correlation was significantly higher within the anterior cluster in FXS animals ($p = 0.003$, Wilcoxon rank sum test), but there were no significant differences between genotypes on the level of correlation within the posterior cluster ($p = 0.180$, Wilcoxon rank sum test). Interestingly, the long-range correlation between the anterior and posterior cluster was stronger in FXS mice ($p = 0.022$, Wilcoxon rank sum test). Thus, these results suggest that the widespread increase in functional connectivity observed in FXS mice is reflected as an anomalously increased level of long-range coherence during slow oscillations, an activity pattern known to be dominated by short-range connectivity [8]. This is also apparent at the short-range level, but only at the most anterior part of the recorded area, putatively corresponding to the prefrontal cortex, a cortical area involved in high-order cognitive functions.

Complex dynamic systems, like the cerebral cortex, require an optimal balance between the level of coherence, which allows the integration of information, and that

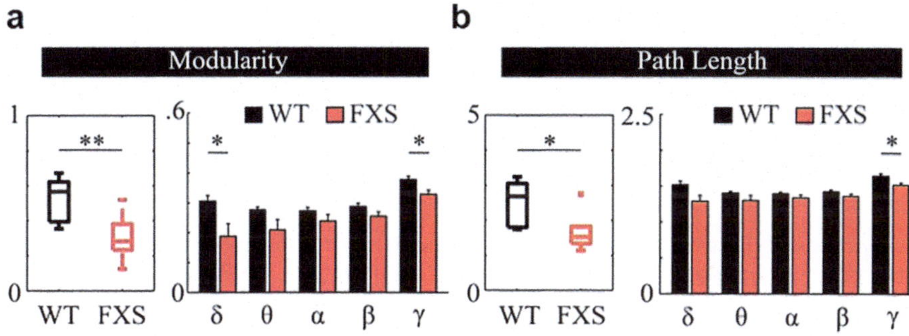

Fig. 32.2 Impaired cortical clustering and abnormal information exchange capacity. Group differences in modularity (**a**) and path length (**b**) calculated on the wideband (left) and frequency-band filtered (right) micro-ECoG signal recorded on the cortical surface of FXS and WT mice. *$p < 0.05$, **$p < 0.01$, Wilcoxon rank sum test. Box plots represent the first and third quartiles, median is depicted by the horizontal line, and extreme values are shown by whiskers. Bar plots represent mean ± SEM

of functional segregation, in which the different elements of the system can express relatively independent dynamics and therefore perform specialized functions [11]. With this in mind and considering the increment in functional connectivity shown by FXS mice, we wondered whether the relative balance between integration and segregation was also affected in these animals. To answer this, we calculated the degree of modularity (segregation) and path length (that reflects the inverse of integration) based on the raw micro-ECoG signal recorded in the same group of FXS and WT mice.

As shown in Fig. 32.2, FXS mice expressed a significantly lower level of modularity (Fig. 32.2a, left, $p = 0.005$, Wilcoxon rank sum test) and shorter path length, revealing larger integration than WT (Fig. 32.2b, left, $p = 0.016$, Wilcoxon rank sum test).

We next asked how network activation patterns were related to the different frequency bands, approaching in this way the mechanisms underlying cognitive impairment reported in these animals. We therefore calculated the degree of modularity and path length at different frequencies, from delta to gamma, in FXS and WT mice. Modularity was consistently lower in FXS mice, reaching significance at delta and gamma bands (Fig. 32.2a, right, $p < 0.05$, Wilcoxon rank sum test). Path length showed a similar trend, with an overall lower magnitude in FXS animals that reached significance at the gamma range (Fig. 32.2b, right, $p = 0.023$, Wilcoxon rank sum test). Overall, these results suggest that the general decrement of cortical segregation (reduced modularity) and anomalously high integration capacity (shorter path length) observed when analyzing the wideband micro-ECoG signal were indeed frequency-dependent and mostly specific to high frequency, gamma-like oscillations. Gamma frequency oscillatory activity has been described to support cognitive function [5], so it is likely that the neurocognitive deficits expressed by FXS mice might be related to an imbalance on the level of cortical integration and segregation in the gamma band.

32.4 Discussion

Here, we extend previous studies in FXS patients during resting-state conditions [12, 13] to the anesthetized FXS mouse model under slow oscillatory activity showing an overall and robust increment of functional connectivity that is both area and scale specific, affecting primarily the prefrontal area of the cortex and its long-range interactions. Our data fit very well with previous findings in FXS mouse models reporting an imbalance of excitation over inhibition [3], which may induce an overexcited state in which the connectivity between different elements of the cortical network lacks proper gating of regulatory mechanisms. Our data also extend previous behavioral and neurophysiological findings in these animals reporting both cognitive deficits and hyperconnectivity at the level of the prefrontal cortex [14], in which we report the highest impairment on the level of functional connectivity.

The balance between the level of cortical integration and segregation has also been reported to be affected in humans suffering from FXS [12, 15]. Here, in line with our findings on increased functional connectivity, we show that the cortical network of FXS mice manifests an abnormal increment of integration capacity (lower path length) alongside low segregation levels (modularity), both especially compromised at high-frequency bands. These results are in agreement with previous findings in FXS mice reporting neural network hyperexcitability to be the consequence of alterations in the inputs to fast-spiking interneurons, which control high-frequency gamma-band neural activity [16]. It thus seems, in light of these results, that the cognitive deficits present in FXS could be associated with a broken equilibrium between integration and segregation at gamma-like frequencies. This would hamper the capacity of different elements of the cortical network to properly integrate information and perform specialized functions.

Overall, our results show that the cortical network of *Fmr1* knockout mice is functionally altered, with anomalously high levels of functional connectivity between different elements of the network that hinder the ability of the cortex to integrate information by functionally specialized neuronal groups. These alterations are mostly present at the level of the prefrontal cortex and at gamma-like frequency bands and may contribute to the neurocognitive deficits manifested by FXS mice, also reported in FXS patients.

Acknowledgments Funded by EU H2020 Research and Innovation Programme, Grant 720270 (HBP SGA1), BFU2014-52467-R (MINECO), BFU2015-68568-P (MINECO/FEDER, UE), FLAGERA-PCIN-2015-162-C02-01, FRAXA Research Foundation, and CERCA (Generalitat de Catalunya). We thank Ernesto Pereda de Pablo (Universidad de La Laguna) for his valuable suggestions for the analysis. We also thank Guglielmo Fortunato (CNR) for the collaboration on the design and fabrication of the electrodes.

References

1. Penagarikano, O., Mulle, J.G., Warren, S.T.: The pathophysiology of fragile x syndrome. Annu. Rev. Genomics Hum. Genet. **8**, 109–129 (2007)
2. Lozano, R., Rosero, C.A., Hagerman, R.J.: Fragile X spectrum disorders. Intractable Rare Dis. Res. **3**, 134–146 (2014)
3. Gibson, J.R., Bartley, A.F., Hays, S.A., Huber, K.M.: Imbalance of neocortical excitation and inhibition and altered UP states reflect network hyperexcitability in the mouse model of fragile X syndrome. J. Neurophysiol. **5**, 2615–2626 (2008)
4. Compte, A., Reig, R., Descalzo, V.F., Harvey, M.A., Puccini, G.D., Sanchez-Vives, M.V.: Spontaneous high-frequency (10–80 Hz) oscillations during up states in the cerebral cortex in vitro. J. Neurosci. **28**, 13828–13844 (2008)
5. Wang, X.J.: Neurophysiological and computational principles of cortical rhythms in cognition. Physiol. Rev. **90**, 1195–1268 (2010)
6. Sporns, O.: Structure and function of complex brain networks. Dialogues Clin. Neurosci. **15**, 247–262 (2013)
7. Gomis-Gonzalez, M., Busquets-Garcia, A., Matute, C., Maldonado, R., Mato, S., Ozaita, A.: Possible therapeutic doses of cannabinoid type 1 receptor antagonist reverses key alterations in fragile X syndrome mouse model. Genes (Basel). **7**, 56 (2016)
8. Bettinardi, R.G., Tort-Colet, N., Ruiz-Mejias, M., Sanchez-Vives, M.V., Deco, G.: Gradual emergence of spontaneous correlated brain activity during fading of general anesthesia in rats. NeuroImage. **114**, 185–198 (2015)
9. Rubinov, M., Sporns, O.: Complex network measures of brain connectivity: uses and interpretations. NeuroImage. **52**, 1059–1069 (2010)
10. Ruiz-Mejias, M., Ciria-Suarez, L., Mattia, M., Sanchez-Vives, M.V.: Slow and fast rhythms generated in the cerebral cortex of the anesthetized mouse. J. Neurophysiol. **106**, 2910–2921 (2011)
11. Zhao, M., Zhou, C., Chen, Y., Hu, B., Wang, W.H.: Complexity versus modularity and heterogeneity in oscillatory networks: combining segregation and integration in neural systems. Phys. Rev. **82**, 046225 (2010)
12. Van der Molen, M., Stam, C., Van der Molen, M.: Resting-state EEG oscillatory dynamics in fragile X syndrome: abnormal functional connectivity and brain network organization. PLoS One. **9**, e88451 (2014)
13. Wang, J., Ethridge, L.E., Mosconi, M.W., White, S.P., Binder, D.K., Pedapati, E.V., Erickson, C.A., Byerly, M.J., Sweeney, J.A.: A resting EEG study of neocortical hyperexcitability and altered functional connectivity in fragile X syndrome. J. Neurodev. Disord. **9**, 11 (2017)
14. Testa-Silva, G., Loebel, A., Giugliano, M., de Kock, C.P., Mansvelder, H.D., Meredith, R.M.: Hyperconnectivity and slow synapses during early development of medial prefrontal cortex in a mouse model for mental retardation and autism. Cereb. Cortex. **22**, 1333–1342 (2012)
15. Bruno, J.L., Hosseini, S.M.H., Saggar, M., Quintin, E.M., Raman, M.M., Reiss, A.L.: Altered brain network segregation in fragile X syndrome revealed by structural connectomics. Cereb. Cortex. **27**, 2249–2259 (2017)
16. Cea-Del Rio, C.A., Huntsman, M.M.: The contribution of inhibitory interneurons to circuit dysfunction in fragile X syndrome. Front. Cell. Neurosci. **8**, 1–7 (2014)

Chapter 33
Multiple Epileptogenic Foci Can Promote Seizure Discharge Onset and Propagation

Denggui Fan and Qingyun Wang

Abstract Clinical electroencephalogram (EEG) of focal seizures shows that the cerebral cortical local neurons are first activated (epileptic foci), followed by rapid synchronous discharges which then rapidly spread to the surrounding normal brain regions. Based on a spatially extended computational model of cortical electric activity with realistic mesoscopic connectivity, we reconstructed the single epileptic focus and multiple epileptic foci by activating the activity level of local neuronal cluster with single-pulse stimulation disturbance, to simulate the physiologically observed synchronous brain activity of epileptogenic focus onsets and propagations. Results show that single epileptic focus with smaller excitatory activity region fails to spread into the surrounding brain regions, except for transient propagation to the neighborhood which eventually disappears with the recession of activity energy. However, as the excitatory activity region corresponding to the single epileptic focus getting larger, the synchronous oscillations of adjacent neuronal populations can not only be successfully initiated but also can propagate to the surrounding farther normal brain regions, due to the local and remote feedforward excitatory connectivities. This can ultimately cause a comprehensive onset of focal epilepsy. In particular, to investigate the effect of multifocal seizures on the propagations of epileptic onsets, we divided the large single focus into three equal multifocal regions. Detailed investigation reveals that compared to the single epileptic focus, multifocal seizure onsets are more easily to promote the generalized epileptic seizures and increase the spreading speed of synchronous oscillations of epileptic seizures.

D. Fan
School of Mathematics and Physics, University of Science and Technology Beijing, Beijing, P.R. China

Q. Wang (✉)
Department of Dynamics and Control, Beihang University, Beijing, P.R. China
e-mail: nmqingyun@163.com

J. M. Delgado-García et al. (eds.), *Advances in Cognitive Neurodynamics (VI)*,
Advances in Cognitive Neurodynamics,
https://doi.org/10.1007/978-981-10-8854-4_33

Keywords Spatially-extended model · Epileptogenic focus · Epileptic seizures · Synchronization

33.1 Introduction

Focal seizures are originated from the spatiotemporally localized pathological brain tissues, i.e., epileptogenic foci [1]. However, the dynamical mechanisms underlying the focal seizures, especially the multifocal seizures [2], remain unclear. Currently, surgical therapy for focal seizures is only applicable to epileptic patients with clearly localized anatomical abnormality of the brain. But the most widely used techniques for localizing the epileptogenic foci is only the visual identification of raw data, lacking the spatiotemporal correlations of different seizure regions which might mask the true epileptogenic foci. Therefore, it is needed to explore the dynamic characteristic and evolutionary property of focal seizures including the unifocal and multifocal seizures.

Electric stimuli are believed to cause the specific state-dependent transitions between non-seizure background state and seizure state [3]. In particular, electric stimuli have been considered as the potential inducement to initiate the epileptic seizures, even though the dynamic mechanism underlying the seizure onsets is still unknown. Our aim is to computationally reproduce the multifocal seizure onsets and evolutions. In particular, we compare the effects of multiple foci to that of single focus on the onsets, evolutions, and spreading speed of focal seizures. In this letter, we address this question by introducing single-pulse stimulation perturbation to a spatially extended computational model of cortical electric activity with realistic mesoscopic connectivity [4]. We then investigate how stimulation perturbations applied on the small and large spatial regions, as well as the multiple spatial regions, can successively initiate and promote the focal seizure onsets and propagations.

33.2 Descriptions of the Model

The computational model used here is an established Wilson-Cowan model [5], which is composed of excitatory (E) pyramidal and inhibitory (I) interneuronal population activities. The governing equations for our E-I model can be described as follows [4],

$$\tau_e \frac{dE}{dt} = -E + \mathbf{Sigm}[C_{e-e} \cdot E + C_{i-e} \cdot I + \mathbf{B}_e + A_s \cdot S(t)] + \mathbf{U}_e(t)$$

$$\tau_i \frac{dI}{dt} = -I + \mathbf{Sigm}[C_{e-i} \cdot E + \mathbf{B}_i] + \mathbf{U}_i(t)$$

where E and I represent fractional firing activities of excitatory and inhibitory neuronal populations, respectively. B_e and B_i represent the basal activity levels in excitatory and inhibitory neural populations, respectively. $A_s \cdot S(t)$ denote the noise input from other unmodelled brain regions to the excitatory neuronal populations, where A_s is the input coupling strength. $C_{p \to q}$ (p, q = e and i) is the synaptic coupling strength between the populations E and I. Sigmoid function Sigm[.] is defined as $S[x] = 1/(1 + \exp(-a(x - \theta)))$, where $a = 1$ and $\theta = 4$ are the steepness and offset of the sigmoid, respectively. In addition, in order to mimic the effect of stimulation on the initiations and evolutions of focal onset seizures, we performed a stimulus control U(t) on the cortical variables, E and I, respectively.

The smallest cortical minicolumn, represented by the single Wilson-Cowan model, is used to display the electric activity with the highest spatial resolution of clinical observations, which is comparable to the model output. Following the work by Wang et al. [4], we investigate a cortical sheet with 150 × 150 minicolumn units, where each minicolumn is assumed to be 50 × 50 μm in size, i.e., the cortical sheet is 7.5 × 7.5 mm in size. Furthermore, based on the cortical connectivity from [4], several types of inter-minicolumn connections are applied (see Fig. 33.1): (1) local/remote excitatory connections $C_{e \to e_{L/R}}$ (L/R denoting local/remote connections), that is, each excitatory population in minicolumn connects to the excitatory populations from close minicolumns or minicolumns at some distance, and (2) local inhibitory connections $C_{e \to i_L}$, i.e., the excitatory population from each minicolumn connects to the inhibitory populations from close minicolumns.

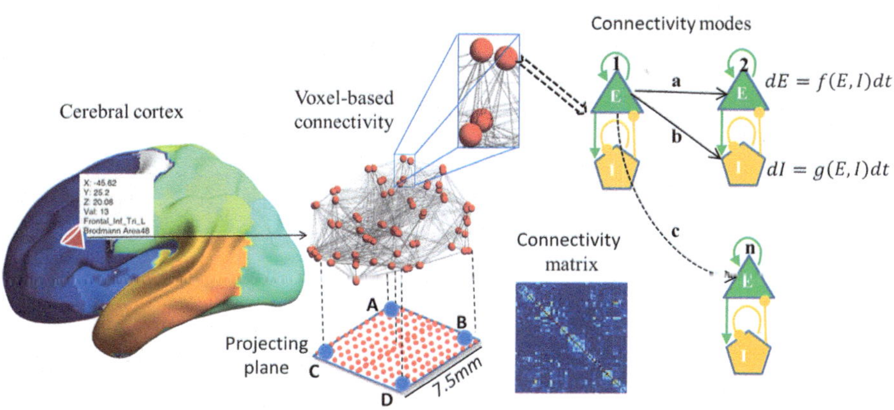

Fig. 33.1 (Color online) Schematic illustration of the structure of the model of a cortical volume, which is projected into a plane sheet. The cortical sheet consists of 150 × 150 minicolumns; each minicolumn is assumed to be 50 × 50 μm in size, i.e., the cortical sheet is 7.5 × 7.5 mm in size. Several connectivity modes within the cortical sheet are used, that is, a, local feedforward excitation; b. local feedforward inhibition; and c (dashed line), remote feedforward excitation. Four blue dots numbered with letters indicate microelectrode positions for the traces, i.e., time series of firing activity of the population, E, in the following figures

33.3 Results

During simulation the basic parameter setting can be seen from [4]. In addition, we apply local single-pulse stimulation to generate epileptic focus. The effect of stimulation amplitude (SA), stimulation fraction (SF), and stimulation density (SD) on the seizure onsets and propagations will be considered. In particular, we set SA = 1.5 and SD = 1 and vary SF from 0.015 to 0.05. Figure 33.2 shows the results of single epileptic focus with SF = 0.015. It is seen from Fig. 33.2a that as T = 0.05 s the local excitatory activity corresponding to epileptic focus spreads to the adjacent region due to the local excitatory connections. At the same time, several remote smaller regions are also activated due to the remote excitatory connections. Nevertheless, with T increasing from 0.05 to 0.1 s and 0.2 s, the activity energy is gradually weakened. Finally, as T = 0.3 s, the whole system restores into the background oscillating state, which can be illustrated by the Fig. 33.2b. However, as shown in Fig. 33.3a, with SF getting larger, the excitatory synchronous oscillations of adjacent neuronal populations and neuronal populations at some distance can be successfully ignited due the local and remote excitatory connections, respectively. In particular, as time goes by, synchronous oscillations gradually propagate to the surrounding farther normal brain regions. When T ≈ 0.5 s, a comprehensive onset of focal epilepsy is eventually caused. Figure 33.3b illustrates that the synchronous oscillations have propagated into the positions indicated by blue dots in Fig. 33.1 at T ≈ 0.4 s.

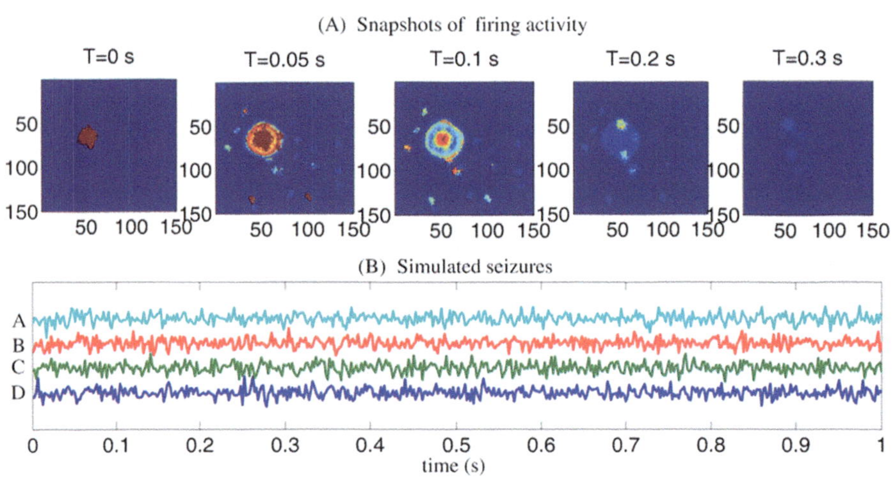

Fig. 33.2 (Color online) Unifocal seizure onset and spread with SF = 0.015, SA = 1.5 and SD = 1: (**a**) Snapshots of firing activity for E population within cortical sheet at different time points. (**b**) Time series of synchronous firing activity of the E population for the 4 units marked with letters in Fig. 33.1

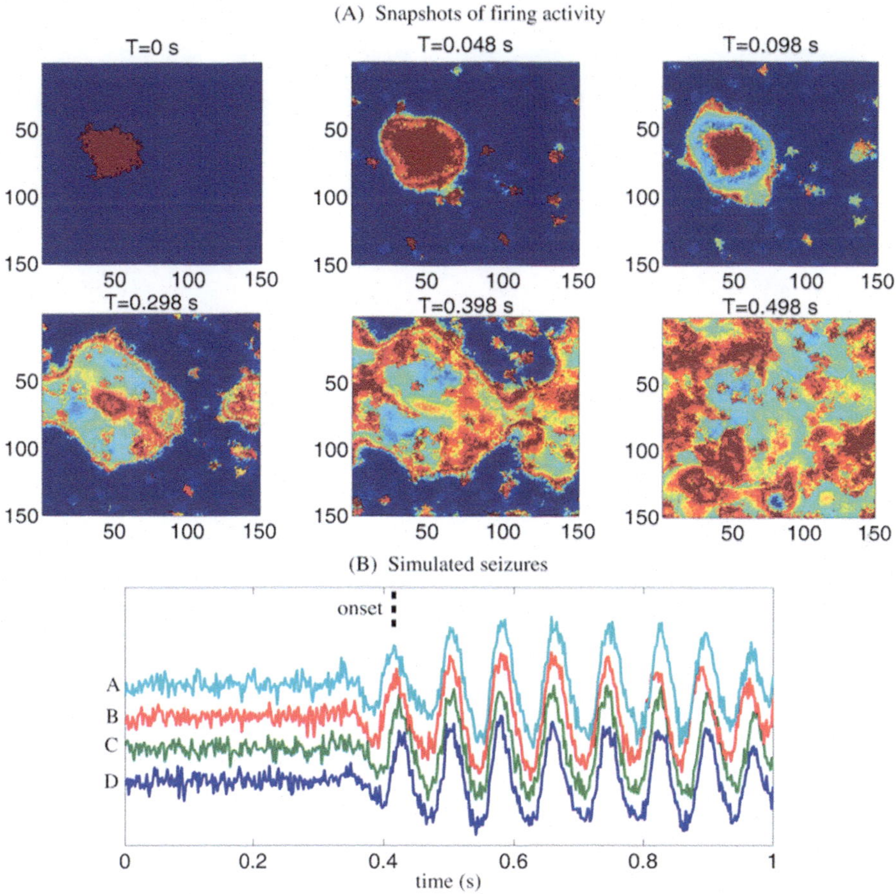

Fig. 33.3 (Color online) Unifocal seizure onset and spread with SF = 0.05, SA = 1.5, and SD = 1 (i.e., T = 0 s): (**a**) snapshots of firing activity for E population within cortical sheet at different time points. (**b**) Time series of synchronous firing activity of the E population for the four units marked with letters in Fig. 33.1

In order to investigate the effect of multifocal seizures on the propagations of epileptic onsets, as shown in Fig. 33.4a (i.e., corresponding to the case of T = 0 s), we disintegrate the large single focus (SF = 0.05) into three equal multifocal regions. Similar to the case of unifocal seizures of Fig. 33.3, each unifocal region of multifocal seizures first, respectively, activates their adjacent and remote neuronal populations through local and remote excitatory connectivities. Then the three enlarged excitatory activity regions can be integrated again with the further diffusion. In particular, as shown in Fig. 33.4b, when T ≈ 0.2 s, the synchronous oscillations have propagated into the positions indicated by blue dots in Fig. 33.1.

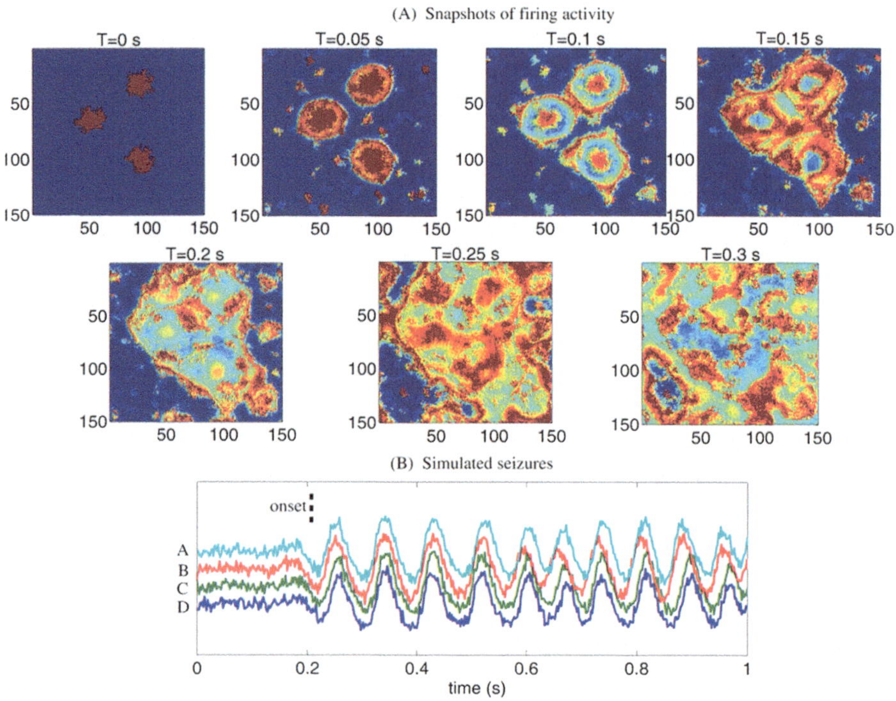

Fig. 33.4 (Color online) Trifocal seizure onset and spread with SF = 0.05/3, SA = 1.5, and SD = 1 for each focus (i.e., T = 0 s). (**a**) Snapshots of firing activity for E population within cortical sheet at different time points. (**b**) Time series of synchronous firing activity of the E population for the four units marked with letters in Fig. 33.1

As T ≈ 0.2 s, the comprehensive onset of focal epilepsy is displayed. However, compared Figs. 33.4b to 33.3b, it is seen that the multifocal seizure onsets are more easily to promote the generalized epileptic seizures and speed up the spread of synchronous oscillations of epileptic seizures than that of unifocal seizures. Physiologically and clinically, multiple foci can be essentially generated by disintegrating a single large focus with single-pulse stimulations; the facilitation of multifocal seizures may interpret the unsuccessful applications of stimulus perturbations on the experimental models and clinical therapy for focal seizures.

Acknowledgements This research is supported by the National Natural Science Foundation of China (Grant Nos. 11325208, 11572015, 11172017, and 11702018), the Project funded by China Postdoctoral Science Foundation (Grant No. 2016M600037), and the Fundamental Research Funds for the Central Universities (FRF-TP-16-068A1).

References

1. Rosenow, F., Luders, H.: Presurgical evaluation of epilepsy. Brain **124**, 1683–1700 (2001)
2. Devinsky, O., Romanelli, P., Orbach, D., Pacia, S., Doyle, W.: Surgical treatment of multifocal epilepsy involving eloquent cortex. Epilepsia **44**, 718–723 (2003)
3. Taylor, P., Wang, Y., Goodfellow, M., Dauwels, J., Moeller, F., Stephani, U., Baier, G.: A computational study of stimulus driven epileptic seizure abatement. PLoS One **9**, e114316 (2014)
4. Wang, Y., Goodfellow, M., Taylor, P.N., Baier, G.: Dynamic mechanisms of neocortical focal seizure onset. PLoS Comput. Biol. **10**, e1003787 (2014)
5. Wilson, H., Cowan, J.: Excitatory and inhibitory interactions in localized populations of model neurons. Biophys. J. **12**, 1–24 (1972)

Chapter 34
An ERP Study Reveals How Training with Dual N-Back Task Affects Risky Decision Making in a Gambling Task in ADHD Patients

Sarah K. Mesrobian, Alessandra Lintas, Manon Jaquerod, Michel Bader, Lorenz Götte, and Alessandro E. P. Villa

Abstract Impaired decision making and working memory (WM) are among the characteristic symptoms of patients affected by attention deficit hyperactivity disorder (ADHD). In order to investigate whether a WM training can affect the attitude toward risky decision making, we designed a study where participants had to perform a probabilistic gambling task. Our study has demonstrated that WM training affects in a different way controls and ADHD patients, who showed an increased tendency toward a risk taking attitude in case of the adaptive variant of the memory task. In ADHD patients, the frontal sites appeared the most affected, whereas global brain activity was likely to be affected in controls. This study shows also the benefits of cognitive training in ADHD patients and healthy subjects.

Keywords ERP · ADHD · Working memory · Decision making · Prefrontal cortex · Cingulate cortex

---

S. K. Mesrobian · M. Jaquerod
NeuroHeuristic Research Group, University of Lausanne, Lausanne, Switzerland

A. Lintas · A. E. P. Villa (✉)
NeuroHeuristic Research Group, University of Lausanne, Lausanne, Switzerland

LABEX, HEC Lausanne, University of Lausanne, Lausanne, Switzerland
e-mail: alessandro.villa@unil.ch

M. Bader
Research Unit of the University Department of Child and Adolescent Psychiatry (SUPEA), CHUV University Hospital and Faculty of Biology and Medicine, University of Lausanne, Lausanne, Switzerland

L. Götte
Institute for Applied Microeconomics, University of Bonn, Bonn, Germany

J. M. Delgado-García et al. (eds.), *Advances in Cognitive Neurodynamics (VI)*,
Advances in Cognitive Neurodynamics,
https://doi.org/10.1007/978-981-10-8854-4_34

34.1 Introduction

Patients with a diagnosis of attention deficit hyperactivity disorder (ADHD) are characterized by working memory (WM) impairment, difficulties in concentrating and maintaining focused attention, high degree of impulsivity, as well as excessive level of activity and talking and a multiple range of associated disorders. Experimental studies have not clarified yet to what extent, if any, ADHD were characterized by impaired performances in risky decision-making tasks. In a double-blind placebo experiment, using the Iowa gambling task (IGT), no significant difference was detected neither in the selection of bad deck nor in the total net score between all groups [1]. Other studies have pointed out a riskier behavior using the Iowa gambling task, with ADHD disclosing a lower total score compared with a control group [2].

During decision making it is important to determine whether the subjects are sensitive to the frequency but blind to the magnitude of a penalty/reward. In order to investigate whether a WM training can provoke a neural response having an influence on risky decision making, we designed a study where participants had to perform a probabilistic gambling task (PGT), modified from the original Gneezy and Potters' neuroeconomic game [3]. The protocol included a pre-training session in the laboratory, 20 days of WM training at home, and a post-training session in the laboratory. During the WM training, half of the participants played the adaptive variant of the memory task, and half played a baseline variant blocked at the first level of difficulty. We analyzed the effect of WM training on the behavior and on event-related potentials (ERPs). Preliminary results have been published in the PhD thesis of SKM [4].

34.2 Experimental Protocol

We selected 127 participants (22 years old ± 0.28 SEM, $N_{ADHD} = 43$ and $N_{CTRL} = 84$) for this study. All participants underwent a short structured diagnostic interview assessing psychiatric disorders. The session started by playing the PGT with an endowment of 20 points at the beginning of each trial and requested to gamble 0, 4, 8, 12, 16, or 20 points with 1/3 chance to win [4]. The outcome of the gambling was either to win four times the chosen amount, with a probability $P_{\text{win}} = 1/3$, or to lose the entire amount with a probability $P_{\text{loose}} = 2/3$ with a uniformly distributed probability. If the participant selected 8 points, the outcome would be $12 = (20-8)$ in case of loss or $44 = (20-8)+(8\times4)$ in case of win. The participants had the possibility to modify their initial choice during 4 s. The overall amount of points held by the participant was displayed every four trials. The whole session was composed of 10 games $\times$ 16 blocks, overall 160 trials. The duration of PGT was approximately 30 min.

Subsequently, the participants exercised the adaptive version of the dual n-back memory task [5], which consisted of 20 blocks of $20+\text{N}$ trials, composed of visual

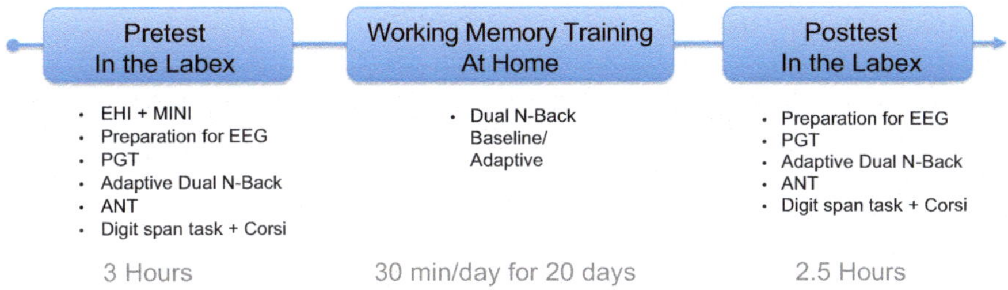

Fig. 34.1 Experimental procedure of the whole study

and auditory stimuli. Participants were asked to detect and to press a key if any of the current stimuli corresponded to the ones presented in the previous trial (Level 1). Participants pressed the "A" key to report the correspondence with a visual target and pressed the "L" key to report for the correspondence with an auditory target. In the adaptive variant, the level of difficulty was adjusted as follows: in case of less than 3 mistakes in both modalities, the difficulty was increased by 1, while 5 errors in any modality decreased the level by 1. Level 2 meant to recognize if any of the current stimuli corresponded to the ones presented two trials earlier, level 3 corresponding to 3 trials earlier, and so on.

During the entire duration of a session, ended with the attentional network test (ANT) [6], EEG was continuously recorded using 64 active Ag/AgCl electrodes and referenced to the linked earlobes [7]. Epochs triggered by pressing the spacebar (event S) and by clicking on the selected amount to gamble (event I) were used for the analysis of ERPs from 500 ms before to 1000 ms after the triggering event. Following the removal of EEG's equipment, the participants performed the WAIS-IV digit span task [8] and the Corsi block-tapping task [9]. The complete protocol of the study (Fig. 34.1) included a pre-training session in the laboratory (with all participants playing the adaptive version of the dual n-back task), 20 days of WM training at home, and a post-training session in the laboratory (with all participants playing again the adaptive version of the dual n-back task). During the WM training, half of the participants of each group (ADHD and controls) played the adaptive variant (trained participants), whereas the other half played the dual n-back blocked at the level 1 for the whole training period (baseline participants). Participants were required to complete at least 18 training sessions within a month. The dual n-back task played during each training session (composed of 20 blocks) was made available to the participants via a customized secured internet website.

34.3 Results

Before WM training, all participants reached level 2 of the dual n-back task. No difference was observed between the baseline and trained participants of either group ($U_{CTRL} = 953.5$, $p > .05$; $U_{ADHD} = 335.5$, $p > .05$) neither between

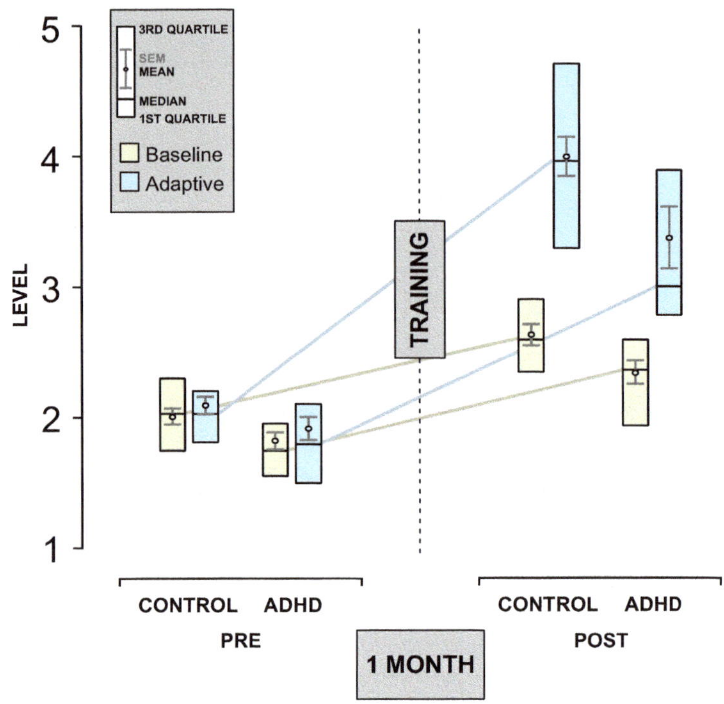

Fig. 34.2 Boxplot of the dual n-back performance of $N = 84$ controls and $N = 43$ ADHD participants before and after the WM training

controls and ADHD ($\chi^2(1) = 0.74$, $p > .05$). After training, the performance on the dual n-back was enhanced in both controls ($Z_{\text{CTRL-base}} = -5.59$, $p < .01$; $Z_{\text{CTRL-train}} = -5.7$, $p < .01$) and ADHD patients ($Z_{\text{ADHD-base}} = -3.67$, $p < .01$; $Z_{\text{ADHD-train}} = -4.23$, $p < .01$. It is interesting to notice that in both groups, even a training with dual 1-back (baseline participants) improved the performance during the post-training session ($\chi^2(1) = 104.2$, $p < .001$, Fig. 34.2).

The WT training did not affect the total gains earned by all participants at the probabilistic gambling task, irrespective of the group, ($\chi^2(1) = 1.8$, $p > .05$). The count of times a participant gambled a small amount (i.e., 0, 4, or 8 points), termed LIR, and the count of times a participant gambled a large amount (i.e., 12, 16, or 20 points), termed HIR, were used to compute an investment risk index $IRi = \frac{HIR-LIR}{HIR+LIR}$. Thus, the value of IRi is centralized within the range $[-1 - +1]$, an index closer to -1 characterizes a participant with risk averse strategy, and an index closer to $+1$ characterizes a risk seeking participant. Hence, each individual was characterized by an investment risk index. WM training affected the participants of the two groups in a very distinct way. Controls tended to centralize their attitude toward a risk neutral attitude after WM training, and risky behavior was not affected by the training protocol ($t_{(70)} = -0.171$, $p > .05$). On the contrary, ADHD patients tended to maintain the same risk-taking behavior before and after WM training for the baseline participants. Moreover, ADHD patients who performed the adaptive training showed an increase in their attitude toward risk ($t_{(40)} = 3.516$, $p < .01$).

The analysis of the ERPs after WM training (Fig. 34.3) showed a N2-P3 complex triggered by the self-paced start of trial (event S). In controls we observed no differences due to the training variant. In ADHD we observed a larger N2 component in baseline, especially toward posterior sites, and a larger P3 component toward the frontal sites after the adaptive training. After the selection of the gambling (event I), the controls showed an increase in P3 amplitude close to central sites after the adaptive training, suggesting a localization toward the cingulate cortex. In the ADHD after the adaptive training, we observed also an important secondary component, P3b, mainly localized toward the frontal sites.

34.4 Discussion

Several hypotheses have been suggested in order to explain the suboptimal decision-making abilities characterizing ADHD patients [10]. The medial prefrontal cortex–posterior cingulate cortex circuitry is involved in the abilities to have a sense of self and to pursuit specific goals. This circuitry has been shown to be less effective in ADHD and might explain goal-setting impairments, planning difficulties, and intention deficits involved in the valuation and intention to take risk processes. Executive functioning deficits in ADHD associated with abnormal activity in the dorsolateral prefrontal cortex postulate that WM deficits prevent to hold informations in mind, resulting in poor economic decision-making capacities in ADHD. The last hypothesis involves dopaminergic dysregulation in ventral frontostriatal loops supporting outcome prediction and evaluation, and learning competences, that have been reported to be less efficient in ADHD individuals. Our study has demonstrated that WM training affects in a different way controls and ADHD patients, who showed an increased tendency toward a risk-taking attitude in case of the adaptive variant of the memory task. The inherent difficulty of this task is likely responsible of an increased activity of the frontal areas, observed in the ERPs after the selection of the gambling. We are undergoing further analyses of these results aimed at providing a more precise localization of the cortical areas that characterize altered brain activity in ADHD patients.

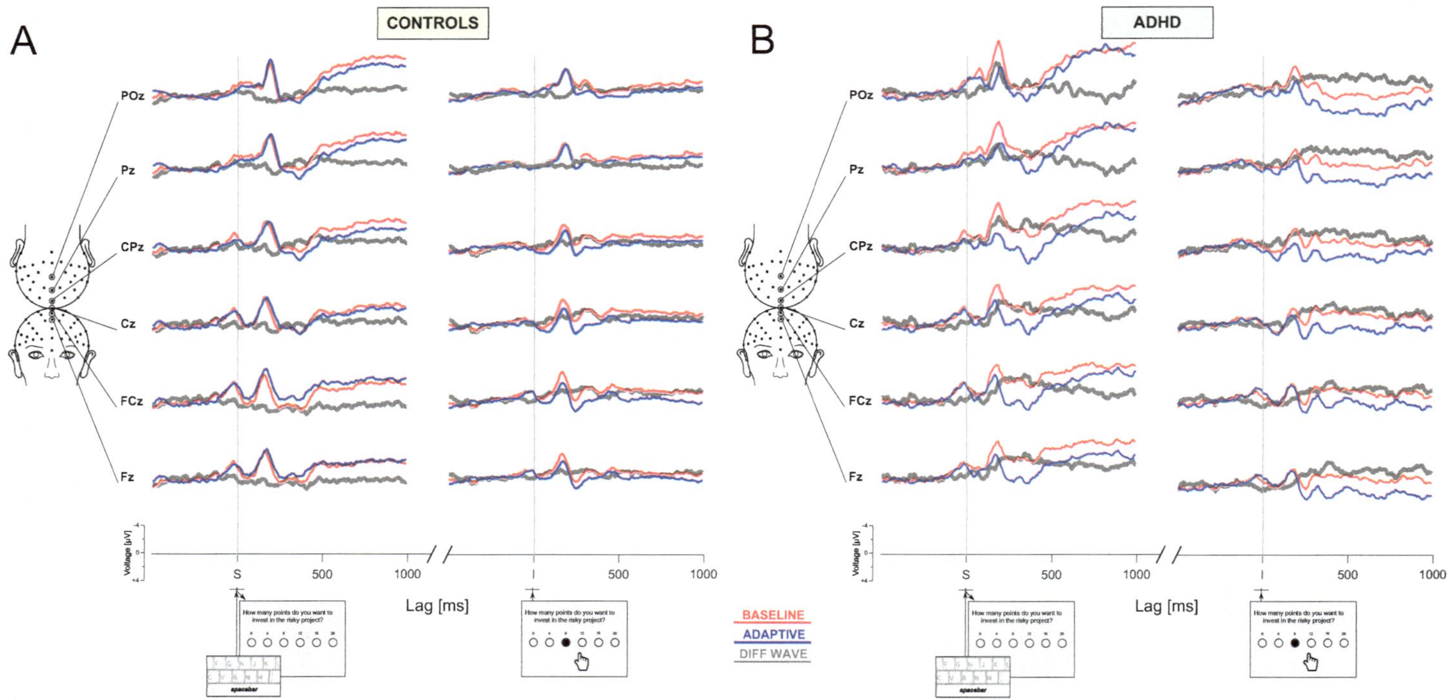

Fig. 34.3 ERPs: Grand average ERPs recorded after WM training at Fz, FCz, Cz, CPz, Pz, and POz sites triggered by pressing the spacebar (event S) and following the selection of the amount to gamble (event I) in control (**a**) and ADHD (**b**) participants. ERPs of participants trained with dual 1-back (the baseline variant, red curves), and trained with the adaptive variant (blue curves) are plotted with the differential curves (dark gray)

Acknowledgements This study was supported by the Swiss National Science Foundation (grant CR 1311–138032).

References

1. Agay, N., Yechiam, E., Carmel, Z., Levkovitz, Y.: Non-specific effects of methylphenidate (Ritalin) on cognitive ability and decision-making of ADHD and healthy adults. Psychopharmacology **210**, 511–519 (2010)
2. Malloy-Diniz, L., Fuentes, D., Leite, W.B., Correa, H., Bechara, A.: Impulsive behavior in adults with AD/HD: characterization of attentional, motor and cognitive impulsiveness. J. Int. Neuropsychol. Soc. **13**, 693–698 (2007)
3. Gneezy, U., Potters, J.: An experiment on risk taking and evaluation periods. Q. J. Econ. **112**, 631–645 (1997)
4. Mesrobian, S.K.: Does working memory training affect decision making? A neuroeconomic study. Ph.D. thesis, University of Lausanne (2015)
5. Jaeggi, S.M., Buschkuehl, M., Jonides, J., Perrig, W.J.: Improving fluid intelligence with training on working memory. PNAS USA **105**, 6829–6833 (2008)
6. Fan, J., McCandliss, B.D., Sommer, T., Raz, A., Posner, M.I.: Testing the efficiency and independence of attentional networks. J. Cogn. Neurosci. **14**, 340–347 (2002)
7. Mesrobian, S.K., Bader, M., Götte, L., Villa, A.E.P., Lintas, A.: Imperfect Decision Making and Risk Taking Are Affected by Personality. Studies in Computation Intelligence, vol. 538, pp. 145–184. Springer, Heidelberg (2015)
8. Wechsler, D.: Wechsler Adult Intelligence Scale–Fourth Edition (WAIS–IV). NCS Pearson, San Antonio (2008)
9. Kessels, R.P., van Zandvoort, M.J., Postma, A., Kappelle, L.J., de Haan, E.H.: The Corsi block-tapping task: standardization and normative data. Appl. Neuropsychol. **7**, 252–258 (2000)
10. Sonuga-Barke, E.J.S., Fairchild, G.: Neuroeconomics of attention-deficit/hyperactivity disorder: differential influences of medial, dorsal, and ventral prefrontal brain networks on suboptimal decision making? Biol. Psychiatry **72**, 126–133 (2012)

Chapter 35
Working Memory Development in Attention Deficit Children and Adolescents

Elena I. Rodríguez-Martínez, Antonio Arjona-Valladares, Francisco J. Ruíz-Martínez, Manuel Morales, Catarina I. Barriga-Paulino, Jaime Gómez-González, and Carlos M. Gómez

Abstract Attentional deficit disorder (ADD, ADHD) is a complex disorder in which attention and working memory (WM) are impaired. The central hypothesis is that WM behavioral performance would be impaired in ADD and would facilitate the classification of control and ADD subjects. One hundred and eighty-one control and 41 ADD children and adolescents (6–17 years old) were behaviorally recorded using the Working Memory Test Battery for Children (WMTB-C), delayed match-to-sample test (DMTS), and oddball tasks. ADD children presented a behavioral impairment in WMTB-C, DMTS, and oddball tasks. ADD obtained lower direct scores in the three subcomponents of the Baddeley's WM model, lower d' values, produced more errors, and presented higher variability in RTs than controls. The discriminant analysis was able to classify correctly around 70% of controls and ADD children. The results suggest that WM is a central core dysfunction in ADD and useful as a diagnostic tool.

Keywords Attention deficit disorder · Working memory impairment · Linear discriminant analysis · Subject's classification

E. I. Rodríguez-Martínez (✉) · A. Arjona-Valladares · F. J. Ruíz-Martínez · M. Morales
C. I. Barriga-Paulino · C. M. Gómez
Human Psychobiology Lab, Experimental Psychology Department, University of Seville, Sevilla, Spain
e-mail: elisroma@us.es

J. Gómez-González
UGC of Mental Health, Virgen Macarena Hospital, Seville, Spain

J. M. Delgado-García et al. (eds.), *Advances in Cognitive Neurodynamics (VI)*, Advances in Cognitive Neurodynamics,
https://doi.org/10.1007/978-981-10-8854-4_35

35.1 Introduction

Brain development is a complex process, involving genetic and environmental factors, as well as their interaction [1]. ADD is a common childhood disorder that involves inattention, hyperactivity, and impulsivity [2].

The relationship of WM impairment and ADD has been well documented in several meta-analyses [3, 4] and suggests that some of the behavioral and curricular impairments of ADD children could be secondary to WM deficits. WM is a central function for performance, which is developing across childhood and adolescent periods. The impairment of WM has been considered a landmark of ADD. Our group has elaborated, using the WMTB-C and the DMTS, an extensive description, how WM develops with age. Particularly, and using the WMTB-C, we have demonstrated an improvement of performance with age of the three subcomponents of Baddeley's model: central executive (CE), phonological loop (PL), and visuospatial sketchpad (VSSP) [5]. By means of DMTS and oddball tasks, an improvement of RTs, a reduction of variability of RTs, an improvement of the index of discriminability (d′), and a decrease of behavioral errors have also been demonstrated with increasing age [6, 7]. Present experiment tries to elucidate possible WM behavioral differences between control and ADD children and adolescents and determine the percentage of subjects that can be correctly classified by means of linear discriminant analysis operating on WM variables. The central hypothesis is that WM behavioral performance would be impaired in ADD, and these variables would permit the classification of control and ADD subjects.

35.2 Methods

35.2.1 *Sample*

The ADD group was extracted from a group of children and adolescents, with ages between 6 and 17 years old, diagnosed with ADD. The ADD group was composed of 41 subjects ($M = 10.8$ years, $SD = 3.4$, 8 females). Additionally the DuPaul questionnaire [8] was administered to the ADD subjects. To guarantee that behavioral recordings in the ADD group were not influenced by medication, the subjects ceased taking medication 48 h prior to testing.

One hundred and eighty-one subjects between 6 and 17 years old participated as a control group in this study ($M = 11.1$, $SD = 3.37$, 87 females). There was not a statistically significant difference between the ages of control and ADD groups ($F_{[1, 220]} = 0.171$, $p = 0.68$). The subjects were recruited from public schools, and the experiments were conducted with informed and written parental consent following the Declaration of Helsinki guidelines. The ethics committee of biomedical research of the autonomous community of Andalucía approved the experimental protocol.

35.2.2 *Experimental Session*

All subjects were assessed in three different tasks to evaluate WM:

35.2.2.1 Working Memory Test Battery for Children (WMTB-C)

The WMTB-C [9] was used. The test consists of nine subtests designed to evaluate the components of WM: PL, VSSP, and CE. This particular test is guided by the Baddeley and Hitch WM model [10]. For the analysis of the WMTB-C variables, the direct scores in these subtests, which are the scores obtained directly from the subtests, were used.

35.2.2.2 Delayed Match-to-Sample Task (DMTS)

The DMTS task was presented in a computer screen. For the control of the stimuli presentation, the E-Prime version 2.0 was used, and for recording the subject's responses, a SRBOX Cedrus was used. One hundred twenty-eight trials organized in four experimental blocks (32 trials each) were presented. One trial was composed of one stimulus to be maintained in memory (S1, 1000 ms), which appeared in the center of the screen; S1 was followed by a 1500 ms delay period (the retention period). Then, two stimuli (S2, 2000 ms) appeared simultaneously, one on the left and the other on the right side of the screen. Subjects were asked to press the corresponding button when S2 matched with S1. The response window was 2000 ms. Auditory feedback was provided, indicating to the subject if a correct or incorrect response was given.

All stimuli were different in each trial. They were obtained from images of children's cartoons. The order of presentation of the target stimulus (the S2 same than S1) was randomized for each subject. A rest period was allowed between blocks. The recorded variables were *RTs* and errors. The three types of errors taken into account were no responses to S2 (omissions), incorrect responses to S2 (commissions), and anticipations. *Means*, *SDs*, and coefficients of variation (*CV*) of *RTs* were obtained from individual subjects, as well as the sum of different types of errors.

35.2.2.3 Oddball Task

The oddball paradigm was composed of a total of 120 trials (in 83 control subjects and 24 ADD subjects, 240 stimuli were presented); 25% of the trials included a novel stimulus, while 75% presented the same repeated stimulus (the standard stimulus). The stimuli were presented in the center of the screen for 700 ms, and a fixed interstimulus interval of 700 ms was used. The standard and novel stimuli were cartoons. Subjects were asked to press the button only when novel stimuli were presented. The order of stimuli presentation was random. The recorded variables were RTs and the three types of errors (omissions, incorrect responses, and anticipations). Means, SDs, and CV of RTs were obtained from individual subjects. The sum of different types of errors was obtained.

35.3 Data Analysis

Behavioral measures obtained from the DMTS and oddball tasks were analyzed. RTs, *Means*, *SD* of *RTs*, *CV* of *RTs*, and the total errors, computed as the sum of the percentage of anticipations, omissions, and false alarms, were calculated with the 2010 version of MATLAB (MathWorks Inc., MA, USA). The *SD* and *CV* of *RTs* permitted to operationalize the intra-subject behavioral variability. For the WMTB-C, the direct scores obtained on each subtest (PL, VSSP, and CE) were the analyzed variables.

The sensitivity index (d') which is a measure of the ability of the subject to discriminate between signal and noise was computed [11], only for the oddball task as:

$$d' = Z\,(\text{hit rate}) - Z\,(\text{false alarm rate}) \tag{35.1}$$

Z (hit rate): Z values of the proportion of hits (responses to targets)
Z (false alarm rate): Z values of the proportion of false alarms (responses to the standards)

High d' positive values are related to high sensitivity, while zero values indicate low sensitivity, and negative values would indicate poor comprehension of the instructions, with the subject responding to standard rather than target stimuli.

As a different number of boys and girls were recruited, a one-way ANOVA was computed to analyze if the gender factor influenced the values of empirical variables in the control group. As gender was not a significant factor for any of these variables in the control group, all the control subjects were included in the subsequent ANOVAs comparing control and ADD subjects.

Linear discriminant analysis was used to classify control and ADD subjects. The direct scores of the three subcomponents of the Baddeley's model, the *RTs*, *SD* of *RTs* and percentage of errors of the DMTS and oddball tasks, and the d' parameters of the oddball task were the variables included in the analysis. Validation was obtained using the leave-one-out method.

35.4 Results

Table 35.1 shows that all recorded behavioral variables were statistically significant when ADD vs. controls were compared. In all cases it was due to a functional impairment in the ADD group with respect to the controls. The ADD showed lower direct scores in the three subcomponents of the WMTB-C, and in the DMTS and oddball tasks, they showed lower RTs, higher number of errors, and higher variability in *RTs*, as indicated by the *SD* and *CV* of *RTs*. Furthermore, the ADD group showed lower d' parameter than controls in the oddball task indicating a lower stimulus discriminability than control subjects.

Table 35.1 Descriptive and statistical comparisons of the behavioral variables (one subject in DMTS and other in oddball task were not able to complete these tasks in the ADD group; all control subjects completed them: WMTB-C, DMTS, and oddball)

Behavioral and statistical measures			*Mean*	*SD*	*DF*	*F*	*p*
WMTB-C	*Direct scores PL*	Controls	117.89	19.54	[1, 219]	17.98	0.001
		ADDs	104.1	15.05			
	Direct scores vs.	Controls	50.96	9.64	[1, 219]	6.16	0.014
		ADDs	46.8	9.78			
	Direct scores EC	Controls	66.94	15.82	[1, 219]	25.14	0.001
		ADDs	53.39	14.69			
Oddball	*Total errors (%)*	Controls	3.33	4.04	[1, 216]	13.04	0.001
		ADDs	6.01	5.03			
	RTs mean	Controls	465.34	64.17	[1, 216]	3.03	0.083
		ADDs	484.74	61.57			
	SD RTs	Controls	67.11	15.92	[1, 216]	13.04	0.001
		ADDs	77.08	15.12			
	CV RTs	Controls	0.14	0.31	[1, 216]	8.35	0.004
		ADDs	0.16	0.32			
	d′	Controls	3.82	0.79	[1, 216]	12.29	0.001
		ADDs	3.32	0.95			
DMTS	*Total errors (%)*	Controls	3	6.16	[1, 215]	19.63	0.001
		ADDs	9.38	14.18			
	RTs mean	Controls	688	193.99	[1, 215]	5.67	0.018
		ADDs	770.93	219.37			
	SD RTs	Controls	202.39	93.22	[1, 215]	15.72	0.001
		ADDs	266.75	90.4			
	CV RTs	Controls	0.28	0.08	[1, 215]	19.89	0.001
		ADDs	0.34	0.07			

Direct scores of the phonological loop (PL), direct scores of the visuospatial sketchpad (VSSP), direct scores of the central executive (CE)

RTs response times, *SD RTs* standard deviation of response times, *CV RTs* coefficient of variation of *RTs*, *d′*, *d* prime

The Box's test for equality of covariance indicates that the null hypothesis of equality of covariance matrices can be accepted ($F_{[55,\ 15570.33]} = 1.659$; $p = 0.002$). The Wilk's lambda ($WL = 0.806$; $\chi^2 = 44.79$; $p = 0.001$; $DF = 10$) indicated a global difference between ADD and control subjects. The linear discriminant analysis permitted to classify correctly 76.7% of control subjects and 69.2% of ADD. When the cross validation method of one-leave-out was applied, 72.1% of control subjects and 64.1% of ADD were correctly classified.

An additional effort was made to eliminate the effect of age by means of recomputing the linear discriminant analysis on the residuals of the empirical variables vs. age. The Box's test indicated equal covariance matrices ($F_{[55,\ 15570.33]} = 1.53$; $p = 0.007$). The Wilk's lambda ($WL = 0.754$; $\chi^2 = 58.7$; $p = 0.001$; $DF = 10$) indicated a global difference between ADD and control subjects. The linear discriminant analysis permitted to classify correctly 81.7% of control subjects and 69.2% of ADD. When the cross validation method of one-leave-out was applied, 78.4% of control subjects and 61.5% of ADD were correctly classified.

35.5 Discussion

ADD children presented a behavioral impairment in the three tasks: WMTB-C, DMTS, and oddball. They obtained lower direct scores in the three subcomponents of the Baddeley's WM model and lower d´ values indicating a lower sensory discrimination between target and distractors, and they produced more errors and presented higher variability in *RTs* than controls. The results indicated that there was a general impairment in all the parameters related to WM in ADD children but also in d′, more related to pure sensory discrimination abilities. The present results add evidence to WM as a main dysfunction in ADD [3, 4]. The impairment of WM is probably causally related to the curricular associated problems [12], given the close association between WM scores and curricular attainments [13].

The discriminant analysis was able to classify correctly around 70% of controls and ADD children. Interestingly, eliminating the effects of age did not improve the efficacy of classification. Other procedures for classifying ADD children have also been successful using structural magnetic resonance [14] and EEG-derived parameters [15]. However, present classification is exclusively based on behavioral data which are much easier to collect at community settings. Given the complexity of the clinical diagnosis, the use of discriminant functions "ready-to-use" which would incorporate the empirical values obtained during the behavioral assessment would optionally help the clinician in obtaining an accurate diagnosis, rather than a determinant criterion for diagnosis. The results, given the extension of behavioral impairment in the WM variables measured in present experiment, suggest that WM is a central core dysfunction in ADD and useful as a diagnostic tool.

Acknowledgements This work was supported by grants of the Spanish Ministry of Economy and Competitiveness [grant numbers PSI2013-47506-R and PSI2016-80059-R] and by a research grant from Janssen-Cilag.

References

1. Shaw, P., Kabani, N.J., Lerch, J.P., Eckstrand, K., Lenroot, R.: Neurodevelopmental trajectories of the human cerebral cortex. J. Neurosci. **28**, 86–94 (2008)
2. Barkley, R.: El desorden de hiperactividad y déficit de atención. Investigación y ciencia. 48 (1998)
3. Martinussen, R., Hayden, J., Hogg-Johnson, S., Tannock, R.: A meta-analysis of working memory impairments in children with attention-deficit/hyperactivity disorder. J. Am. Acad. Child Adolesc. Psychiatry. **44**, 4 (2005)
4. Frazier, T.W., Demaree, H.A., Youngstrom, E.A.: Meta-analysis of intellectual and neuropsychological test performance in attention-deficit/hyperactivity disorder. Neuropsychology. **18**, 543–555 (2004)
5. Rodriguez-Martinez, E.I., Barriga-Paulino, C.I., Rojas-Benjumea, M.A.: Co-maturation between the spontaneous electroencephalogram and working memory. Neurosci. Lett. **550**, 134–138 (2013)
6. Rojas-Benjumea, M.A., Barriga-Paulino, C.I., Rodríguez-Martínez, E.I., Gómez, C.M.: Development of behavioral parameters and ERPs in a novel target visual detection paradigm in children, adolescents and young adults. Behav. Brain Funct. **11**, 22 (2015)
7. Barriga-Paulino, C.I., Rodríguez-Martínez, E.I., Rojas-Benjumea, M.A., Gómez, C.M.: Principal component analysis of working memory variables during child and adolescent development. Span. J. Psycho. **19**, e62: 1–e62:13 ((2016)
8. DuPaul, G.J., Power, T.J., Anastopoulos, A.D., Reid, R.A.D.H.D.: Rating Scale-IV: Checklists, Norms, and Clinical Interpretation. Guilford, New York (1998)
9. Gathercole, S., Pickering, S.: Working Memory Test Battery for Children (WMTB-C). Pearson Education Ltd, London (2001)
10. Baddeley, A.D., Hitch, G.: Working memory. In: Bower, G.A. (ed.) Recent Advances in Learning and Motivation, vol. 8, pp. 47–90. Academic, New York (1974)
11. Stanislaw, H., Todorov, N.: Calculation of signal detection theory measures. Behav. Res. Meth. Ins. C. **31**, 137–149 (1999)
12. Alloway, T.P., Gathercole, S., Elliot, J.: Examining the link between working memory behavior and academic attainment in children with ADHD. Dev. Med. Child Neurol. **52**, 632–636 (2010)
13. Alloway, T.P.: Working memory but not IQ predicts subsequent learning in children with learning difficulties. Eur. J. Psych. Assess. **25**, 92–98 (2009)
14. Colby, J.B., Rudie, J.D., Brown, J.A., Douglas, P.K., Cohen, M.S., Shehzad, Z.: Insights into multimodal imaging classification of ADHD. Front. Syst. Neurosci. **6**, 59–52 (2012)
15. Poil, S., Bollmann, S., Ghisleni, C., O'Gorman, R.L., Klaver, P., Ball, J., Eich-Höchli, D., Brandeis, D., Michels, L.: Age dependent electroencephalographic changes in attention-deficit/hyperactivity disorder (ADHD). Clin. Neurophysiol. **125**, 1626–1638 (2014)

Chapter 36
Spectral Power and Maturational Frequency-Coupling Differences Between Attention Deficit and Control Children and Adolescents

Elena I. Rodríguez-Martínez, Brenda Y. Angulo-Ruíz, Antonio Arjona-Valladares, Francisco J. Ruíz-Martinez, Jaime Gómez-González, and Carlos M. Gómez

Abstract Present report explores the possibility that attentional deficit disorder (ADD) subjects would show a different spectral power values compared with healthy subjects and possibly a different theta-beta maturational frequency coupling. Open-eyes, resting-state EEG was recorded in a sample of 36 controls and 36 ADD subjects (6–17 years old). The power spectral density (PSD) from 0 to 46 Hz was computed. ANOVAs to compare spectral power between control and ADD subjects were obtained. PSD correlations between the whole ranges of frequencies were calculated to observe possible differences in the co-maturation of the different brain rhythms in both groups of subjects. An increase in delta power in ADD subjects with respect to control subjects was obtained, indicating a predominance of slow waves in ADD subjects. While control subjects presented a significant correlation between low-frequency rhythms and beta rhythm, ADD subjects presented a lower maturational power-to-power frequency coupling between these rhythms.

Keywords Attention deficit · EEG resting state · Spectral power · Delta rhythm · Power-to-power frequency coupling

E. I. Rodríguez-Martínez · B. Y. Angulo-Ruíz · A. Arjona-Valladares · F. J. Ruíz-Martinez
C. M. Gómez (✉)
Human Psychobiology Lab, Experimental Psychology Department, University of Sevilla, Sevilla, Spain
e-mail: cgomez@us.es

J. Gómez-González
UGC of Mental Health, Virgen Macarena Hospital, Sevilla, Spain

J. M. Delgado-García et al. (eds.), *Advances in Cognitive Neurodynamics (VI)*,
Advances in Cognitive Neurodynamics,
https://doi.org/10.1007/978-981-10-8854-4_36

36.1 Introduction

From a neuropsychological point of view, two theories try to account for the different neuropsychological profile of ADD children: The maturational lag model proposes that ADD children are developmentally delayed for their age [1] and they would perform in behavior and EEG parameters in a way that would be normal in younger children. In this sense, Matsuura et al. [2] demonstrated an increase in slow frequencies in frontal areas in ADD children, which is typical of young children. Also the increase in theta rhythms in ADD subjects also support the maturational lag model [3].

Several EEG studies in ADD children have found that they have a greater activity of slow waves, mainly in theta band (5–7 Hz) and in theta-beta and theta-alpha ratios, as well as a decrease of beta power when compared to control subjects [4]. The analysis of theta-alpha and theta-beta ratios has proven to be a very useful measurement tool both for maturative EEG changes in children with no disorder [5] and for differences in children with inattentive and combined ADD (Clarke et al.) [6, 7].

We have previously found a correlation between theta and beta spectral power along development in control subjects, indicating a co-maturation between these two rhythms [8]. If a reduction of theta-beta frequency coupling would appear in ADD subjects, it would imply a decoupling in the co-maturation of both rhythms and would suggest a differential developmental trajectory in ADD children and adolescents with respect to controls.

We hypothesized that ADD subjects would show a different spectral power values compared with healthy subjects and possibly a different theta-beta maturational coupling. A last point to be addressed is that most EEG resting state results in ADD children have been obtained in closed-eyes condition, while the present report is based on open eyes. Present results would permit to test how general or specific to the closed-eyes condition are the already described increase in low-frequency rhythms power and decrease in high-frequency rhythms power in ADD with respect to controls.

36.2 Methods

36.2.1 Sample

The ADD group was extracted from a group of children and adolescents, with ages between 6 and 17 years old, diagnosed with ADD. The ADD group was composed of 36 subjects ($M = 10.8$ years, $SD = 3.4$, 8 females). To guarantee that EEG recordings in the ADD group were not influenced by medication, subjects ceased taking medication 48 hours prior to testing.

Control subjects were selected to match by age and gender the selected ADD subjects. The experiment was conducted with the informed and written parental consent following the Declaration of Helsinki guidelines. The ethics committee of biomedical research of the autonomous community of Andalucía approved the experimental protocol.

36.2.2 *Experimental Session*

The EEG was recorded during 3 min of spontaneous activity in an open-eyes condition. Subjects were asked to look at the screen, to blink as little as possible, and to be focused on a cross presented at the center of the screen.

Recordings were obtained from 32 scalp sites from the 10–20 international system, and an average reference was used.

Eye movements were recorded by two electrodes at the outer canthus of each eye for horizontal movements, and electrodes placed above and below the left eye for vertical movements. Impedance was maintained below 10 kilo-ohms.

Data were recorded in direct current mode at a sampling rate of 512 Hz, with a 20.000 amplification gain using an analog digital acquisition and analysis board (ANT).

36.2.3 *Data Analysis*

EEG recordings were analyzed with the EEGLAB and Matlab 2010a software packages. To eliminate alternant current power line interference, eye movements, blink and muscle artifacts on the EEG, an independent components analysis was performed. These components were discarded, and the EEG signal was reconstructed. The segmented epochs lasted 2000 ms. All epochs for which the EEG exceeded ±120 microvolts in any channel were discarded. The PSD of individual epochs was computed by means of the fast Fourier transform (FFT). The PSD was computed in windows of 2 s. The EEGLAB function spectopo was applied to the EEG data, and the PSD results were expressed in logarithms for the statistical analysis. For spectrum display purposes, the PSD values were presented eliminating the logarithm in order to show up the differences in spectral power between ADD and control subjects. Subjects with less than ten trials after the rejection protocol were not considered for further analysis. For the ADD group, the mean number of accepted trials was 57.5 ($SD = 26.7$) and 70.3 ($SD = 21.7$) for controls, with a range between 10–90 for ADD and 12–90 for controls. There was an statistical significant difference in the number of accepted epochs due to a higher number of accepted epochs in the control subjects group ($F_{[1,70]} = 5.018$, $p = 0.28$).

The PSD obtained from individual trials was averaged in each individual subject. The frequency resolution of PSD was 1 Hz. The EEG frequencies in the range from 1 to 46 Hz were selected for analysis.

36.2.4 Statistical Analysis

Using the Statistical Package for the Social Sciences (SPSS) 20.0, a mixed-model analysis of variance (ANOVA) was applied to the logarithm of the PSD of absolute power to compare the PSD of the ADD and control groups. Six independent ANOVAs were computed for the six different brain rhythms considered: delta (1–2 Hz), theta (4–7 Hz), alpha (8–11 Hz), low beta (13–20 Hz), high beta (21–30 Hz), and gamma (31–46 Hz). The between-subject factor was the subjects group (control or ADD), and the within-subject factor was the electrodes (30 electrodes). The main purpose of the ANOVA in the present report was to observe if there were differences in spectral power between ADD and control subjects.

36.2.4.1 Inter-frequency Correlations

The PSD of each frequency in each electrode was correlated against the other frequencies in the same electrode in order to observe whether there were patterns of co-variation between PSD of different frequencies and if these patterns were different in ADD and control subjects. Correlations were computed by means of the Spearman's rank correlation coefficient. The two-tailed statistical significance of the correlations between different frequencies was estimated, taking into account the number of comparisons. Bonferroni correction for multiple correlations was applied to the inter-frequency correlation matrices. Statistical significance values for correlation matrices were obtained considering the number of correlations in each single electrode. As 46×46 correlations were made per single electrode, but the matrix was symmetric, the number of correlation was $[(46 \times 46)-46)/2 = 1035]$. The Bonferroni corrected for multiple comparisons p-value for type I error was $p = (0.05/1035) = 0.000048$. This p-value was applied as a threshold for statistical significance.

36.3 Results

Figure 36.1 shows the PSD for ADD and control groups in the midline electrodes for the 1–46 Hz frequency range. An increase of PSD can be observed in the ADD group with respect to controls in delta and alpha bands. However, the ANOVA showed that only delta band reached a significant effect for the between subjects factor, due to a higher PSD in the ADD group with respect to the control group ($F_{[1,70]} = 4.66$, $p = 0.034$). No significant interactions appeared between the subjects group and the electrodes factor.

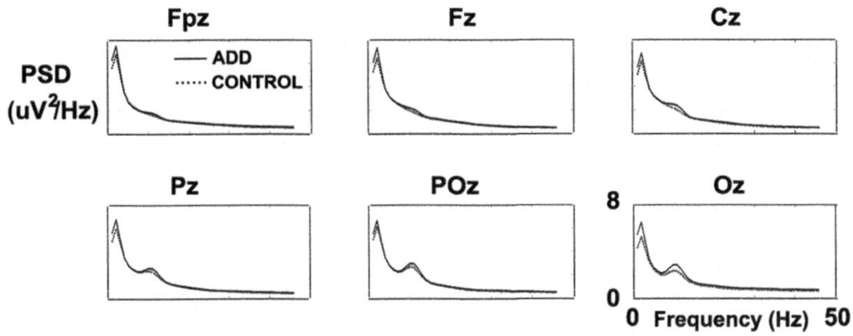

Fig. 36.1 ADD and control groups power spectral density (PSD) in the 1–46 Hz frequency range for six electrodes. Notice the increase of power in ADD with respect to controls in the delta and alpha band. However, only delta reached statistical significance (see results)

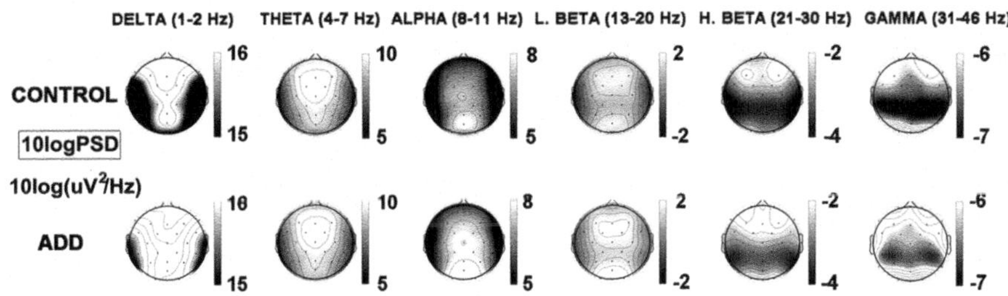

Fig. 36.2 Topographical maps of PSD log in ADD and control groups. Observe the similarity between topographies of both groups

Figure 36.2 shows the topographical maps of the log PSD in both groups. Topographies were very similar between ADD and control groups. In order to check the degree of similarity of the ADD and control group topographies, the Spearman correlation between the topographical maps of both groups in each frequency range was computed. These correlations were very high, indicating a high similarity between the topographies of both groups (delta correlation $= 0.96$, $p < 10^{-7}$; theta correlation $= 0.98$, $p < 10^{-7}$; alpha correlation $= 0.96$, $p < 10^{-7}$; low-beta correlation $= 0.95$, $p < 10^{-7}$; high-beta correlation $= 0.85$, $p < 6.37 \times 10^{-7}$; gamma correlation $= 0.89$, $p < 3.07 \times 10^{-7}$).

The inter-frequency correlations (Fig. 36.3) showed that both groups presented a co-maturation between low-frequency rhythms (delta and theta) and beta rhythm (mainly low beta). These inter-frequency correlations are interpreted as a power-to-power maturational frequency coupling of the theta and beta rhythms. However, the ADD group presented a lower range of frequencies in which a significant correlation between slow frequencies and beta range frequencies was obtained, suggesting a lower intensity in the frequency coupling of these rhythms for the ADD with respect to control group.

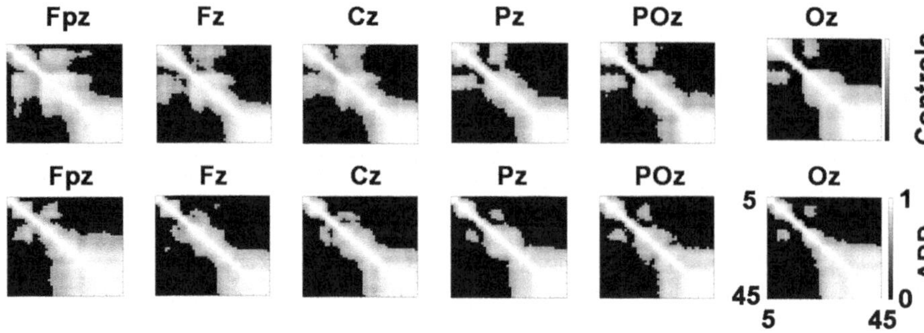

Fig. 36.3 Inter-frequency Spearman correlations for the six electrodes displayed in Fig. 36.1. A significant correlation between low-frequency rhythms and beta rhythms was obtained

36.4 Discussion

The increase of low-frequency rhythms in ADD with respect to controls suggests a developmental delay in ADD children, given that power of brain rhythm is decaying with age in normal subjects [9]. The higher PSD in ADD can be interpreted as a delay in the maturational processes underlying power reduction with age. Synaptic pruning has been suggested as the more plausible mechanism for power reduction with age. In this sense, Whitford et al. [10] have found a parallel reduction in cortical thickness and slow wave maturation in normal children, suggesting a direct link between neuroanatomical and neurophysiological age-related processes. As ADD children have shown a delay in the cortical thickness reduction that occurs normally during development [11], the delay in decrease of delta power in ADD children would be possibly due to a decrease in the pruning pace. The results obtained of delta power reduction in ADD children support the maturational lag model [2, 3].

However, the diminished maturational frequency coupling between low-frequency rhythms and beta rhythm suggest a differential pattern of development in ADD children with respect to controls. The power-to-power coupling decrease during maturation in ADD children could be partially related to the behavioral and/or cognitive impairments in ADD subjects. It must be mentioned that the maturational theta-beta correlation obtained in present report in control subjects has been obtained from an independent sample to that reported by Rodriguez-Martínez et al. [8], indicating the replicability and robustness of this developmental biomarker.

The classic increase of theta power and decrease in beta in ADHD children [4] has not been obtained in present report. However this pattern has been consistently obtained in closed-eyes condition and not in open-eyes condition [12]. The fact that the open-eyes condition did not obtain the same results with closed-eyes condition suggests that the classical pattern of theta increase and beta decrease would be highly dependent on the recording conditions and not constitute a standard model of oscillatory activity in ADD children.

Therefore, both neurobiological theories for ADD can be supported from present data, the maturational lag is coherent with the higher amplitude of the delta rhythms in ADD children, but the reduced theta-beta power-to-power frequency coupling is suggestive of a developmental deviation model for ADD.

Acknowledgments This work was supported by grants of the Spanish Ministry of Economy and Competitiveness [grant numbers PSI2013-47506-R and PSI2016-80059-R] and by a research grant from Janssen-Cilag.

References

1. Kinsbourne, M.: Minimal brain dysfunction as a neurodevelopmental lag. Ann. N. Y. Acad. Sci. **205**, 268–273 (1973)
2. Matsuura, M., Okubo, Y., Toru, M., et al.: A cross-national EEG study of children with emotional and behavioral problems: a WHO collaborative study in the Western Pacific Region. Biol. Psychiatry. **34**, 59–65 (1993)
3. Hermens, D.F., Soei, E.X., Clarke, S.D., Kohn, M.R., Gordon, E., Williams, L.M., Resting, E.E.G.: Theta activity predicts cognitive performance in attention-deficit hyperactivity disorder. Pediatr. Neurol. **32**, 248–256 (2005)
4. Barry, R.J., Clarke, A.R., Johnstone, S.J., McCarthy, R., Selikowitz, M.: Electroencephalogram θ/β ratio and arousal in attention-deficit/hyperactivity disorder. evidence of independent processes. Biol. Psychiatry. **66**, 398–401 (2009)
5. Matousek, M., Petersen, I.: Frequency analysis of the EEG in normal children and normal adolescents. In: Kellaway, P., Petersen, I. (eds.) Automation of Clinical Electroencephalography, pp. 75–102. Raven Press, New York (1973)
6. Clarke, A.R., Barry, R.J., McCarthy, R., Selikowitz, M.: EEG differences in two subtypes of attention-deficit/hyperactivity disorder. Psychophysiology. **38**, 212–221 (2001)
7. Clarke, A.R., Barry, R.J., Dupuy, F.E., Heckel, L.D., McCarthy, R., Selikowitz, M., Johnstone, S.J.: Behavioural differences between EEG-defined subgroups of children with attention-deficit/hyperactivity disorder. Clin. Neurophysiol. **122**, 1333–1341 (2011)
8. Rodríguez-Martínez, E.I., Barriga-Paulino, C.I., Rojas-Benjumea, M.A., Gómez, C.M.: Co-maturation of theta and low-beta rhythms during child development. Brain Topogr. **28**, 250–260 (2015)
9. Rodríguez-Martínez, E.I., Barriga-Paulino, C.I., Zapata, M.I., Chinchilla, C., López-Jiménez, A.M., Gómez, C.M.: Narrow band quantitative and multivariate electroencephalogram analysis of peri-adolescent period. BMC Neurosci. **13**, 104 (2012)
10. Whitford, T.J., Rennie, C.J., Grieve, S.M., Clark, C.R., Gordon, E., Williams, L.M.: Brain maturation in adolescence: concurrent changes in neuroanatomy and neurophysiology. Hum. Brain Mapp. **28**, 228–237 (2007)
11. Narr, K.L., Woods, R.P., Lin, J., Kim, J., Philips, O.R., Del'Homme, M., Caplan, R., Toga, A.W., McCracken, J.T., Levitt, J.G.: Widespread cortical thinning is a robust anatomical marker for attention-deficit/hyperactivity disorder. J. Am. Acad. Child Adolesc. Psychiatry. **48**,\break 1014–1022 (2009)
12. Arns, M., Conners, C., Kraemer, H.C.: A decade of EEG theta/beta ratio research in ADHD: a meta-analysis. J. Atten. Disord. **17**, 374–383 (2013)

Chapter 37
Event-Related Potentials During a Delayed Match-to-Sample Test to Evaluate Working Memory Development in Control and Attention Deficit Children and Adolescents

Antonio Arjona-Valladares, Elena I. Rodríguez-Martínez, Francisco J. Ruíz-Martínez, Jaime Gómez-González, and Carlos M. Gómez

Abstract Working memory (WM) impairment is a core feature of attention deficit disorder (ADD). Differences in event-related potentials (ERPs) amplitude with respect to controls would permit to characterize neurophysiologically WM processing in ADD. Thirty-two ADD-diagnosed subjects and thirty-eight controls (6–17 years old) were recorded during a visual delayed match-to-sample test (DMTS). ERP amplitudes of ADD and control subjects were compared during the encoding, retention, and matching phases. Cluster mass permutation test, controlling for multiple comparisons in time and space (electrodes), showed statistically significant lower amplitude for the negative slow wave (NSW) in ADD children compared to the control group during the retention period. Electrodes CP6, P4, and P8 showed these significant differences. The lower amplitude of the NSW in ADD children suggests a dysfunctional activation of the WM, possibly related to the difficulty to focus attention during the coding and retention phases.

Keywords Attention deficit disorder · Working memory · Delayed match-to-sample test · Event-related potentials · Negative slow wave · Retention phase

A. Arjona-Valladares (✉) · E. I. Rodríguez-Martínez · F. J. Ruíz-Martínez · C. M. Gómez
Human Psychobiology Lab, Experimental Psychology Department, University of Seville, Sevilla, Spain
e-mail: aarjona@us.es

J. Gómez-González
UGC of Mental Health, Virgen Macarena Hospital, Seville, Spain

J. M. Delgado-García et al. (eds.), *Advances in Cognitive Neurodynamics (VI)*,
Advances in Cognitive Neurodynamics,
https://doi.org/10.1007/978-981-10-8854-4_37

37.1 Introduction

Attention deficit disorder (ADD) is a developmental disorder affecting people from children to adults and characterized by inattention, hyperactivity, and/or impulsivity. Three different phenotypes have been proposed as characterizing ADD [1]: (i) shortened delay gradient related to the inability to expect delayed rewards, (ii) difficulties in temporal processing, and (iii) working memory (WM) deficits which would project on executive functions as impulse control and attention focusing. There is a consensus, suggested by meta-analysis studies [2, 3], about WM impairment as one of the central symptoms in ADD.

The neurophysiology of ADD has been studied during sensory stimulation inducing ERPs, and the results suggest that ADD subjects present problems in numerous capacities [4]. With respect to WM paradigms, Gomarus et al. [5] reported that ADD subjects show no influence of task load in the amplitude of the search related negativity (an important component for selecting the relevant target in the matching period), while control subjects show an increase of amplitude with task load. Besides, Kim et al. [6] have found a reduced P3 in young ADD adults during the encoding phase in a DMTS task, suggesting a deficit for attentional allocation of processing resources, which would be related to the WM-associated problems.

A neurophysiological WM model has been proposed by Fuster [7] based on the recording of single neurons during delayed match-to-sample tests (DMTS). The model proposes a reciprocal interaction between prefrontal cortex and posterior sensory neurons that would permit the maintenance of the remembered item in WM. Using the same DMTS paradigm, we have shown the human neurophysiological correlates for coding, retention, and matching phases in children, adolescents, and adults [8]. During the coding and matching phases, a sequence of visual event-related potentials (ERPs) appeared, and a slow wave was present during the retention phase.

Differences in ERPs amplitude of ADD subjects with respect to controls, using the DMTS task, would permit to characterize these possible WM processing impairments. A decrease in ERPs amplitude, particularly in the slow wave during the retention phase, is expected in ADD subjects with respect to controls to justify the WM behavioral impairment.

37.2 Methods

37.2.1 Sample and Experimental Session

The ADD group (32 subjects, 25 males, and 7 females) was extracted from a group of children and adolescents between 6 and 17 years of age (*mean* = 11.21 years old, *SD* = 3.55) diagnosed by an expert psychiatric clinical service, following a comprehensive structured interview, and the DePaul questionnaire [9] was administered to the parents. Subjects ceased taking medication 48 hours prior to testing. Control subjects were recruited from public schools and selected to match by age and gender with the ADD group. The experiments were conducted with the informed and written parental consent, following the Declaration of Helsinki guidelines. The ethics committee of biomedical research of the autonomous community of Andalucía approved the experimental protocol.

The DMTS task was presented in a computer screen. The stimulus presentation and response recording were computer-controlled (E-Prime 2.0). 128 trials, organized in 4 experimental blocks (32 trials each), were presented in the DMTS task. Every trial was composed of one stimulus (S1) to be maintained in memory (1000 ms of duration), which appeared in the center of the screen, followed by a 1500 ms delay period. Then, two stimuli (S2) appeared, one on the left and the other on the right side of the screen (2000 ms of duration). One of the S2 stimuli corresponded to the same image than S1, and subjects were asked to press to the corresponding button (left or right) where the S2 matched the S1. The target S2 appeared randomly on the left or right side of the screen, and the response time was 2000 ms. Auditory feedback was provided, indicating to the subject if a correct or incorrect response was given.

37.2.2 EEG Recording and Statistical Analysis

The EEG was recorded during the DMTS task. Recordings were obtained from 32 scalp sites using the 10–20 international system and an average reference. Data were recorded in direct current mode at 512 Hz with a 20.000 amplification gain. After registration, the data were filtered with a low-pass filter of 100 Hz and a high-pass filter of 0.05 Hz.

The RTs and the percentage incorrect responses, anticipation errors (responses to S1, responses given in the delay period, or responses with durations less than 180 ms after S2 appearance), omissions (no responses to S2), and total errors of ADD and controls subjects were independently compared by means of one-way ANOVA, with group subjects as the within-subjects factor.

The ERPs of ADD and control subjects during the encoding, retention, and matching phases were compared by means of the cluster mass permutation test, which controls for multiple comparisons in time and space (electrodes) [10]. Two different periods were selected for the comparisons between ADD and controls: a first period of 2500 ms from the S1 onset (baseline of 100 ms before the S1) and a second period of 1000 ms from the S2 onset (baseline of 100 ms before the S2). The first period was selected to analyze the encoding and retention phases, and the second period was selected to include the matching phase, in which the influence of the slow wave would vanish.

37.3 Results

37.3.1 Analysis of Reaction Times and Errors

For the RTs, despite the mean of the ADD group was higher than the control subjects, ANOVA showed that the difference was not significant. Instead, the percentage of errors presented significant differences for (i) incorrect responses ($F_{[1,69]} = 11.16$, $p < 0\ 0.001$), due to a higher percentage of incorrect responses in the ADD group; (ii) anticipation errors ($F_{[1,69]} = 6.2$, $p < 0.015$), due to a higher percentage of hasty responses in the ADD group; (iii) omissions ($F_{[1,69]} = 8.38$, $p < 0.005$), due to a higher percentage of trials with no-response in the ADD group; and (iv) total errors ($F_{[1,69]} = 12.77$, $p < 0.001$). Overall, behavioral results showed that the performance in the DMTS task was more accurate in the control group compared to the ADD subjects.

37.3.2 Analysis of Event-Related Potentials

A cluster mass permutation test was performed in all the electrodes of the scalp (30 channels) comparing controls vs. ADD subjects. Results showed significant differences located on the right side of the posterior area (electrodes CP6, P4, and P8) during the whole retention phase and the last part of the encoding phase (Figs. 37.1 and 37.2). This differences were due to a reduced amplitude of the negative slow wave (NSW) in the posterior area (specially on the right side) of the ADD subjects compared to the control group, which would indicate a possible impairment of the ADD subjects to retain information in the WM.

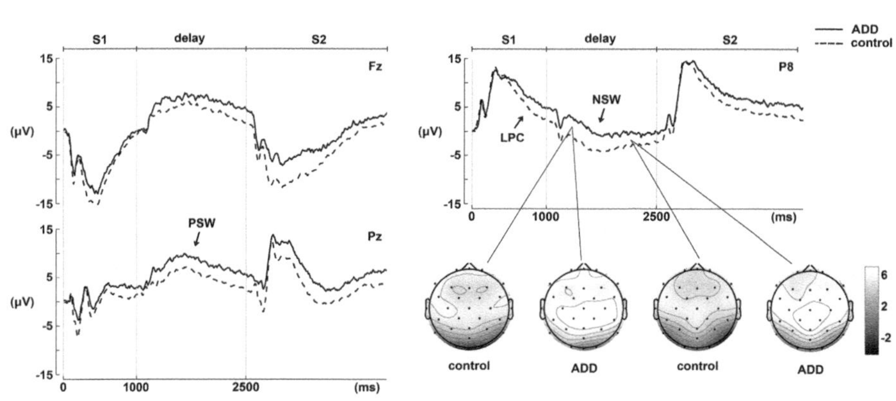

Fig. 37.1 ERPs during the encoding, retention, and matching phases. Fz, Pz, and P8 electrodes are displayed. Notice that, in the waves and topographies of P8 electrode, the amplitude of the negative slow wave (NSW) in posterior sites is lower in ADD than in controls. PSW, positive slow wave. LPC, late-positive component

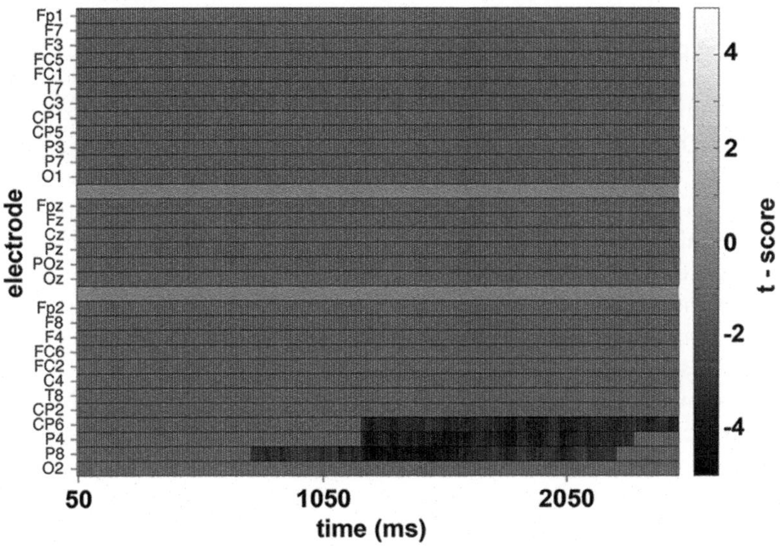

Fig. 37.2 Cluster mass permutation statistical analysis of the comparison between control and ADD subjects during the S1 and delay periods. The black-labeled areas refer to time and electrodes in which there were statistically significant differences between both groups

37.4 Discussion

The results showed an impaired behavioral performance in the DMTS task in ADD subjects with respect to controls and a statistically significant reduction of the NSW amplitude in the ADD group (in right posterior sites) with respect to controls. Thereby, present study support that WM behavioral impairment is a possible central cognitive dysfunction in ADD subjects [11].

The behavioral results obtained in present report show a lower performance in ADD children, with an increased number of omissions, incorrect responses, and anticipations during the visual DMTS task. The impairment of WM during DMTS tasks has been previously described in ADD subjects [12]. In fact, a deficit in visual memory has been suggested as a behavioral endophenotype of ADD [13]. The WM deficit would be able to explain the behavioral inhibition because information should be first activated in WM before a decision to inhibit responses can be taken [14]. The ERPs obtained in control subjects in present report were quite similar to those reported in a different children and adolescents sample [8, 15], including a sequence of anterior and posterior visually induced ERPs during the encoding and matching phases, the visual offset potential after the S1, and a slow wave during the retention period extending from anterior to posterior sites. This slow wave presented a positive polarity in anterior sites and a negative polarity in posterior sites.

When the cluster mass permutation test was applied to the comparison of voltage amplitude between controls and ADD subjects, only the NSW was statistically significant. The significant cluster appeared in right posterior sites during the end of the encoding phase and the whole retention phase. The slow wave, that characterizes the retention phase, starts a few hundreds of milliseconds after the presentation of the S1 component [15]. Therefore, the slow-wave component would explain the reduced amplitude of the late-positive component (LPC) in control subjects with respect to ADD, due to the scalp superposition of both components. On the other hand, present results have not demonstrated the decrease in P3 (named as LPC in present report) during the encoding phase of a DMTS task, obtained by Kim et al. [6] and explained as diminished allocation of attentional resources in ADD. The former cited results were obtained in ADD adults, while present results are obtained in children and adolescents, so the differences in ERPs might be associated to the already known dissimilarities in behavior between ADD adults and children [16].

As conclusion, present results support that WM is impaired in ADD children and adolescents. The WM impairment observed behaviorally would be related to the amplitude reduction in the posterior slow wave (a macroscopic reflection of the sustained activity of neurons in posterior cortices during the retention period of DMTS task [7, 15]) in ADD subjects.

Acknowledgments This work was supported by grants of the Spanish Ministry of Economy and Competitiveness [grant numbers PSI2013-47506-R and PSI2016-80059-R] and by a research grant from Janssen-Cilag.

References

1. Castellanos, F.X., Tannock, R.: Neuroscience of attention deficit/hyperactivity disorder: the search for endophenotypes. Nat. Rev. Neurosci. **3**, 617–628 (2002)
2. Martinussen, R., Hayden, J., Hogg-Johnson, S., Tannock, R.: A meta-analysis of working memory impairments in children with attention-deficit/hyperactivity disorder. J. Am. Acad. Child Adolesc. Psychiatry. **44**, 377–384 (2005)
3. Frazier, T.W., Demaree, H.A., Youngstrom, E.A.: Meta-analysis of intellectual and neuropsychological test performance in attention-deficit/hyperactivity disorder. Neuropsychology. **18**, 543–555 (2004)
4. Johnstone, S.J., Barry, R.J., Clarke, A.R.: Ten years on: a follow-up review of ERP research in attention deficit/hyperactivity disorder. Clin. Neurophysiol. **124**, 644–657 (2013)
5. Gomarus, H.K., Wijers, A.A., Minderaa, R.B., Althaus, M.: ERP correlates of selective attention and working memory capacities in children with ADHD and/or PDD-NOS. Clin. Neurophysiol. **120**, 60–72 (2009)
6. Kim, S., Liu, Z., Glizer, D., Tannock, R., et al.: Adult ADHD and working memory: neural evidence of impaired encoding. Clin. Neurophysiol. **125**, 1596–1603 (2014)
7. Fuster, J.M.: The Prefrontal Cortex. Lippincott-Raven, Philadelphia (1997)
8. Barriga-Paulino, C.I., Rodríguez-Martínez, E.I., Arjona, A., Gómez, C.M.: Developmental trajectories of event related potentials related to working memory. Neuropsychologia. **95**, 215–226 (2017)
9. DuPaul, G.J., Power, T.J., Anastopoulos, A.D., Reid, R.A.D.H.D.: Rating Scale-IV: Checklists, Norms, and Clinical Interpretation. Guilford, New York (1998)
10. Groppe, D.M., Urbach, T.P., Kutasa, M.: Mass univariate analysis of event-related brain potentials/fields I: a critical tutorial review. Psychophysiology. **1**, 1–15 (2011)

11. Rapport, M.D., Chung, K.M., Shore, G., Denney, C.B., Isaacs, P.: Upgrading the science and technology of assessments and diagnosis: laboratory and clinic–based assessment of children with ADHD. J. Clin. Child Psychol. **29**, 555–568 (2000)
12. Kempton, S., Vance, A., Maruff, P., Luk, E., Costin, J., Pantelis, C.: Executive function and attention deficit hyperactivity disorder: stimulant medication and better executive function performance in children. Psychol. Med. **29**, 527–538 (1999)
13. Shang, C.Y., Gau, S.S.: Visual memory as a potential cognitive endophenotype of attention deficit hyperactivity disorder. Psychol. Med. **41**, 2603–2614 (2011)
14. Rapport, M.D., Chung, K.M., Shore, G., Isaacs, P.: A conceptual model of child psychopathology: implications for understanding attention–deficit/hyperactivity disorder (ADHD) and treatment efficacy. J. Clin. Child Psychol. Spec. Edn. **30**, 48–58 (2001)
15. Barriga-Paulino, C.I., Rodriguez-Martinez, E.I., Rojas-Benjumea, M.A., Gómez, C.M.: Slow wave maturation on a visual working memory task. Brain Cognit. **88**, 43–54 (2014)
16. DuPaul, G.J., Weyandt, L., O'Dell, S., Varejao, M.: College students with ADHD: current status and future directions. J. Atten. Disord. **13**, 234–250 (2009)

Chapter 38
Postnatal Development of Sleep-Wake Cycle in Wild-Type Mice

Ángeles Prados-Pardo, Sandra Yaneth Prieto-Soler, and Eduardo Domínguez-del-Toro

Abstract Many neurological disorders affecting considerably the population, such as attention deficit hyperactivity disorder (ADHD), epilepsy, or Alzheimer's disease concern a nucleus of the brainstem: the locus coeruleus. This fact motivates the study of the early development of the noradrenergic system, its normal functions, and how different lesions are involved in diverse diseases. In fact, neuronal connections are still forming in the brain of the neonate, while brainstem neurons project rostrally to the developing cortex. Lesions affecting the maturation of the noradrenergic system at first may go unnoticed. In this chapter we present how the respiratory, circulatory, and sleep parameters, all controlled by the noradrenergic system, evolve during early postnatal development in two mice strains (CD-1 and C57). Oxygen consumption and heart rate increase from P7 to P14 in both strains. C57 mice reduce strongly time spent in REM sleep and increase time spent in NREM sleep from P7 to P14.

Keywords ECG · EMG · REM sleep · Sleep-wake cycle · Postnatal development · Locus coeruleus · Mice

38.1 Introduction

One of the challenges for the actual neuroscience is to understand how the brain develops the behavior and which are the neuronal mechanisms underlying the acquisition of new motor tasks by different structures of the central nervous system. During last years, it has been reinforced the theory that the activation of pontine noradrenergic system, from neurons sited at the locus coeruleus, facilitates the reorganization of cortical neuronal networks, affecting the elaboration of adaptive

Á. Prados-Pardo · S. Y. Prieto-Soler · E. Domínguez-del-Toro (✉)
Divisi6n de Neurociencias, Universidad Pablo de Olavide, Sevilla, Spain
e-mail: edomtor@upo.es

J. M. Delgado-García et al. (eds.), *Advances in Cognitive Neurodynamics (VI)*, Advances in Cognitive Neurodynamics,
https://doi.org/10.1007/978-981-10-8854-4_38

and cognitive responses. When trying to explain that function, it becomes essential the knowledge of how these neuronal systems are organized during postnatal development, both morphologically and functionally.

With this it could be possible to explain certain congenital anomalies that are associated to cognitive deficits. In this sense, it is necessary to study of molecular and physiological mechanisms involved in the organization of the proper development of the adult pontine noradrenergic system, necessary for the acquisition of new motor tasks, and with special emphasis on its activity during the sleep-wake cycle. Mice appear as good experimental models, as many mutants may suffer, in different developmental moments, selective impairments in those territories leading to the locus coeruleus, causing both early embryonic neuronal death and neuronal degeneration in this nucleus during postnatal development.

New researching lines give an important role to the noradrenergic system in learning processes during conditioning or during execution in different mazes, by using, for example, adrenergic agonist or antagonist effects on the amygdala fear conditioning. This reinforces the idea of the importance of adrenergic receptors in the acquisition of motor responses. A reduced activity of noradrenergic neurons due to chronic postnatal treatments has been shown to affect behavior in adult rats, also to delay motor development, and to block short-term memory in young mice [1, 2].

Learning and memory consolidation depend on the sleep-wake cycle. And this cycle is also related to the activity of noradrenergic neurons, which are active during wakefulness, partially active during NREM sleep and silent during REM sleep. Ear2 mutant mice with a great reduction in neurons at the locus coeruleus at birth present sleep problems when they are adults. The reduced size of this nucleus may affect adult cycle directly or be a consequence of developmental deficiencies due to a lack of proper postnatal organization of brain structures [3].

Postnatal sleep is supposed to be involved in functional organization and maturation of connections between brainstem and cortical areas. Because of that some studies focused on the research on postnatal development of behavioral stages of rats, showing a high percentage of REM sleep (near to 90%) at birth, which decreases drastically around the second postnatal week, near postnatal day 14 (P14), together with an expansion of slow sleep (NREM) and wakefulness. During those two postnatal weeks, duration of time spent in each stage increases. The high proportion of REM sleep and the reduced duration of each stage are indicators of the immaturity of the central nervous system and also good indicators of brain development and cortical maturation [4, 5].

Chronic administration of antihypertensive drugs such as clonidine (an alpha adrenergic agonist) during postnatal development in rats changes the sleep-wake patterns and has been proposed as a REM deprivation model [1]. Consequently, in adults, those rats exposed to clonidine during first developmental stages show a high percentage of REM sleep along with an increase in motor activity during this state. Furthermore, treatment with clonidine did not change NA levels in the brain. Studies on rats after treating them postnatally with clonidine demonstrate neuronal death in the rostral region of the brainstem, very interesting, as sleep is involved in prevention of apoptosis.

As mentioned, some of these treatments generate deficits in short-term declarative memory, and this is an essential point, as both sleep phases are involved in acquisition of different memories. REM sleep is necessary for implicit memory, whereas NREM sleep is necessary for declarative memory. It is important to remember that the function of each stage during sleep is to facilitate consolidation of different memories [6].

Apart from rats, acquiring knowledge of sleep-wake cycle parameters during postnatal wild-type mice development is essential in order to study the effect of these alterations on the developmental pattern of the sleep-wake cycle and their consequences on memory and learning tests in mutants. Ear2 mutant mice, lacking 70% neurons at LC, exhibit a higher fragmentation of different sleep phases together with explicit memory problems, meaning a great immaturity in cortical system.

38.2 Recordings in Mice During Postnatal Sleep-Wake Cycle

With perspective, the objective of these recordings consists on the characterization of the dynamics of sleep-wake cycle during postnatal development in mice and their relationship with learning, looking for the precise moment in which acquisition of different memories is possible. The approximation is electrophysiological as many phenotypes may go unnoticed in mutants. A pharmacological approximation, by administering agonist and antagonist drugs, can be used.

We have set up the technique (see Fig. 38.1) for recording electrocardiographic (ECG) and electromyographic (EMG) activity (also integrated EMG), in order to identify the different states in sleep-wake cycle.

In Fig. 38.2a we can observe an electrocardiogram recording from a P7 CD-1 mouse from which we can differentiate the pulses and analyze heart rate. Before

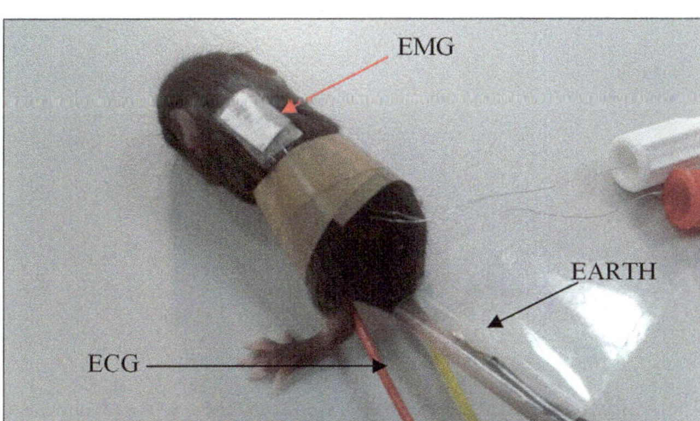

Fig. 38.1 Electrode location in a C57 P10 mouse. Nuchal electrodes for the EMG, red and yellow for the ECG, and black ones as earth

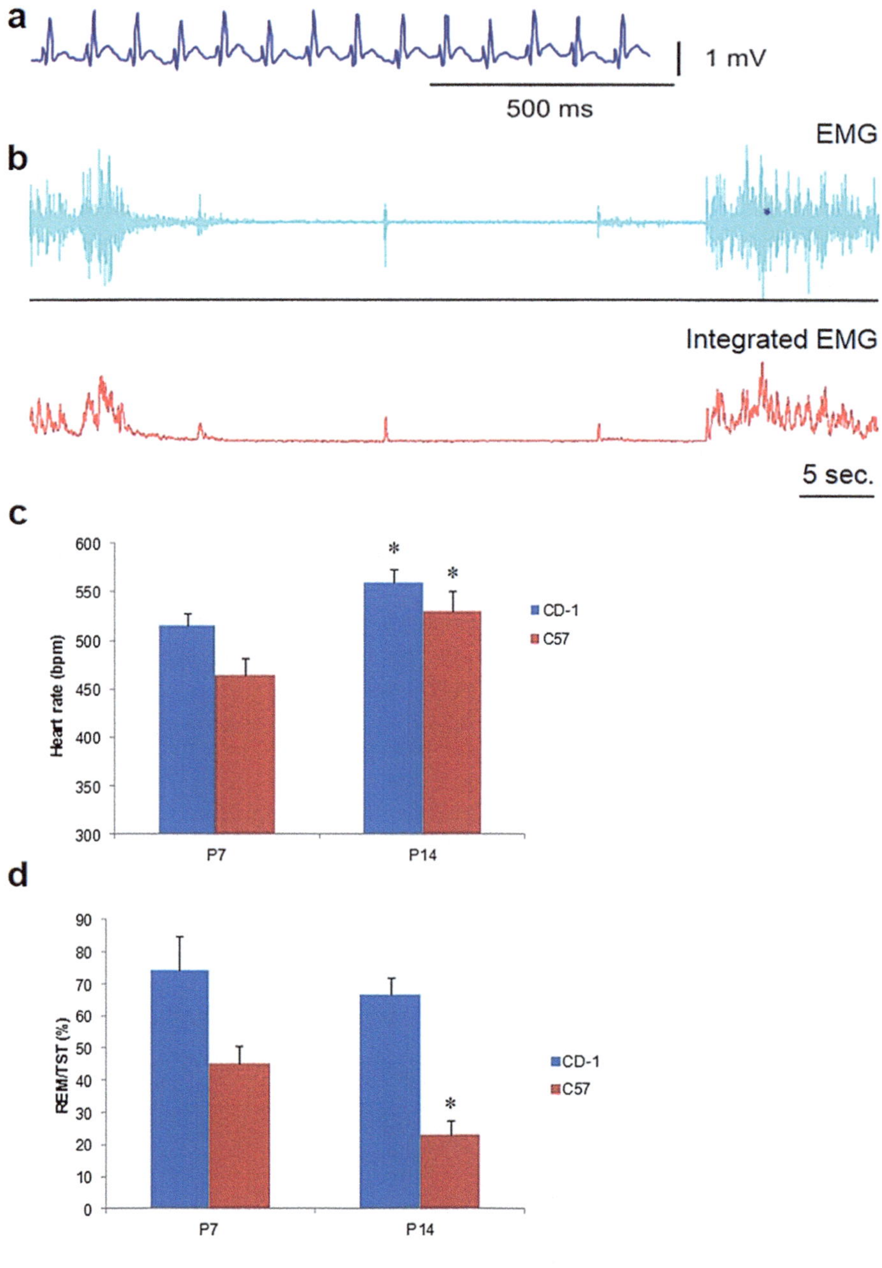

Fig. 38.2 (**a**) Electrocardiogram recording from a P7 CD-1 neonate. (**b**) Nuchal electromyogram (EMG) recording and integrated signal (integrated EMG) from a P7 C57 mouse. (**c**) Heart rate (b.p.m.) evolution from P7 to P14 in both strains. (**d**) Maturation of REM (as percentage of total sleep time- TST)

P14, EMG activity corresponds well with REM (atony), NREM (tone), and wakefulness (high activity) stages (see Fig. 38.2b). In both strains, oxygen consumption (not shown) and heart rate (see Fig. 38.2c) tend to increase with age, from P7 to P14 (CD-1 mice, from 515 ± 12 to 559 ± 14 b.p.m., $F_{[1, 13]} = 4.7899$, $p < 0.05$; C57 mice, from 464 ± 17 to 530 ± 20 b.p.m., $F_{[1, 11]} = 5.3872$, $p < 0.05$). EMG recordings at different postnatal ages show a significant increase in quiet sleep (NREM) and a significant decrease in REM sleep between P7 and P14 in both CD-1 and C57 mice (see Fig. 38.2d), whereas wakefulness shows an increasing tendency.

This corresponds well with sleep developmental parameters described in rats. During postnatal development, P7 mice exhibit a higher fragmentation of different sleep phases and a higher percentage of REM sleep, meaning a great immaturity in cortical system. At P14, duration of each stage increases as percentage of REM sleep decreases, coincident with cortical maturation.

This reduction of REM sleep is higher in C57 mice (from 45 ± 5 to 23 ± 4%REM/TST; $F_{[1, 6]} = 10.1487$, $p < 0.05$) compared to CD-1 mice (from 74 ± 10 to 66 ± 5%REM/TST; $F_{[1, 5]} = 0.3908$, $p = 0.5657$). This means that postnatal maturation of sleep-wake cycle is different in both mice strains, and this is something to have into account.

From now on, we assume that sleep parameters will be altered in mutant mice models as their noradrenergic system is damaged. Given that the noradrenergic system activation is related to wake state, we expect an increase in sleep (especially in REM sleep which is related to noradrenergic system inactivation).

We also expect a worse performance in attention and memory tasks because of the noradrenergic system damage and the sleep disturbances. With this we want to understand the relationship between sleep phases maturation and acquisition of different memories and how these are affected in different animal models.

Degeneration of locus coeruleus neurons occurs in adult Ts65Dn and Alzheimer's mice, models showing severe cognitive impairments and sleep problems. Those malfunctions can appear in the adulthood, or perhaps they are present from the very beginning. With this recording technique, we can now investigate it.

References

1. Mirmiram, M., Brenner, E., Van der Gugten, J., Swaab, D.F.: Neurochemical and electrophysiological disturbances mediate developmental behavioral alterations produced by medicines. Neurobehav. Toxicol. Teratol. **6**, 677–683 (1985)
2. Calvino-Núñez, C., Domínguez-del-Toro, E.: Clonidine treatment delays postnatal motor development and blocks short-term memory in young mice. PLoS One. **9**, e114869 (2014)
3. Warnecke, M., Oster, H., Revelli, J.P., Álvarez Bolado, G., Eichele, G.: Abnormal development of the locus coeruleus in Ear2 (Nr2f6)-deficient mice impairs the functionality of the forebrain clock and affects nociception. Genes Dev. **19**, 614–625 (2005)

4. Blumberg, M.S., Seelke, A.M.: The form and function of infant sleep: from muscle to neocortex. In: Blumberg, M.S., Freeman, J.H., Robinson, S.R. (eds.) Oxford Handbook of Developmental Behavioral Neuroscience, pp. 406–438. Oxford University Press, New York (2010)
5. Cirelli, C., Tononi, G.: Cortical development, EEG rhythms, and the sleep/wake cycle. Biol. Psychiatry. **77**, 1071–1078 (2015)
6. Li, W., Ma, L., Yang, G., Gan, W.-B.: REM sleep selectively prunes and maintains new synapses in development and learning. Nat. Neurosci. **20**, 427–437 (2017)

Chapter 39
Complexity of Heart Rate As a Value of Behavioral Complexity

Anastasiia Bakhchina

Abstract Autonomic nervous system is the main way for the brain-body coordination, of which mode can be evaluated by dynamics of heart rate variability (HRV). HRV analysis is used for evaluation of different psychological states (stress, arousal, cognitive control, etc.), which can be considered as characteristics of behavior that formed at different stages of ontogeny. We investigated whether HRV differs between the early-formed behavior and the later-formed behavior. Heart rate was recorded in 33 healthy subjects (mathematical specialists). Participants performed two tests which included sentences with mathematical terms and sentences with common current used words. They had to add one missing word in each sentence. SampEn as a measure quantifying the complexity of time series was used to analyze HRV. SampEn was significantly higher in the mathematical test performance, when participants actualized the later-formed behavior.

Keywords Autonomic nervous system · Complexity of heart rate · Different stages of ontogeny · Behavioral complexity · Sample entropy

39.1 Introduction

Physiological supporting of behavior includes activation of different linked neurons groups and optimization of physiological processes. As a rule traditionally in the conserved approach, internal bodily states are ignored in the searching of the neural basis of behavior. Such mental functions as perceptions, thoughts, feelings, etc. are for the most part considered in isolation from the physiological state of the

A. Bakhchina (✉)
Department of Neural Basis of Mind Named of V.B. Shvirkov, Institute of Psychology of RAS, Moscow, Russia

J. M. Delgado-García et al. (eds.), *Advances in Cognitive Neurodynamics (VI)*, Advances in Cognitive Neurodynamics,
https://doi.org/10.1007/978-981-10-8854-4_39

body. A mechanistic understanding of distinct interoceptive pathways, which can influence brain functions, leads to the impossibility of considering the behavior at the whole organism level [1]. Therefore the current main task is forming of the system approach for psychophysiology describing of behavior.

39.1.1 Autonomic Nerve System in Supporting of Behavior

Studies of autonomic psychophysiology are beginning to have a big part in the current field of neuroscience. As an example, the fundamental association between bodily changes and emotions was founded by James and Lange in the nineteenth century. Since that time a lot of studies with electrical stimulation in animals have demonstrated the coupling of visceral responses to cortical regions, which include cingulate, insular [2], visual [3], and somatosensory [4] regions. It means that not only the nucleus of the solitary tract, ventrolateral medulla, parabrachial nucleus and hypothalamus but also many cortex regions take part in the brain-body cooperation (processing of visceral information).

Experimental researches into the mechanisms through which visceral afferent information is represented within the brain haven't shown clearly how visceral signals shape human behavior yet. The majority of visceral signals that shape behavior are unnoticed despite there is anatomical and experimental information about the representation and influence of the visceral state in brain processes.

In this way, describing principles through which human behavior and experience are colored by internal bodily signals, Benarroch [5] showed the central autonomic network (CAN), which included different structures of central and autonomic nervous systems. The main statement of the theory is that the CAN is an integral component of an internal regulation system through which the brain controls visceromotor, neuroendocrine, pain, and behavioral responses essential for survival (for goal-directed behavior supporting).

39.1.2 Heart Rate Variability As a Tool for Measurement of ANS Activity

Heart rate variability (HRV) is the variation over time of heart beat intervals (the periods between consecutive heartbeats), which depends on such physiological processes as autonomic neural regulation, thermoregulation, breathing, etc. [6]. HRV is thought to reflect the heart's ability to changing behavior and can be considered as an indicator of central-peripheral neural feedback and central nervous system – autonomic nervous system integration. Therefore HRV was used in the current study as noninvasive tool for assessing the activities of the autonomic nervous system.

39.1.3 Complexity of Behavior

It is shown that HRV associated with a diverse range of processes, including affective and attention regulation, cognitive functions (such as working memory, sustained attention, behavioral inhibition, general mental flexibility) [7]. These processes can be considered as characteristics of behavior formed at different stages of ontogeny.

From the system-evolutionary theory [8] perspective, a new behavior is subserved by co-activation of specialized neuron groups that had emerged in learning. The result of learning is a system that is a set of brain and body elements activity for providing efficient interaction with environment [9]. Formation of new systems during development results in growing complexity and differentiation of organism-environment relations [10].

Therefore we investigated whether HRV in the early-formed behavior ("old") differs from HRV in the later-formed ("new") behavior. Basing on the fact that usually "old" behavior is less complicated than "new" behavior [11], we hypothesized that heart rate complexity would be higher at "new" behavior performing.

39.2 Methods

39.2.1 Subjects

The experiment was approved by the Ethics Committee of the Institute of Psychology of Russian Academy of Science. Prior to the experiment, all subjects signed an informed consent form stating that participation was voluntary and that they could withdraw from the study at any moment.

Thirty-three healthy subjects (28 men, $M = 27.78$ years, range: 23–37 years) participated.

Participants did not suffer from any self-reported respiratory, cardiac diseases, epilepsy, psychiatric disorder, or any minor or major illness. All participants were professional mathematicians and had work experience ($M = 4.84$ years).

39.2.2 Task

The experimental task was to add one missing word in the sentence. The sentences ($N = 64$) were divided into two groups by age of acquisition of words. The first group of sentences ($N = 32$) included sentences with mathematical terms. Subjects had known these terms from University studying (from the age of 18–19 years). The example is "A normal is a vector that is perpendicular to a given object." These sentences included later-acquired words and made actual a "new" behavior.

The second group of sentences ($N = 32$) included sentences with commonly used words. Subjects had known these words from childhood (from the age of 5–6 years). The example is "Plasticine is a material for modeling figures." These sentences included early-acquired words and made actual an "old" behavior. The sentences in both groups were equal in the linguistic estimations, such as the quantity of words, syllables, letters, and Fog's index. The sentences were performed individually on a standard computer randomly without repetition in each group.

39.2.3 Physiologic Monitoring

The ECG was obtained using the wireless device HxM BT by Zephyr Technology. The plastic electrodes were filled with electrolyte and placed on the thorax across the heart. The signal was sampled at 400 Hz. The interbeat intervals (IBI) were extracted from ECG through the threshold algorithm.

The time domain index of HRV used in our analyses was the mean (HR, ms) and standard deviation (SDNN, ms) of IBI. These indexes closely reflect all nervous regulatory inputs to the heart.

For estimation of heart rate complexity, we used sample entropy (SampEn) as a set of measures of system complexity reporting on similarity in time series. SampEn (m, r, N, where N is the length of the time series, m is the length of sequences to be compared, r is the tolerance for accepting matches) is precisely the negative natural logarithm of the conditional probability that two sequences similar for m points remain similar at the next point, where self-matches are not included in calculating the probability [12]. Thus a low value of SampEn reflects a high degree of regularity. SampEn is independent on the record length and displays relative consistency under circumstances. The parameters m and r were fixed: $m = 2$, $r = 0.5*\text{SDNN}$.

39.2.4 Statistical Analyses

Normality of variables was tested in Shapiro-Wilk's test (W-test). HRV data of two conditions (performing tasks with sentences with mathematical or commonly used words) were tested in Wilcoxon signed-rank test. Statistical analyses for all measures were accomplished with Statistica 10.

39.3 Results

We compared time domain indexes (HR and SDNN) and nonlinear index (SampEn) of HRV between two periods: performing the task with mathematical words and performing the task with commonly used words, using nonparametric Wilcoxon signed-rank test. Heart rate (HR) and standard deviation of heart rate (SDNN) values did not significantly differ between two conditions (Tables 39.1 and 39.2).

Table 39.1 Description statistics (median, lower and upper quartiles) and the results of Shapiro-Wilk's test of HRV parameters in task performing with sentences with mathematical words (MW) and sentences with commonly used words (CW)

Statistics	SamEn	SampEn	SDNN	SDNN	HR	HR
	CW	MW	CW	MW	CW	MW
Median	0.65	0.72	56.72	58.92	789.64	781.35
Lower quartile	0.51	0.62	44.67	45.48	734.11	708.86
Upper quartile	0.77	0.79	76.48	76.88	922.31	907.13
W-test	$W = 0.92$,	$W = 0.97$,	$W = 0.94$,	$W = 0.89$,	$W = 0.97$,	$W = 0.78$,
	$p = 0.04$	$p = 0.68$	$p = 0.15$	$p = 0.01$	$p = 0.21$	$p = 0.02$

Table 39.2 The distributions of the medians of HRV parameters in two type's tasks performing were compared using Wilcoxon signed-rank test

Variables	T	Z	p
SampEn	82.00	2.37	0.01*
SDNN	128.00	1.21	0.22
HR	132.00	0.65	0.49

*Significant level

Heart rate complexity (SampEn) was significantly higher in the performing the task with mathematical terms than performing the task with commonly used words (Tables 39.1 and 39.2).

39.4 Conclusions

The aim of the current study was to examine the relationship between the complexity of behavior and complexity of heart rate. It was shown that the early-formed behavior, which realized less differentiated organism-environment relationships, was corresponded with less complexity of heart rate than the latter-formed behavior, which realized more differentiated organism-environment relationships. We can suppose that neuronal supporting of the latter-formed behavior includes more neuronal systems. In this case, the central autonomic network has to realize more nonstationary activity which demands many different changes in the activity of the heart and other parts of the organism. It leads to lack of regularity of activity and to the increase of complexity.

It is important that HR was the same in both conditions. It means that these different modes of heart activity cannot be explained through the different intensity of cognitive load, which demands different quantity of internal recourses, during early- and latter-formed behavior.

The main output of the study is that the behavioral complexity is reflected not only in the brain activity but also in the body activity.

Acknowledgments The search is supported by grant RFBR N16-36-60044 mol_a_dk.

References

1. Critchley, H.D., Harrison, N.A.: Visceral influences on brain and behavior. Neuron. **77**, 624–638 (2013)
2. Bagaev, V., Aleksandrov, V.: Visceral-related area in the rat insular cortex. Auton. Neurosci. Basic. Clin. **125**, 16–21 (2006)
3. Pigarev, I.N., Bagaev, V.A., Levichkina, E.V., Fedorov, G.O., Busigina, I.I.: Cortical visual areas process intestinal information during slow-wave sleep. Neurogastroenterol. Motil. **25**, e169–e268 (2013)
4. Brüggemann, J., Shi, T., Apkarian, A.V.: Viscero-somatic neurons in the primary somatosensory cortex (SI) of the squirrel monkey. Brain Res. **756**, 297–300 (1997)
5. Benarroch, E.E.: The central autonomic network—functional organization, dysfunction, and perspective. Mayo Clin. Proc. **68**, 988–1001 (1993)
6. Acharya, U.R., Joseph, K.P., Choo, N.K., Jasjit, M.L., Suri, S.: Heart rate variability: a review. Med. Bio. Eng. Comput. **44**, 1031–1051 (2006)
7. Matthews, S.C., Paulus, M.P., Simmons, A.N., Nelesen, R.A., Dimsdale, J.E.: Functional subdivisions within anterior cingulate cortex and their relationship to autonomic nervous system function. NeuroImage. **22**, 1151–1156 (2004)
8. Shvyrkov, V.B.: Behavior specialization and the system-selection hypothesis of learning. In: Human Memory and Cognitive Capabilities, pp. 599–611. Elsevier, Amsterdam (1986)
9. Anokhin, P.K.: Biology and neurophysiology of the conditioned reflex and its role in adaptive behavior. In: Corson, S.A. (Scientific and Transl. Ed.): International Series of Monographs in Cerebrovisceral and Behavioral Physiology and Conditioned Reflexes, Vol. 3. Pergamon Press, New York (1974) 1–592
10. Edelman, G.M.: Naturalizing consciousness: a theoretical framework. Proc. Natl. Acad. Sci. U. S. A. **100**, 5520–5524 (2003)
11. Lewin, K.: Action research and minority problems. In: Lewin, G.W. (ed.) Resolving Social Conflicts. Harper & Row, New York (1946)
12. Richman, J.S., Moorman, J.R.: Physiological time-series analysis using approximate entropy and sample entropy. Am. J. Physiol. Heart Circ. Physiol. **278**, H2039–H2049 (2000)

Chapter 40
Neural Generators of the N2 Component for Abstinent Heroin Addicts in a Dot-Probe Task

Hongqian Li, Qinglin Zhao, Bin Hu, Yu Zhou, and Quanying Liu

Abstract Target-elicited N2 component of event-related potential (ERP) has been considered to be involved in target detection in the attentional processes. We aim to link the target-elicited N2 in a dot-probe task and the drug-related attention bias in heroin dependence and further estimate the brain regions involved in the generation of the target-elicited N2. We recorded 64-channel electroencephalograms (EEG) from 17 abstinent heroin addicts (AHAs) and 17 healthy controls (HCs) during the dot-probe visual task. Individual N2 sources were localized using exact low-resolution electromagnetic tomography (eLORETA). Compared to HCs, AHAs generated larger N2 amplitude in both congruent and incongruent conditions, suggesting that target detection processing in AHAs might require more attention resources. Moreover, N2 component was mainly generated in the Brodmann areas (BAs) 7, 23, 24, 31, 30, 32, and 40, implying that the frontoparietal cortex played a critical role in target detection processes.

Keywords ERP · N2 component · Target detection · eLORETA · Dot-probe task

40.1 Introduction

Attention can be captured by relevant signals derived from task demands such as target stimuli [1, 2]. The late cognitive event-related potentials (ERPs), such as N2 and P3, have been thought to reflect high-order cognitive processes involved in target detection, for instance, the selective attention processing, conflict processing,

H. Li · Q. Zhao · B. Hu · Y. Zhou
Laboratory of Ubiquitous Awareness and Intelligent Solutions, Lanzhou University, Lanzhou, China

Q. Liu (✉)
Department of Computing and Mathematical Sciences, California Institute of Technology, Pasaden, USA

J. M. Delgado-García et al. (eds.), *Advances in Cognitive Neurodynamics (VI)*,
Advances in Cognitive Neurodynamics,
https://doi.org/10.1007/978-981-10-8854-4_40

and stimulus categorizing [3–6]. The N2 components can be divided into a more anterior subcomponent with a frontocentral scalp distribution, namely, N2b, and a more posterior subcomponent known as N2c [1]. The target-elicited posterior N2 component is still largely unknown [7]. In this study, we mainly explored the posterior N2 component that is related with target detection in AHAs using the dot-probe task. Previous study has showed that the posterior N2 component is sensitive to task difficulty and observed in visual tasks [8]. Moreover, some studies have applied the distributed source methods to estimate N2 sources in smokers, obesity, and drug addiction [9–11].

In present study, we used eLORETA method to localize and then compare brain sources of N2 in AHAs and HCs during a visual dot-probe task. Behavior results showed that AHAs responded faster than healthy controls (HCs) to target (dot) stimuli that replace drug-related compared to neutral cues. Furthermore, we proved that the brain sources for target-elicited N2 were mainly in frontal-parietal areas such as anterior cingulate cortex, posterior cingulate cortex, precuneus, and superior parietal lobule.

40.2 Materials and Methods

40.2.1 Subjects

Thirty-four participants (all males) were involved in the study, including 17 abstinent heroin addicts (AHAs) and 17 healthy controls (HCs). The AHAs (age, 36.76 ± 6.01) were recruited from the Gansu Compulsory Isolated Detoxification Center in China, meeting the criteria of Diagnosis and Statistics of Mental Disorder 5th edition (DSM-V) for heroin dependence. The HCs (age, 36.65 ± 6.12) were recruited from the local community with no history of alcohol or drug abuse. These two groups showed no significant difference in the age ($t_{[32]} = 0.057, p = 0.955$ for unpaired t-test) but a significant difference in educational level (3.06 ± 2.82 years for AHAs, 7.65 ± 3.30 for HCs; $t_{[32]} = -4.363$, $p < 0.01$ for unpaired t-test). All the subjects were right-handed, had normal or corrected-to-normal visual acuity, and had no history of neurological problems. None of the subjects were taking any psychotropic, neurological, or psychiatric medications at the time of experiment. All participants were given written informed consent before participating in the study, which had been approved by the local ethics committee.

40.2.2 Procedure

The experiment was performed, in accordance with the approved guidelines, in a quiet, air-conditioned laboratory with dimly natural light. The participants seated comfortably 80 centimeters in front of a 21-inch computer screen. The dot-probe

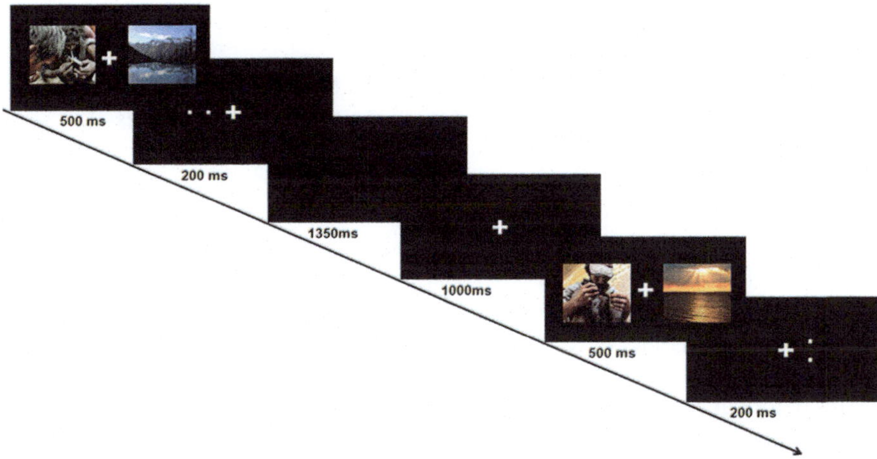

Fig. 40.1 Experimental paradigm. Each trial began with a fixation cross with 1000 ms duration. Immediately following offset of the fixation cross, a pair of pictures, including one drug-related cue and one neutral picture, was presented bilaterally for 500 ms. Next, a target stimulus occurred and lasted for 200 ms, which was either a horizontal pair of dots or a vertical pair. In the end, the trial ended with a 1350 ms intertrial interval

experimental paradigm was shown in Fig. 40.1. Specifically, each trial began with a fixation cross (1 cm × 1 cm) in the center of the screen for 1000 ms. Immediately following offset of the fixation cross, a pair of images was presented for 500 ms. Each pair contained one picture of the heroin-related cues and a neutral image. In each pair, one of the pictures appeared to the left of the fixation cross and one appeared to the right, with 15 cm center distance between the stimuli.

When the images disappeared, the target stimuli, either a horizontal pair of dots or a vertical pair, immediately presented in the place of one of the images. The pair of dots had 5 mm center distance, with 1 mm radius of each dot. The subjects were required to judge the vertical and horizontal and to press the response key as soon as possible. All subjects are instructed to use the middle finger and the index finger of the right hand to press the button.

There are two conditions: the target stimulus replaced the drug-related cue, which was defined as the "congruent" occasion, and replaced the control cue on the other occasion, which was defined as the "incongruent" occasion.

Participants had one practice block before the real experiment. Practice block was exactly the same as the experimental blocks and was given until the subject was able to reach a mean accuracy of 90% in the task. After the practice, the subject started the real experiment with three blocks. Each block has 80 trials that last about 6 min. The subject had 1–2 min breaks between blocks to reduce the fatigue.

40.2.3 EEG Recording and ERP Analysis

EEG signals were recorded during the task using a 64-channel electrode cap (Brain Products, Gilching, Germany) with the International 10/20 system sites. The scalp impedance of each sensor was kept below 10 kΩ, as suggested by the manufacturer. The EEG signals were recorded at a sampling rate of 5000 Hz with vertex as reference and amplified with an analog band-pass filter of 0.01–100 Hz.

For the off-line analysis, the raw EEG signals were resampled at 1000 Hz and then band-pass filtered at 1–40 Hz with an IIR filter, and then independent component analysis (ICA) was performed to remove the ocular and muscle artifacts. The cleaned EEG signals were re-referenced to the grand average. The target-locked data were segmented into epochs 700 ms before the dot stimulus onset, which is 200 ms before image onset, to 1000 ms after dot stimulus onset. Due to the carryover effects of image stimulus, the average voltage of 200 ms to 0 ms before dot stimulus onset is biased, thus not suitable for the baseline. Therefore, the target-locked data were baseline corrected using the average voltage of 700 ms to 500 ms before dot stimulus onset.

In the present study, we examined the target-elicited (posterior) N2 component. The N2 peak was defined as the largest negative deflection in the target stimulus waveforms within the 220–300 time window. N2 amplitude and latency measures were calculated at the Pz electrodes.

40.2.4 N2 Source Localization

In the present report, N2 source localization was performed a current density analysis on the scalp-recorded electrical activity using the sLORETA and eLORETA software package [12, 13] using the standard electrode positions of the EEG cap (Montreal method: ten-twenty electrode system). The eLORETA solutions were computed with the sLORETA and eLORETA software package for each time period within a 80-ms time window in each condition. We then compared N2 time window with their respective 80ms baseline period (-160 to -80 ms as the "baseline window"). The eLORETA images represented the electrical neural activity of each voxel in the neuroanatomic Montreal Neurological Institute (MNI) space as amplitude of the computed current source density ($\mu A/mm^2$). The brain sources were restrained in the cortical gray matter, resulting in 6239 cortical gray matter voxels at 5 mm resolution [12, 13].

40.3 Results

40.3.1 Behavioral Results

A 2×2 ANOVA on the RTs showed no significant main effect for group and condition, respectively, (group: $F_{[1, 32]} = 0.414$, $p = 0.525$; condition: $F_{[1, 32]} = 0.265$, $p = 0.610$), whereas it showed a significant interaction between the position of drug cue and the group ($F_{[1, 32]} = 14.47$, $p = 0.001$). Moreover, by simple effect test, main effect of condition is significant for both AHA ($F_{[1, 32]} = 9.89$, $p = 0.004$) and HC ($F_{[1, 43]} = 4.99$, $p = 0.033$) group. AHAs tended to have a significant quicker response to targets preceded by drug-related cues compared to targets preceded by neutral images, whereas the opposite pattern was observed in HC.

40.3.2 N2 Amplitude and Latency

Table 40.1 shows the mean ± standard deviation (*SD*) of N2 amplitudes in Pz for target stimuli in the congruent and incongruent condition. Grand averages of ERP waveforms for each condition on Pz are shown in Fig. 40.2. The 2 (group: AHA vs. HC) × 2 (condition: congruent vs. incongruent) repeated measures ANOVA of N2 amplitude revealed a significant main effect of group ($F_{[1, 32]} = 4.358$, $p = 0.045$) and condition had no main effect ($F_{[1, 32]} = 0.364$, $p = 0.550$). Moreover, there was no significant "group × condition" interaction ($F_{[1, 32]} = 0.687$, $p = 0.413$). N2 latency had a main effect of condition ($F_{[1, 32]} = 25.57$, $p < 0.001$), but no main effect of group ($F_{[1, 32]} = 0.10$, $p = 0.754$) and no significant "group × condition" interaction ($F_{[1, 32]} = 0.428$, $p = 0.518$).

40.3.3 Localization of N2 Cortical Generators

The results of eLORETA showed that the sources for the target-elicited N2 component were in the frontal and parietal lobe, such as cingulate cortex and precuneus. In congruent condition, N2 generators were mainly observed in the posterior cingulate cortex and superior parietal lobule (SPL) in AHAs, whereas the

Table 40.1 Target-elicited (posterior) N2 amplitude and latencies (mean ± standard deviation) at the Pz electrode

	Target-elicited N2 (congruent)		Target-elicited N2 (incongruent)	
	AHA	HC	AHA	HC
Amplitude (uV)	−1.46 ± 0.27	−0.67 ± 0.30	−1.35 ± 0.23	−0.50 ± 0.20
Latency (ms)	270.65 ± 5.22	266.53 ± 6.77	257.86 ± 4.40	252.88 ± 6.08

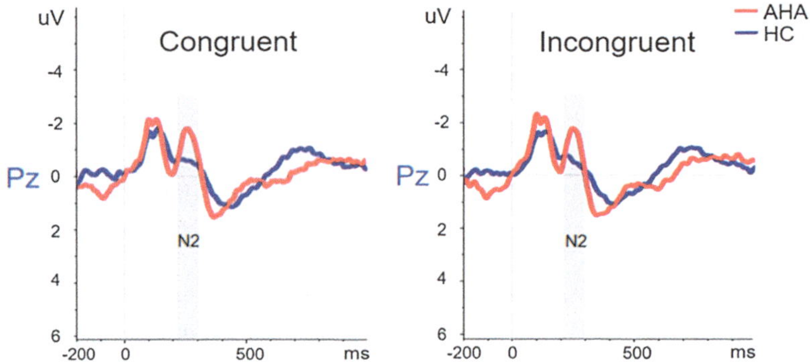

Fig. 40.2 N2 waveform and Scalp topography. Grand average N2 waveforms from the middle electrode (Pz) for target stimuli (congruent and incongruent)

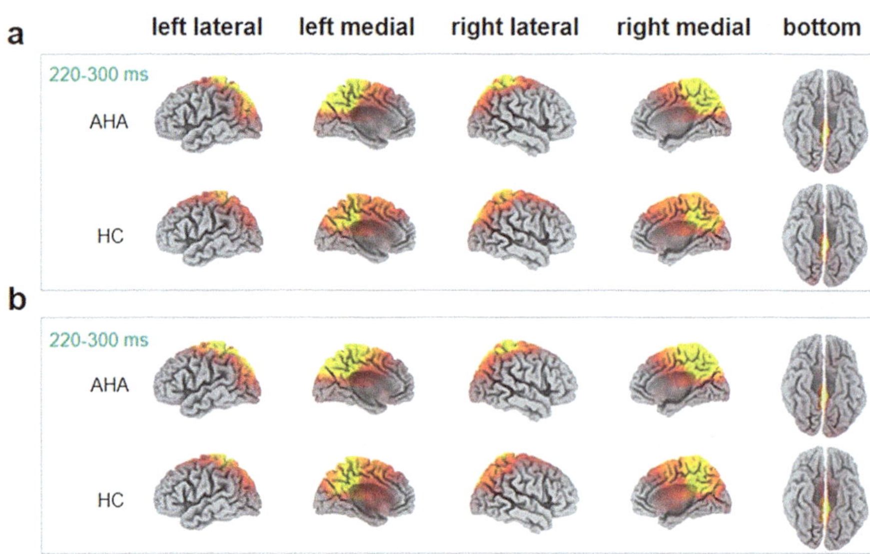

Fig. 40.3 Source localization of target-elicited N2 generators. Statistical maps (from nonparametric *t*-tests) of the gray matter current density for the target-elicited N2 ($p < 0.001$) within the 220–300 ms in the congruent (**a**) and incongruent (**b**) condition for AHA and HC groups, respectively. The MNI coordinate regions for each condition are reported in Table 40.2

estimated sources for N2 were located in the cingulate cortex, including anterior cingulate cortex (ACC) and posterior cingulate cortex (PCC), and SPL in HCs (see Fig. 40.3a). In the incongruent condition, N2 were generated by the cingulate cortex and parietal lobe for both AHA and HC group (see Fig. 40.3b).

Table 40.2 Location of target-elicited N2 generators in congruent and incongruent condition for AHA and HC group

Stimuli	Condition	Time period (ms)	Group	Brodmann areas	*t*-value
Target	CON	220–300	AHA	2, 3, **4**, 5, **6**, 7, 18, 19, **22**, 23, 31, 39, 40	3.509
			HC	**1**, 2, 3, 5, 7, **8**, **13**, **17**, 18, 19, 23, **24**, **30**, 31, **32**, 39, 40	1.983
	INCON	220–300	AHA	2, 3, 4, 5, 6, 7, 8, 13, 18, 19, 23, 24, 31, 32, 39, 40	3.425
			HC	**1**, 2, 3, 4, 5, 6, 7, 8, 13, **17**, 18, 19, **22**, 23, 24, **25**, **29**, **30**, 31, 32, 39, 40	1.954

Clusters surviving FWE-corrected threshold $p < 0.001$. The **bold** BAs indicate different activated brain regions between the two groups. Abbreviations: *CON* congruent, *INCON* incongruent

40.4 Discussion

Behavioral results showed that the faster RTs in congruent compared to incongruent in AHAs, which suggested an attentional bias to heroin cues in AHAs. The main objective of the present study was to use eLORETA to identify brain areas involved in the generation of the target-elicited N2 components during a visual dot-probe task, which might add new electrophysiological evidence of attentional bias for heroin-related cues in AHAs. Our results showed that although the target-elicited N2 components for both groups generally shared a common network, certain differences were observed between two groups in two conditions. The amplitude analysis confirmed the frontal-central topography of N2 component for AHA group, whereas for N2 mainly distributed in the central-parietal regions for HC group. Although N2 latency showed no differences between two groups, N2 amplitude was higher in AHAs than HCs, suggesting that AHA had attentional bias for heroin-related cues [14].

In congruent condition, source location results showed that PCC was activated both in AHAs and HCs, while ACC was only activated in HCs, suggesting that heroin-related cues might affect the subsequent target detection resulting in detection of AHA for target stimuli maintained only where the previous cues appeared. PCC has been considered to be associated with arousal, regulating the focus of attention, and conscious awareness [15], indicating that the attention of AHAs was consciously focused on the heroin-related cues.

In incongruent condition, both ACC and PCC activated for two groups. Previous studies have confirmed the critical role of the ACC in cognitive control and conflict monitoring [16, 17] and PCC has been thought to be indirectly involved in manipulating the attention concentration [18], suggesting that the target response phase was involved in conflict monitoring and target-locked attention.

Besides, the BAs 7, 18, 19, 23, and 31 were activated in target-elicited N2 time window for both conditions and both groups. These BAs have been considered to be connected with the motor areas such as BA 4, 6, and 8, referring to the primary motor cortex on the cingulate gyrus, the somatosensory cortex, and the superior parietal lobule [1, 19]. Target stimuli evoked that subject psychologically has to produce in advance a motor response. Hence, our result implied that the BAs 7, 18, 19, 23, and 31 might belong to a common parietal-occipital network for detecting targets that induce a motor response.

40.5 Conclusion

We have investigated sources of target-elicited N2 components in a dot-probe task. ACC and PCC interacted with the frontal-parietal attention networks in target detection and conflict monitoring cognitive processing and BAs 7, 18, 19, 23, and 31 of the parietal-occipital network involved in target detection processing as well.

Acknowledgments This work was supported by the National Basic Research Program of China (973 Program) (No.2014CB744600), the Program of International S&T Cooperation of MOST (No.2013DFA11140), the National Natural Science Foundation of China (grant No.61210010, No.61632014), the National key foundation for developing scientific instruments (No.61627808), and the Program of Beijing Municipal Science and Technology Commission (No.Z171100000117005).

Reference

1. Bocquillon, P., et al.: The spatiotemporal dynamics of early attention processes: a high-resolution electroencephalographic study of N2 subcomponent sources. Neuroscience. **271**, 9–22 (2014)
2. Kastner, S., Ungerleider, L.G.: Mechanisms of visual attention in the human cortex. Annu. Rev. Neurosci. **23**, 315 (2000)
3. Berg, I.V.D.: Cue-reactivity and Attentional Bias in Smokers and Non-smokers: an ERP study. Erasmus University, Rotterdam (2007)
4. Hoeksma, M.R., et al.: Abnormal selective attention normalizes P3 Amplitudes in PDD. J. Autism Dev. Disord. **36**, 643–654 (2006)
5. Azizian, A., et al.: Electrophysiological correlates of categorization: P300 amplitude as index of target similarity. Biol. Psychol. **71**, 278–288 (2006)
6. Jr, R.J., Donchin, E.: P300 and Stimulus Categorization: two plus one is not so different from one plus one. Psychophysiology. **17**, 167–178 (1980)
7. Näätänen, R., Picton, T.W.: N2 and automatic versus controlled processes. Electroencephalogr. Clin. Neurophysiol. **38**, 169–186 (1986)
8. Folstein, J.R., Petten, C.V.: Influence of cognitive control and mismatch on the N2 component of the ERP: a review. Psychophysiology. **45**, 152–170 (2008)

9. Anderer, P., et al.: Non-invasive localization of P300 sources in normal aging and age-associated memory impairment. Neurobiol. Aging. **24**, 463–479 (2003)
10. Kropotov, J.D.P., Decomposing, V.A.: N2 NOGO wave of event-related potentials into independent components. Neuroreport. **20**, 1592 (2009)
11. Parvaz, M.A., et al.: Neuroimaging for drug addiction and related behaviors. Rev. Neurosci. **22**, 609–624 (2011)
12. Pascual-Marqui, R.D.: Standardized low-resolution brain electromagnetic tomography (sLORETA): technical details. Methods Find. Exp. Clin. Pharmacol. **24**, 5–12 (2002)
13. Pascual-Marqui, R.D., et al.: Assessing interactions in the brain with exact low-resolution electromagnetic tomography. Philos. Trans. R. Soc. A Math. Phys. Eng. Sci. **369**, 3768–3784 (1952)
14. Thai, N., Taberthomas, B.C., Pérezedgar, K.E.: Neural correlates of attention biases, behavioral inhibition, and social anxiety in children: an ERP study. Dev. Cogn. Neurosci. **19**, 200 (2016)
15. Dillen, K.N., et al.: Aberrant functional connectivity differentiates retrosplenial cortex from posterior cingulate cortex in prodromal Alzheimer's disease. Neurobiol. Aging. **44**, 114 (2016)
16. Braver, T.S., et al.: Anterior Cingulate Cortex and Response Conflict: effects of Frequency, Inhibition and Errors. Cereb. Cortex. **11**, 825–836 (2001)
17. Carter, C.S., Botvinick, M.M., Cohen, J.D.: The contribution of the anterior cingulate cortex to executive processes in cognition. Rev. Neurosci. **10**, 49 (1999)
18. Leech, R., et al.: Fractionating the default mode network: distinct contributions of the ventral and dorsal posterior cingulate cortex to cognitive control. J. Neurosci. **31**, 3217–3224 (2011)
19. Margulies, D.S., et al.: Precuneus shares intrinsic functional architecture in humans and monkeys. Proc. Natl. Acad. Sci. U. S. A. **106**, 20069–20074 (2009)

Part V
Oscillation, Synchronization, Neural Plasticity, and Coordination Dynamics from Neural to Social Systems

Chapter 41
Changes in Phase Synchronization of EEG During Development of Symbolic Communication Systems

Masayuki Fujiwara, Takashi Hashimoto, Guanhong Li, Jiro Okuda, Takeshi Konno, Kazuyuki Samejima, and Junya Morita

Abstract To identify neural synchrony changes during the establishment of symbolic communication systems, we analyzed Phase Locking Value, a phase synchronization index reflective of the cognitive process of finding meaning in visual stimuli. Hyper-scanning electroencephalograms were recorded during a symbolic communication task. Good and bad performance pairs were studied to find features of phase synchronization during the establishment of a communication system. During the initial phase of the establishment process, the good performance pair showed synchronization in the 150–300-ms and 450–750-ms latency periods and desynchronization at 300–450 ms in the gamma band after receipt of the partner's message. The synchronization at around 500 ms strengthened and lasted longer during the last phase when a shared communication system was established. This pattern was not observed in the bad pair. These results suggest that phase synchrony around 500 ms and later is involved in the cognitive process of finding meaning in symbolic messages.

Keywords Phase synchronization · Symbolic communication · Phase locking value · Experimental semiotics · Hyper-scanning EEG

M. Fujiwara (✉) · T. Hashimoto (✉) · G. Li
Japan Advanced Institute of Science and Technology, Nomi, Ishikawa, Japan
e-mail: m-fujiw@jaist.ac.jp; hash@jaist.ac.jp

J. Okuda
Department of Intelligent Systems, Faculty of Computer Science and Engineering, Kyoto Sangyo University, Kita-ku, Kyoto, Japan

T. Konno
Kanazawa Institute of Technology, Nonoichi, Ishikawa, Japan

K. Samejima
Tamagawa University, Tokyo, Japan

J. Morita
Shizuoka University, Shizuoka, Japan

J. M. Delgado-García et al. (eds.), *Advances in Cognitive Neurodynamics (VI)*, Advances in Cognitive Neurodynamics,
https://doi.org/10.1007/978-981-10-8854-4_41

41.1 Introduction

The neural synchrony viewpoint has attracted attention in the study of human coordination behavior. For example, it is shown that neural synchrony increases through coordination behavior by tapping [1] and fingertip movement [2, 3]. These studies contribute to the clarification of the neural underpinning of human social coordination from the viewpoint of synchronization within the brain and in interbrain systems.

Despite the neural synchrony viewpoint providing better understanding of coordination within the brain and in interbrain systems, a neural mechanism of coordination via symbolic communication has not been well studied using this viewpoint. Recent research in the development of symbolic communication systems has advanced the field of experimental semiotics [4, 5]. Electroencephalographic (EEG) recordings during the development of symbolic communication systems were obtained in an experiment within the framework of experimental semiotics [6]. We believe that investigating this experiment from the viewpoint of neural synchrony can help elucidate the neural underpinnings of symbolic communication.

We aimed to identify changes in neural synchrony during the establishment of symbolic communication systems. In the present article, we analyzed Phase Locking Value (PLV) [7] obtained during hyper-scanning EEGs recorded during a symbolic communication task [8] to understand the role of neural synchronization in symbolic communication. PLV is an index of phase synchronization, which is known to reflect the cognitive process of finding meaning in visual stimuli [9].

41.2 Materials and Methods

41.2.1 Symbolic Communication Task

41.2.1.1 Participants

To study differences in phase synchrony between successful and unsuccessful communications, we selected two pairs of participants (four males, age, mean = 23.3, standard deviation = 2.77, range = 20–23 years). The pairs had good or bad performance in typical behavior and were selected from 40 neurologically normal participants (20 pairs) who had undergone hyper-scanning EEG recording [6] during the symbolic communication task [6, 8]. Performance was measured in accordance with a Bayesian inference model constructed on the basis of behavioral data obtained during the task [10]. The good performance pair constantly succeeded after several trials in the task, while the bad performance pair succeeded to a much lesser extent.

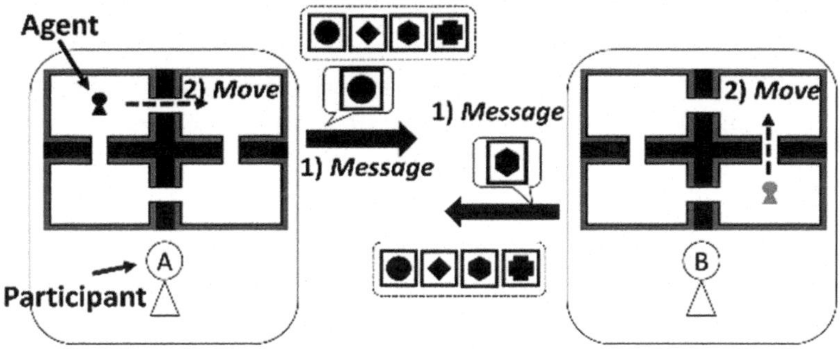

Fig. 41.1 Schematic view of the symbolic communication task

41.2.1.2 Procedure

In this experiment, we used a coordination game wherein symbolic messages were exchanged [8]. This was called a symbolic communication task and was based on an experiment performed in an experimental semiotic framework [4]. In this task, which is schematically described in Fig. 41.1, a pair of participants played a coordination game on computer screens. The aim of the game was to bring their agents, which were initially located randomly in one of four different rooms (2×2 configuration), into the same room by moving them once horizontally or vertically. Before moving the agents, each participant sent his partner a message by choosing one geometric figure from four options (●◆⬢✚), whose meanings were neither predefined nor initially shared. Diagonal movement was prohibited. This demanded a mutual understanding of the denotations (mapping between the figures and the rooms) and the connotations (informing initial room/destination) of symbolic messages for better performance [8]. The game was repeated 60 times (trials).

EEG Recording and Preprocessing. We recorded EEGs using a sampling rate of 1000 Hz (BrainAmp MRplus, Brain Products GmbH, Germany) at 32 scalp sites according to the International 10–20 system and its extension using active Ag/AgCl electrodes (actiCAP 32Ch Standard-2). A reference electrode was located at AFz. To remove ocular artifacts, four electrodes were used to record vertical and horizontal eye movements.

The scalp EEGs were recorded using remaining 28 electrodes. Re-referencing with common averages and filtering in the range of 1–70 Hz were performed on the data. The EEGs were segmented into epochs (from −500 ms to 2000 ms) and independent component analysis rejection using SemiAutomatic Selection of Independent Components for Artifact correction [11] was performed. We used complex Morlet wavelet transforms of the signals for each EEG.

41.2.2 Phase Locking Value and Z-Score

Phase Locking Value (PLV) [7] was used to quantify the degree of phase synchrony between two signals in several trials. $\mathrm{PLV}_t \in [0, 1]$ is defined as:

$$\mathrm{PLV}_t = \frac{1}{N}\left|\sum_{n=1}^{N} e^{j\theta(t,n)}\right|, \tag{41.1}$$

where t, n, N, and j denote the time, the trial, the number of trials, and the imaginary unit, respectively. $\theta(t,n) = \phi_1(t,n) - \phi_2(t,n)$ is the phase difference between the two signals at time t and in trial n, where ϕ_1 and ϕ_2 are transient phases of the two signals.

To study changes in phase synchrony from baseline, we used the Z-score of the PLV, which is defined as $\mathrm{PLVz}_t = \{\mathrm{PLV}_t - \mathrm{mean}(\mathrm{PLV}_{\mathrm{baseline}})\}/\mathrm{std}(\mathrm{PLV}_{\mathrm{baseline}})$, where $\mathrm{mean}(\mathrm{PLV}_{\mathrm{baseline}})$ and $\mathrm{std}(\mathrm{PLV}_{\mathrm{baseline}})$ indicate the mean and the standard deviation of PLV_t during a baseline period, respectively. Here the baseline period was from -200 ms to 0 ms, when the message from the partner was displayed on the screen.

We analyzed gamma-phase synchrony at 40 Hz after the partner's message was received. Gamma-phase synchrony was viewed as a mechanism subserving large-scale cognitive integration [12, 13].

41.3 Results

Figure 41.2 shows the transitions of the mean PLVz of all EEG electrodes at a frequency of 40 Hz. The starting point (0 ms) is the time when the partner's message is displayed on the screen. In the initial five trials, the good performance pair showed increases in PLVz during the latency periods of 150–300 ms and 450–750 ms and a reduction during the period of 300–450 ms, as indicated in Fig. 41.2a. In the last five trials (Fig. 41.2b), the later increase became stronger and continued for a longer period. In contrast, as Fig. 41.2c–d depict, the bad performance pair did not show such patterns of PLVz changes.

We observed PLV networks (inter-electrode links) whose degrees of synchronization significantly increased from when compared to the previous latency period shown in Fig. 41.3. We defined Phase Locking Statistics (PLS) [13] for the differences in PLVz compared to the previous latency period (every 150 ms) and show links with PLS < 0.05. Consistent with the transition in Figs. 41.2a and 41.3a show dense synchrony networks during the 150–300-ms and 450–750-ms periods. These networks include frontoparietal synchronizations.

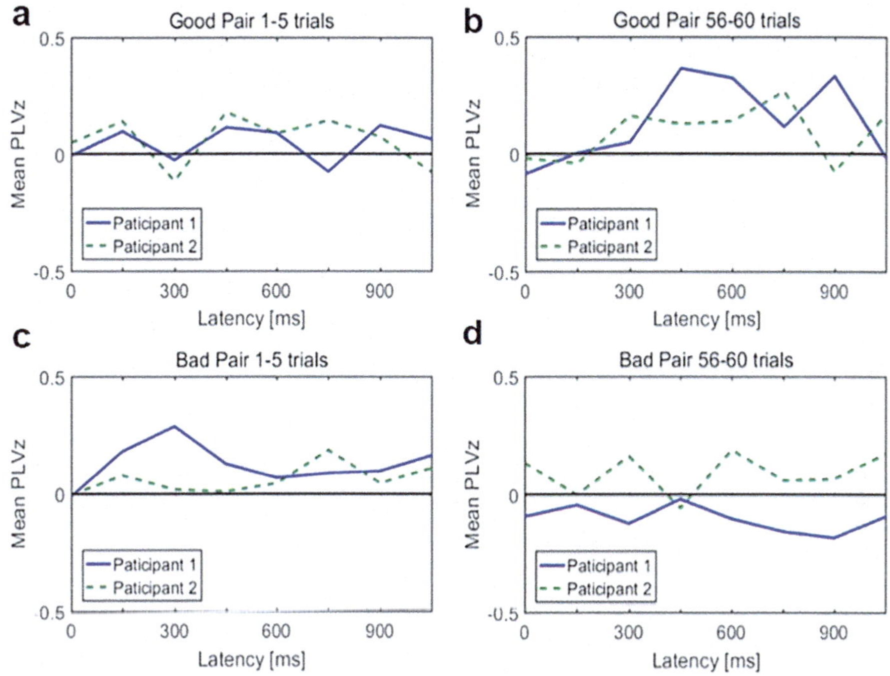

Fig. 41.2 Transitions of PLVz at 40 Hz for the good (**a** and **b**), and bad (**c** and **d**) performance pairs. The vertical and horizontal axes show PLVz averaged over all inter-electrodes and the latencies following the receipt of the partner's message (indicating the beginning times of the 150-ms periods in the graphs), respectively. Graphs (**a**) and (**c**) were obtained during trials $n = 1$–5; and graphs (**b**) and (**d**) were obtained during trials $n = 56$–60

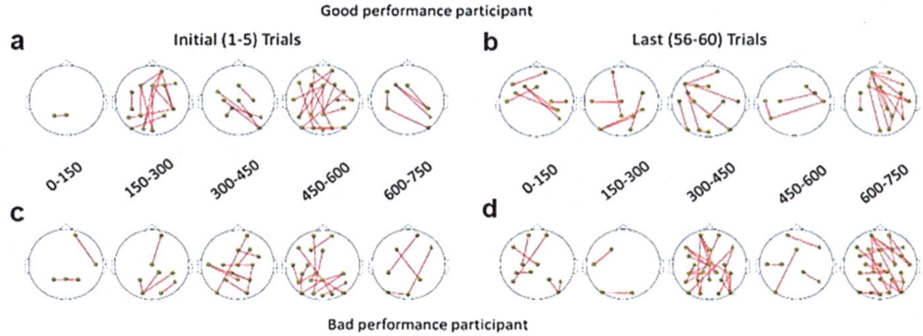

Fig. 41.3 Examples of transitions of the PLVz network at 40 Hz in the good performance pair (**a** and **b**) and the bad performance pair (**c** and **d**) (participant 2 for each case). Electrodes with significant increases in PLVz compared to the previous latency period are linked. Figures in (**a**) and (**c**) were obtained during the trials $n = 1$–5. Figures in (**b**) and (**d**) were obtained during trials $n = 56$–60

41.4 Discussion and Conclusion

In the initial five trials, the two participants in the good performance pair did not understand the received symbols well, as indicted by the behavioral data in the symbolic communication task [10]. Specifically, a communication system was not established. Therefore, the initial trials can be considered a phase wherein the participants attempted to understand the symbolic messages from their partners. In this phase, the good performance pair showed synchronization during two latency periods (150–300 and 450–750 ms) after the partner's message was received (Fig. 41.3a).

There was desynchronization between these two periods. In previous studies, similar synchronization and desynchronization patterns were observed and are considered to reflect switching between cognitive processes, such as visual cognition and motor responses [12, 13]. In the last five trials, where the good performance pair successfully established a shared communication system, strong synchrony continued for a longer duration. Our results thus may suggest that the two latency periods (150–300 and 450–750 ms) are involved in two cognitive processes: recognition of figures and consideration of the meanings of the figures as symbols.

Long-range frontoparietal phase synchronizations were observed in the two latency periods (Fig. 41.3a). The frontoparietal synchronization network is considered to be related to visual working memory [14] and cognitive control [15]. Similar patterns, i.e., synchronizations with frontoparietal connections and desynchronization between synchronization periods, were also found in the last trials of the bad performance pair (Figs. 41.2d and 41.3d) during late latency periods. According to estimations based on the behavioral data [10], this participant seems to still attempt to understand symbolic messages in the last trials. The frontoparietal phase synchronizations found here may play major roles in the understanding of symbolic messages, where the association between visual inputs and working memory for previous instances are related.

We suggest that the gamma-phase synchrony at around 500 ms and later, and especially the frontoparietal synchrony, may be involved in the cognitive process of finding meaning in symbolic messages. We should confirm how robust this phenomenon is in other pairs of participants and in other symbolic communication tasks. Here we show that the neural synchrony viewpoint and analysis of phase synchronization are effective for the examination of cognitive processes and the neural underpinnings of symbolic communication.

Acknowledgments This work was supported by JSPS KAKENHI Grant Number JP26240037 and JP17J06623.

References

1. Kawasaki, M., Kitajo, K., Yamaguchi, Y.: Inter-subject's brain synchronizations for coordination of tapping rhythms. IEICE Tech. Rep. **112**, 73–78 (2012) (in Japanese)

2. Tognoli, E., Lagarde, J., DeGuzman, G.C., Kelso, J.A.S.: The Phi Complex as a neuromarker of human social coordination. Proc. Natl. Acad. Sci. U. S. A. **104**, 8190–8195 (2007)
3. Yun, K., Watanabe, K., Shimojo, S.: Interpersonal body and neural synchronization as a marker of implicit social interaction. Sci. Rep. **2**, 1–8 (2012)
4. Galantucci, B.: An experimental study of the emergence of human communication systems. Cogn. Sci. **29**, 737–767 (2005)
5. Galantucci, B., Garrod, S.: Experimental semiotics: a review. Front. Hum. Neurosci. **5**, 1–15 (2011)
6. Li, G., Konno, T., Okuda, J., Hashimoto, T.: An EEG Study of Human Mirror Neuron System Activities During Abstract Symbolic Communication Advances in Cognitive Neurodynamics, Vol. V, pp. 565–571. Springer, Berlin/Heidelberg/New York (2016)
7. Lachaux, J.P., Rodriguez, E., Martinerie, J., Varela, F.J.: Measuring phase synchrony in brain signals. Hum. Brain Mapp. **8**, 194–208 (1999)
8. Konno, T., Morita, J., Hashimoto, T.: Symbol Communication Systems Integrate Implicit Information in Coordination Tasks Advances in Cognitive Neurodynamics, Vol. IV, pp. 453–459. Springer, Berlin/Heidelberg/New York (2013)
9. Castelhano, J., Rebola, J., Leitão, B., Rodriguez, E., Castelo-Branco, M.: To perceive or not perceive: the role of gamma-band activity in signaling object percepts. PLoS One. **8**, e66363 (2013)
10. Samejima, K., Konno, T., Li, A., Okuda, J., Morita, J., Hashimoto, T.: Statistical inference of meaning by a generative model of signal communication in the "coordination game". IPSJ SIG Tech. Rep. 1–6 (2016) (in Japanese)
11. Chaumon, M., Bishop, D.V.M., Busch, N.A.: A practical guide to the selection of independent components of the electroencephalogram for artifact correction. J. Neurosci. Methods. **250**, 47–63 (2015)
12. Rodriguez, E., George, N., Lachaux, J.P., Martinerie, J., Renault, B., Varela, F.J.: Perception's shadow: long-distance synchronization of human brain activity. Nature. **397**, 430–433 (1999)
13. Varela, F., Lachaux, J.-P., Rodriguez, E., Martinerie, J.: The brainweb: phase synchronization and large-scale integration. Nat. Rev. Neurosci. **2**, 229–239 (2001)
14. Salazar, R., Dotson, N., Bressler, S., Gray, C.: Content specific fronto-parietal synchronization during visual working memory. Science. **338**, 1097–1100 (2012)
15. Zanto, T.P., Gazzaley, A.: Fronto-parietal network: flexible hub of cognitive control. Trends Cogn. Sci. **17**, 602–303 (2013)

Chapter 42
Effect of Spike-Timing-Dependent Plasticity on Stochastic Spike Synchronization in an Excitatory Neuronal Population

Sang-Yoon Kim and Woochang Lim

Abstract We consider an excitatory population composed of subthreshold neurons which exhibit noise-induced spikings. This neuronal population has adaptive dynamic synaptic strengths governed by the spike-timing-dependent plasticity (STDP). In the absence of STDP, stochastic spike synchronization (SSS) between noise-induced spikings of subthreshold neurons was previously found to occur over a large range of intermediate noise intensities. Here, we investigate the effect of STDP on the SSS by varying the noise intensity. A "Matthew" effect in synaptic plasticity is found to occur due to a positive feedback process. Good synchronization gets better via long-term potentiation (LTP) of synaptic strengths, while bad synchronization gets worse via long-term depression (LTD). Emergence of LTP and LTD of synaptic strengths is investigated through microscopic studies based on both the distributions of time delays between the pre- and the postsynaptic spike times and the pair correlations between the pre- and the postsynaptic IISRs (instantaneous individual spike rates).

Keywords LTD · LTP · Spike-timing-dependent plasticity · Stochastic spike synchronization · Synaptic strength

42.1 Introduction

Recently, much attention has been paid to brain rhythms which emerge via population synchronization between individual firings in neural circuits [1]. These synchronized rhythms are associated with sensory and cognitive processes in the brain. Population synchronization has been intensively investigated in neural

S.-Y. Kim · W. Lim (✉)
Institute for Computational Neuroscience and Department of Science Education, Daegu National University of Education, Daegu, Korea
e-mail: sykim@icn.re.kr; wclim@icn.re.kr

J. M. Delgado-García et al. (eds.), *Advances in Cognitive Neurodynamics (VI)*,
Advances in Cognitive Neurodynamics,
https://doi.org/10.1007/978-981-10-8854-4_42

circuits composed of spontaneously firing suprathreshold neurons exhibiting regular discharges like clock oscillators [2]. In contrast to the case of suprathreshold neurons, the case of subthreshold neurons (which cannot fire spontaneously) has received little attention. The subthreshold neurons can fire only with the help of noise. Here we are interested in stochastic spike synchronization (SSS) (i.e., noise-induced population synchronization) between complex noise-induced firings of subthreshold neurons which exhibit irregular discharges like Geiger counters. Recently, such SSS has been found to occur in an intermediate range of noise intensity via competition between the constructive and the destructive roles of noise [3–6]. In the previous works on SSS, synaptic coupling strengths are static. However, in real brains synaptic strengths may vary (i.e., they can be potentiated or depressed) to adapt to the environment. These adjustments of synapses are called the synaptic plasticity which provides the basis for learning, memory, and development [7]. Regarding the synaptic plasticity, we consider a Hebbian spike-timing-dependent plasticity (STDP) [8–10]. For the STDP, the synaptic strengths vary via a Hebbian plasticity rule depending on the relative time difference between the pre- and the postsynaptic spike times. When a presynaptic spike precedes a postsynaptic spike, long-term potentiation (LTP) occurs; otherwise, long-term depression (LTD) appears. In this paper, we consider an excitatory population of subthreshold neurons and investigate the effect of STDP on the SSS by varying the noise intensity.

42.2 Excitatory Small-World Network of Subthreshold RS Izhikevich Neurons with Synaptic Plasticity

We consider the Watts-Strogatz small-world network (SWN) which interpolates between a regular lattice with high clustering (corresponding to the case of $p = 0$) and a random graph with short average path length (corresponding to the case of $p = 1$) via random uniform rewiring with the probability p [11]. Here, we consider the case that the rewiring probability p is 0.15 and the average number of synaptic inputs per neuron $M_{\rm syn}$ is 20. As an element in our neural system, we choose the regular spiking (RS) Izhikevich neuron model with the same parameters as those in Refs. [12, 13]. Each ith RS Izhikevich neuron is stimulated by using the DC current $I_{DC,i}$ and an independent Gaussian white noise ξ_i whose intensity is controlled by using the parameter D. We also consider a subthreshold case (where only noise-induced firings occur) such that the value of $I_{DC,i}$ is chosen via uniform random sampling in the range of [3.55, 3.65]. The synaptic current into the ith neuron is modeled in terms of the delayed double-exponential functions [14], and the coupling strength of the synapse from the jth presynaptic neuron to the ith postsynaptic neuron is J_{ij}. For the excitatory synapse (involving the AMPA receptors), we use the same synaptic delay, synaptic rise time, synaptic decay time, and synaptic reversal potential as those in [15]. Here, we consider a Hebbian STDP for the synaptic strengths $\{J_{ij}\}$. Initial synaptic strengths are normally distributed with the mean $J_0 (= 0.2)$ and the standard deviation $\sigma = 0.02$. With increasing time the synaptic strength for each synapse is updated with an additive nearest spike pair-based STDP rule:

$$J_{ij} \to J_{ij} + \delta \, \Delta J_{ij}(\Delta t_{ij}), \tag{42.1}$$

where δ ($= 0.005$) is the update rate and ΔJ_{ij} is the synaptic modification depending on the relative time difference Δt_{ij} ($= t_i^{(\text{post})} - t_j^{(\text{pre})}$) between the nearest spike onsets of the postsynaptic neuron i and the presynaptic neuron j. We use an asymmetric time window for the synaptic modification $\Delta J_{ij}(\Delta t_{ij})$ [9]:

$$\Delta J_{ij} = \begin{cases} A_+ \, e^{-\Delta t_{ij}/\tau_+} & \text{for } \Delta t_{ij} > 0 \\ A_- \, e^{\Delta t_{ij}/\tau_-} & \text{for } \Delta t_{ij} < 0 \end{cases}, \tag{42.2}$$

where $A_+ = 1.0$, $A_- = 0.7$, $\tau_+ = 35$ ms, $\tau_- = 70$ ms, and $\Delta J_{ij}(\Delta t_{ij} = 0) = 0$. Numerical integration of the governing equations of motion is done using the Heun method [16] (with the time step $\Delta t = 0.01$ ms).

42.3 Effect of the STDP on the Stochastic Spike Synchronization

We consider an excitatory Watts-Strogatz SWN of subthreshold RS Izhikevich neurons (exhibiting noise-induced spikings). In the absence of STDP, SSS is found to occur in an intermediate range of noise intensity D. As D passes a lower critical value D_l^* ($\simeq 0.225$), SSS occurs via the constructive role of noise, while it disappears through the destructive role of noise when passing a higher critical value D_r^* ($\simeq 0.846$). We take into consideration the STDP, and study its effect on the SSS by varying D, the results of which are well shown in Fig. 42.1. Figure 42.1a shows the time evolution of $\langle J_{ij} \rangle$ for various values of D (chosen in the range of SSS in the absence of STDP), where $\langle J_{ij} \rangle$ represents the population-averaged synaptic strength over all synapses. Initial average strengths at $t = 0$ are J_0 ($= 0.2$), independently of D. However, with increasing t, $\langle J_{ij} \rangle$ varies (i.e., potentiated or depressed) depending on D. After a sufficiently long time ($\sim$2000 s), $\langle J_{ij} \rangle$ seems to approach its saturated limit value $\langle J_{ij}^* \rangle$. We note that LTP occurs for $D = 0.27$, 0.3, 0.5, and 0.7, while LTD appears for $D = 0.25$ and 0.77. A plot of population-averaged limit values of synaptic strengths $\langle\langle J_{ij}^* \rangle\rangle_r$ vs. D is given in Fig. 42.1b; $\langle \cdots \rangle_r$ represents an average over 20 realizations. Here, the horizontal dotted line denotes the initial values of synaptic strengths J_0 ($= 0.2$), and the lower and the upper threshold values $\widetilde{D}_l$ ($\simeq 0.253$) and $\widetilde{D}_r$ ($\simeq 0.717$) for LTP/LTD are denoted by solid circles. Hence, LTP occurs in the range of ($\widetilde{D}_l$, $\widetilde{D}_r$); otherwise, LTD appears. Population spike synchronization may be well visualized in the raster plot of spikes.

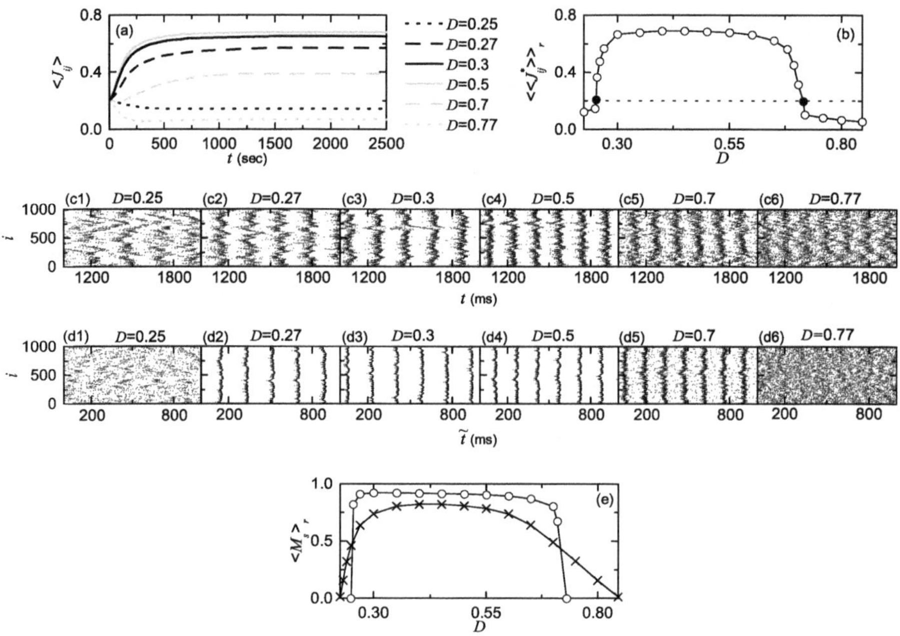

Fig. 42.1 Effect of the STDP on the SSS. (**a**) Time evolution of population-averaged synaptic weight $\langle J_{ij} \rangle$ for various values of D. (**b**) Plot of population-averaged limit value of synaptic weights vs. D. Raster plots of spikes in the absence of STDP [(**c1**)–(**c6**)] and in the presence of STDP [(**d1**)–(**d6**)] [$t = t^*$ (saturation time = 2000 s) $+\tilde{t}$]. (**e**) Plot of the statistical-mechanical spiking measure $\langle M_s \rangle_r$ (represented by open circles) versus D; for comparison, $\langle M_s \rangle_r$ in the absence of the STDP is also shown in crosses

Figure 42.1c1–c6 [Fig. 42.1d1–d6] show the raster plots of spikes in the absence (presence) of STDP. In the case of LTP for $D = 0.27$, 0.3, 0.5 and 0.7, the degree of synchronization is increased. In contrast, for the case of LTD of $D = 0.25$ and 0.77, the population states become unsynchronized. Furthermore, the degree of population synchronization may be quantitatively measured in terms of the realistic statistical-mechanical spiking measure, introduced by considering the occupation pattern and the pacing pattern of the spikes in the spiking stripes in the raster plots of spikes [17]. Figure 42.1e shows plots of $\langle M_s \rangle_r$ in the presence (open circles) and the absence (crosses) of STDP. A "Matthew effect" in the synaptic plasticity seems to occur in the following way. Good synchronization gets better via LTP of synaptic strengths, while bad synchronization gets worse via LTD. As a result, a steplike rapid transition to SSS occurs by varying the noise intensity, in contrast to the relatively smooth transition in the absence of STDP.

We also make an intensive investigation on emergence of LTP and LTD of synaptic strengths via microscopic studies based on both the distributions of time delays $\{\Delta t_{ij}\}$ between the pre- and the postsynaptic spike times and the pair-correlations between the pre- and the postsynaptic instantaneous individual spike

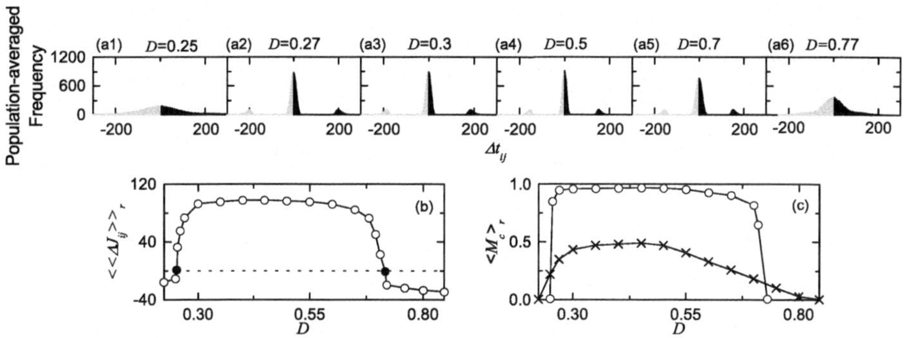

Fig. 42.2 Microscopic investigation on emergence of LTP and LTD via STDP. (**a1**)–(**a6**) Population-averaged histograms for the distributions of time delay Δt_{ij} during the time interval from $t = 0$ to the saturation time ($t = 2000$ s) for various values of D. (**b**) Plot of the population-averaged synaptic modification $\langle\langle \Delta J_{ij} \rangle\rangle_r$ vs. D. (**c**) Plot of the microscopic correlation measure $\langle M_c \rangle_r$ (represented by open circles) versus D in the limiting saturated case; for comparison, $\langle M_c \rangle_r$ in the absence of the STDP is also shown in crosses

rates (IISR) (which is given in Eq. (11) of Ref. [14]). These results are given in Fig. 42.2. Figure 42.2a1–a6 show population-averaged histograms $H(\Delta t_{ij})$ for the distributions of time delay Δt_{ij} during the time interval from $t = 0$ to the saturation time ($t = 2000$ s) for various values of D: for each synaptic pair, its histogram for the distribution of Δt_{ij} is obtained, and then we get the population-averaged histogram via averaging over all synaptic pairs. Here, black and gray regions represent LTP and LTD, respectively. In the case of LTP ($D = 0.27$, 0.3, 0.5, and 0.7), 3 peaks appear: one main central peaks and two minor left and right peaks. When the pre- and the postsynaptic spike times appear in the same spiking stripe in the raster plot of spikes, its time delay Δt_{ij} lies in the main peak; LTP/LTD may occur depending on the sign of Δt_{ij}. On the other hand, time delays Δt_{ij} lie in the minor peaks when the pre- and the postsynaptic spike times appear in the different nearest-neighboring spiking stripes. If the presynaptic stripe precedes the postsynaptic stripe (causality), then its time delay Δt_{ij} lies in the right minor peak (LTP); otherwise, it lies in the left minor peak (LTD). However, in the case of LTD of $D = 0.25$, and 0.77, the population states become desynchronized due to overlap of spiking stripes in the raster plot of spikes. Consequently, the main peak in the histogram becomes merged with the left and the right minor peaks, and then only one broadened main peak appears, in contrast to the case of LTP. The population-averaged synaptic modification $\langle\langle \Delta J_{ij} \rangle\rangle_r$ may be directly obtained from the above histogram $H(\Delta t_{ij})$:

$$\langle\langle \Delta J_{ij} \rangle\rangle_r \simeq \sum_{bins} H(\Delta t_{ij}) \cdot \Delta J_{ij}(\Delta t_{ij}). \qquad (42.3)$$

Figure 42.2b shows a plot of $\langle\langle \Delta J_{ij} \rangle\rangle_r$ [obtained from $H(\Delta t_{ij})$] vs. D. Then, population-averaged limit values of synaptic strengths $\langle\langle J^*_{ij} \rangle\rangle_r$ are given by $J_0 + \delta \, \langle\langle \Delta J_{ij} \rangle\rangle_r$, which agree well with the directly obtained values in Fig. 42.1b. Finally, we study the effect of STDP on the microscopic pair correlation $C_{ij}(\tau)$ between the pre- and the postsynaptic IISRs for the $\langle ij \rangle$ synaptic pair. Then, the microscopic correlation measure M_c, representing the average "in-phase" degree between the pre- and the postsynaptic pairs, is given by the average value of $C_{ij}(0)$ at the zero-time lag for all synaptic pairs. Figure 42.2c shows plots of $\langle M_c \rangle_r$ in the presence (open circles) and the absence (crosses) of STDP. Like the case of M_s, a "Matthew" effect in M_c also occurs: good pair correlation gets better via LTP, while bad pair correlation gets worse via LTD. Hence, a steplike transition occurs, in contrast to the case without STDP. Enhancement in M_c results in the increase in the average in-phase degree between the pre- and the postsynaptic pairs. Then, widths of spiking stripes in the raster plot of spikes decrease, which leads to narrowed distribution of time delays Δt_{ij}. Consequently, LTP may occur. In contrast, for the case of suppression of M_c, the distribution of Δt_{ij} becomes widened, which may lead to occurrence of LTD.

Acknowledgements This research was supported by the Basic Science Research Program through the National Research Foundation of Korea (NRF) funded by the Ministry of Education (Grant No. 20162007688).

References

1. Buzsáki, G.: Rhythms of the Brain. Oxford University Press, New York (2006)
2. Wang, X.-J.: Neurophysiological and computational principles of cortical rhythms in cognition. Physiol. Rev. **90**, 1195–1268 (2010)
3. Wang, Y., Chik, D.T.W., Wang, Z.D.: Coherence resonance and noise-induced synchronization in globally coupled Hodgkin-Huxley neurons. Phys. Rev. E **61**, 740–746 (2000)
4. Lim, W., Kim, S.-Y.: Characterization of stochastic spiking coherence in coupled neurons. J. Korean Phys. Soc. **51**, 1427–1431 (2007)
5. Lim, W., Kim, S.-Y.: Statistical-mechanical measure of stochastic spiking coherence in a population of inhibitory subthreshold neuron. J. Comput. Neurosci. **31**, 667–677 (2011)
6. Kim, S.-Y., Lim, W.: Coupling-induced population synchronization in an excitatory population of subthreshold Izhikevich neurons. Cogn. Neurodyn. **7**, 495–503 (2013)
7. Abbott, L.F., Nelson, S.B.: Synaptic plasticity: taming the beast. Nat. Neurosci. **3**, 1178–1183 (2000)
8. Bi, G.-Q., Poo, M.-M.: Synaptic modifications in cultured hippocampal neurons: dependence on spike timing, synaptic strength, and postsynaptic cell type. J. Neurosci. **18**, 10464–10472 (1998)
9. Song, S., Miller, K.D., Abbott, L.F.: Competitive Hebbian learning through spike-timing-dependent plasticity synaptic plasticity. Nat. Neurosci. **3**, 919–926 (2000)
10. Bi, G.-Q., Poo, M.-M.: Synaptic modification by correlated activity: Hebbs postulate revisited. Annu. Rev. Neurosci. **24**, 139–166 (2001)
11. Watts, D.J., Strogatz, S.H.: Collective dynamics of 'Small-World' networks. Nature **393**, 440–442 (1998)

12. Izhikevich, E.M.: Simple model of spiking neurons. IEEE Trans. Neural Netw. **14**, 1569–1572 (2003)
13. Izhikevich, E.M.: Which model to use for cortical spiking neurons? IEEE Trans. Neural Netw. **15**, 1063–1070 (2004)
14. Kim, S.-Y., Lim, W.: Effect of small-world connectivity on fast sparsely synchronized cortical rhythms. Phys. A **421**, 109–123 (2015)
15. Brunel, N., Wang, X.-J.: What determines the frequency of fast network oscillations with irregular neural discharges? I. Synaptic dynamics and excitation-inhibition balance. J. Neurophysiol. **90**, 415–430 (2003)
16. San Miguel, M., Toral, R.: Stochastic effects in physical systems. In: Martinez, J., Tiemann, R., Tirapegui, E. (eds.) Instabilities and Nonequilibrium Structures VI, pp. 35–130. Kluwer Academic, Dordrecht (2000)
17. Kim, S.-Y., Lim, W.: Realistic thermodynamic and statistical-mechanical measures for neural synchronization. J. Neurosci. Methods **226**, 161–170 (2014)

Chapter 43
Alpha Phase Is Regulated by Gamma Power in Mouse Hippocampus

Tao Zhang, Xiaxia Xu, and Zhuo Yang

Abstract In this study, the PAC_CMI algorithm was validated by simulated data. Afterward, it was used to analyze local field potentials, obtained from mouse's hippocampal DG region. Male mice were divided into two groups: enrich environment (EE, $n = 6$) and society isolation (SI, $n = 6$). Modulation index (MI) was used to detect the PAC. It shows that there is a significant PAC between alpha and gamma rhythms in the hippocampal DG region in both groups. However, the value of MI was bigger in the EE group than in SI group. The analysis of stimulation data showed that PAC_CMI worked reliably. PAC_CMI results exhibited that gamma rhythm directionally drove alpha rhythm in both mice groups. Additionally, the strength of directional driving was considerably higher in the EE group than that in the SI group. The data suggest that such a directional driving is associated with certain cognitive functions.

Keywords Gamma · Alpha · Directional phase-amplitude coupling · Conditional mutual information

43.1 Introduction

The cognitive function of hippocampal DG, which has been indicated to take charge of the spatial information, is closely connected with rhythmic neural activity. The neural oscillation at both theta and gamma rhythms in the hippocampus is often focused on investigated by researchers [1]. On the other hand, a previous study showed that there also existed *alpha rhythm* (8–13 Hz) in the hippocampal DG [2]. Although alpha rhythm has been reported to play an important role in

T. Zhang (✉) · X. Xu
College of Life Sciences, Nankai University, Tianjin, People's Republic of China
e-mail: zhangtao@nankai.edu.cn

Z. Yang
College of Medicine, Nankai University, Tianjin, People's Republic of China

J. M. Delgado-García et al. (eds.), *Advances in Cognitive Neurodynamics (VI)*,
Advances in Cognitive Neurodynamics,
https://doi.org/10.1007/978-981-10-8854-4_43

cognitive function through inhibiting task-irrelevant brain areas, a number of studies link alpha rhythm directly to cognitive functions. These cognitions are possibly associated with alpha-gamma cross-frequency phase-amplitude coupling (PAC) [3]. Furthermore, it is well known that the neurons in hippocampal DG area are granule neurons. In brain regions with granule neurons, such as the striatum and the visual cortex, there existed prominent PAC between alpha and gamma rhythms. Therefore, we suppose that there is alpha-gamma PAC in mouse's hippocampal DG area.

It is well known that there are several analytic algorithms to be used to measure PAC, such as modulation index (MI) [4] and phase locking value (PAC_PLV). However, given that there is no time-delay element in these algorithms, they are not involved in assessing the direction of interaction between different frequencies. That is to say, there might be no approaches, by which the directional interaction between alpha and gamma can be effectively measured. In other words, based upon these analytic algorithms, it is impossible to determine whether alpha phase drives gamma amplitude or reversely. Recently, a study reported a directional coupling between alpha and gamma in ECOG data in cortex, which was measured by phase-slope index (PSI) [5]. In our previous studies, the conditional mutual information (CMI) was successfully applied in assessing the directionality and strength of coupling between different rhythms [6]. However, CMI is well known to be applied to measure the directional coupling between two different brain regions at the same rhythm. It is possible that there is an alteration of cognitive function associated with the change of directional coupling between alpha and gamma rhythms. Therefore, the measurement of directional coupling could be applied in assessing cognitive deficits in animal models, such as society isolation animal model (SI).

In the present study, local field potential (LFP) signals, obtained from the hippocampus of mice, were recorded by in vivo electrophysiological techniques. PAC was evaluated by MI to ensure that there was a robust alpha phase-gamma amplitude coupling. Finally, the PAC_CMI was applied in LFP data, and the comparison was performed between two animal groups, which were the enrich environment (EE) and the society isolation (SI).

43.2 Materials and Methods

Twelve male C57 mice were randomly divided in two groups. One group, in which the mice were *raised in* an *enriched environment, was named* EE ($n = 6$). Another, in which the mice were *raised in* a society isolation, was called SI ($n = 6$). The animals in the EE group were housed together in one big cage ($60 \times 40 \times 35$ cm) containing several toys. The mice in the SI group were caged individually in standard cages ($36 \times 18 \times 14$ cm) and stayed 24 h in dark environment. All the animals in both groups were raised at constant temperature (21–24 °C) with free access to food and water.

After 2 months, an electrophysiological experiment was performed to obtain the LFP data. Briefly, animals were anesthetized by 30% urethane and fixed in a stereotaxic frame. A stainless steel electrode was implanted into the hippocampal DG (0.5 mm posterior to the bregma, 3.2 mm lateral to midline, 1.5–2.0 mm ventral below the dura). Ground and reference electrodes were placed symmetrically over the skull. LFP signals in the hippocampal DG region were collected at a sampling frequency of 1 K Hz for 5 min.

Conditional mutual information (CMI) has also been generally used to evaluate the directional coupling between two brain regions at an identical rhythm. In our study, the CMI was applied in assessing the directional PAC between two frequency bands at identical site in the hippocampal DG area. Here, the method was abbreviated to PAC_CMI.

A_{fam} represents the amplitude of the filtered high-frequency signal. ϕ_{amp} and ϕ_{fph} signify the phase of A_{fam} and the phase of the filtered low-frequency signal, respectively. To estimate the information that theτ-future of the process ϕ_{amp} contained within the process ϕ_{fph}, $I(\phi_{fph}; \Delta_\tau \phi_{amp} | \phi_{amp})$ was calculated. Then, ϕ_{fph} to ϕ_{amp} directional coupling is defined as:

$$I\left(\phi_{fph}; \Delta_\tau \phi_{amp_r} | \phi_{amp_r}\right) = H\left(\phi_{fph} | \phi_{amp}\right) + H\left(\Delta_\tau \phi_{amp} | \phi_{amp}\right) \quad H\left(\phi_{fph}, \Delta_\tau \phi_{amp} | \phi_{amp}\right) \tag{43.1}$$

With the phase increments $\Delta_\tau \phi_{amp} = \phi_{amp}(t + \tau) - \phi_{amp}(t)$.

Similarly, the directional coupling from ϕ_{amp} to ϕ_{fph} could be calculated in the symmetrical form:

$$I\left(\phi_{amp_r}; \Delta_\tau \phi_{fph} | \phi_{fph}\right) = H\left(\phi_{amp_r} | \phi_{fph}\right) + H\left(\Delta_\tau \phi_{fph} | \phi_{fph}\right) - H\left(\phi_{amp_r}, \Delta_\tau \phi_{fph} | \phi_{fph}\right) \tag{43.2}$$

Finally, the directional index is defined as:

$$D = \frac{I\left(\phi_{fph}; \Delta_\tau \phi_{amp_r} | \phi_{amp_r}\right) - I\left(\phi_{amp_r}; \Delta_\tau \phi_{fph} | \phi_{fph}\right)}{I\left(\phi_{amp_r}; \Delta_\tau \phi_{fph} | \phi_{fph}\right) + I\left(\phi_{fph}; \Delta_\tau \phi_{amp_r} | \phi_{amp_r}\right)} \tag{43.3}$$

In this study, a sliding window (length = 24 s) with 50% overlap and $\tau = 100$ ms was used. Both stimulated and experimental LFP data were analyzed using MATLAB R2011a (MathWorks). SPSS software was used to perform the statistical analyses. One-way ANOVA was applied to compare the mean MI and PAC_CMI values between the EE group and the SI group with the significant level setting at 0.05. All the figures were plotted in Origin 8.5.

43.3 Results

Power spectrum was analyzed by multi-taper spectral estimation. A sliding window with a length of 20 s with 50% overlap was applied. Figure 43.1a displayed an example of the time-frequency power spectrum in 1–8 Hz with the original traces of LFPs in one EE mouse and another SI mouse, respectively. Observable stripes at about 4 Hz can be seen, suggesting there is a relatively stable brain state. Figure 43.1b showed that there were corresponding power spectra in [1–100] Hz. Two peaks in theta band (3–5 Hz) and alpha band (about 8–10 Hz, indicated by the arrow) embedded in the 1/f power spectrum were quite visible.

MI method was used to measure the PAC between the low-frequency bands (1–20 Hz, step = 1 Hz) and the gamma frequency bands (30–100 Hz, step = 1 Hz) in the hippocampal DG region. Examples of MI results in one EE mouse and one SI mouse were showed in Fig. 43.2a. There is an observable PAC between alpha (8–15 Hz) and high-gamma (70–100 Hz) frequency bands in either an EE mouse or a SI animal. One-way ANOVA showed that the strength of alpha-gamma PAC in the EE group was significantly higher than that in the SI group (1.91 ± 0.31 vs. 0.60 ± 0.19, $p < 0.01$, Fig. 43.2b).

Furthermore, Fig. 43.2c showed an example of alpha (8–15 Hz, the red line) nesting gamma (70–80 Hz, the blue line) rhythms. There were combined alpha phase-gamma frequency bands (Fig. 43.2d), revealing that the gamma rhythm was nested in the alpha cycle. Furthermore, it was found that gamma oscillation was mainly nested in the ascending branch of the alpha circle (Fig. 43.2c–d).

Figure 43.3a is an example, in which a clear PAC phenomenon has been detected by PAC_CMI. In both EE and SI mice, $D_{alpha->gamma} > 0$ occurred at the low-gamma frequency band (about [30–50] Hz). Interestingly, $D_{gamma->alpha} < 0$ happened at the high-gamma frequency band (about [50–100] Hz). Figure 43.3b–c showed the statistical PAC_CMI results in both groups. The significant difference was detected by one-way ANOVA. Figure 43.3b showed that there were no

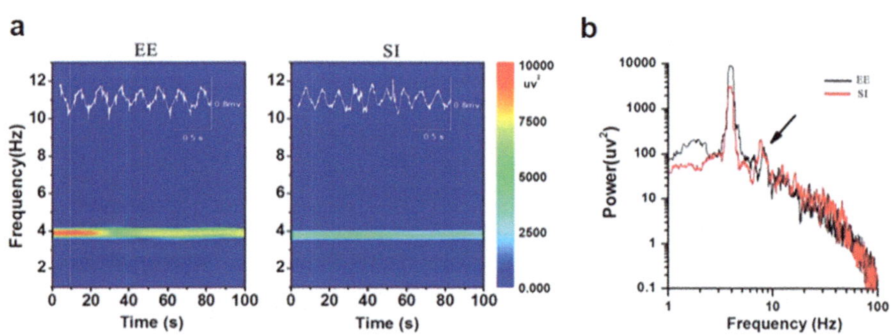

Fig. 43.1 (**a**) Examples of original LFP traces and the time-frequency power spectrum in 1–8 Hz in an EE mouse and another SI mouse. (**b**) The corresponding log-transformed power in an EE mouse and another SI mouse. The arrow indicates the alpha oscillation

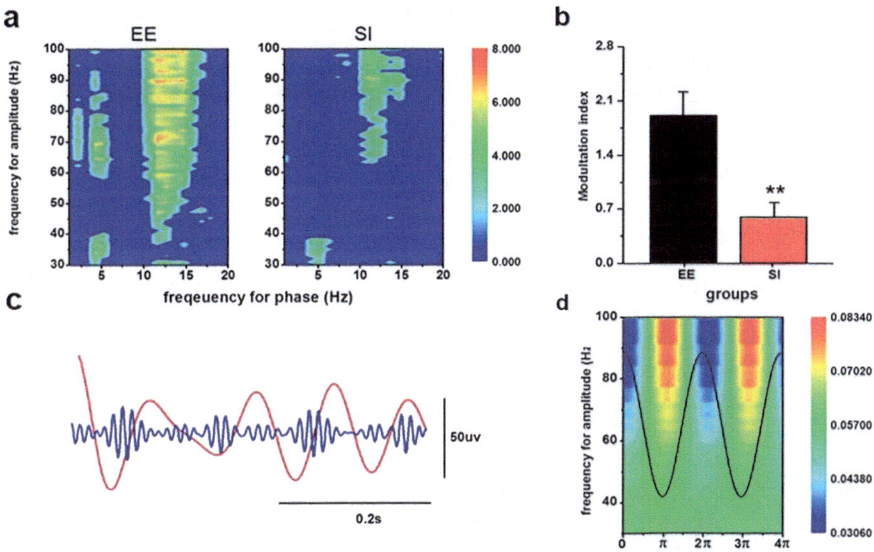

Fig. 43.2 Coupling between alpha and gamma. (**a**) Reprehensive examples of cross-frequency PAC. The highlighted area shows that there is a strong PAC between alpha rhythm and the gamma rhythm. The higher MI value, the stronger PAC. (**b**) Statistical PAC_MI results. $^{**}p < 0.01$ between EE and SI. (**c**) An example of the gamma rhythms is coupled with the ascending branch of the filtered alpha. (**d**) Alpha phase-gamma frequency bands

significant differences of the strength of alpha driving low-gamma rhythm between both groups. However, there were significant differences of the strength of high-gamma rhythm driving alpha rhythm between these two groups (0.47 ± 0.11 vs. 0.17 ± 0.02, $p < 0.01$, Fig. 43.3c).

43.4 Discussion

Our data showed that gamma rhythm could significantly drive alpha rhythm in the hippocampal DG area. Although this result may conflict to our general knowledge that alpha rhythm should control the gamma rhythms, there exist several facts that probably explain our results. (1) From the point of view of the rhythm origination, alpha rhythm is reported to originate from the superficial layers that exercise control over granular layers in the alpha band. Gamma rhythm was found to be produced in granular layer that control superficial layers also in the alpha band. Therefore, the direction between alpha and gamma rhythms might be layer specific [5]. In our study, the electrode was located in granular layer of the hippocampal DG. The explanation could be verified by laminar LFP recordings in the hippocampal DG region. (2) From the point of view of the cognitive function related to alpha rhythm,

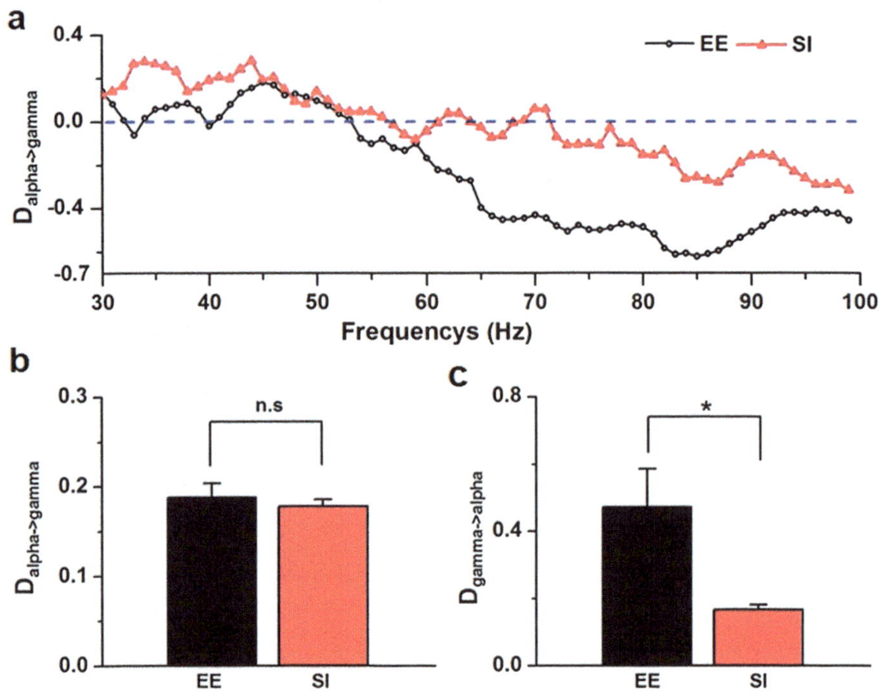

Fig. 43.3 Directional PAC in hippocampal DG region. (**a**) Example of the directional PAC between gamma frequency bands (30–100 Hz, step = 1 Hz) and the alpha rhythm (8–13 Hz). (**b**–**c**) Statistical results of the directional PAC between alpha-low-gamma rhythm (30–50 Hz, **b**) and alpha-high-gamma rhythm (50–100 Hz, **c**). $^*p < 0.05$ between EE and SI

strong alpha could usually be observed in the human scalp electroencephalogram (EEG) during relaxed wakefulness without higher cognitive load. During execution of several cognitive tasks, some task-relevant brain areas showed decreased alpha activity. Thus, alpha rhythm was believed to be an idling rhythm. As we know theta rhythm was more prominent than alpha rhythm in the hippocampus. Furthermore, previous studies showed that the hippocampal theta rhythm played an important role in spatial cognitive function. Accordingly, it may infer that in the hippocampal DG, alpha rhythm is also an idling rhythm [7]. The phenomenon that gamma directionally drives alpha feasibly reflects the active control of the cognition-relevant gamma oscillation on the cognition-irrelevant alpha oscillation. The directional coupling probably makes sure an immediate control of the cognition-irrelevant activity in the default network of hippocampal DG. Such an assumption is consistent with the fact that alpha and gamma power is anticorrelated in the cortical networks [8]. In conclusion, the strength of gamma driving alpha was significantly higher in the EE group than that in the SI group, suggesting that the directional PAC was associated with certain cognitive functions.

Acknowledgments This work was supported by grants from the National Natural Science Foundation of China (31771148 & 11232005 to TZ).

References

1. Xu, X., Liu, C., Li, Z., Tao, Z.: Effects of hydrogen sulfide on modulation of theta–gamma coupling in hippocampus in vascular dementia rats. Brain Topogr. **28**, 1–16 (2015)
2. Nerad, L., Bilkey, D.K.: Ten- to 12-Hz EEG oscillation in the rat hippocampus and rhinal cortex that is modulated by environmental familiarity. J. Neurophysiol. **93**, 1246–1254 (2005)
3. Roux, F., Wibral, M., Singer, W., Aru, J., Uhlhaas, P.J.: The phase of thalamic alpha activity modulates cortical gamma-band activity: evidence from resting-state MEG recordings. J. Neurosci. **33**, 17827–17835 (2013)
4. Tort, A.B., Komorowski, R., Eichenbaum, H., Kopell, N.: Measuring phase-amplitude coupling between neuronal oscillations of different frequencies. J. Neurophysiol. **104**, 1195–1210 (2010)
5. Jiang, H., Bahramisharif, A., van Gerven, M.A., Jensen, O.: Measuring directionality between neuronal oscillations of different frequencies. NeuroImage. **118**, 359–367 (2015)
6. Zheng, C., Zhang, T.: Synaptic plasticity-related neural oscillations on hippocampus-prefrontal cortex pathway in depression. Neuroscience. **292**, 170–180 (2015)
7. Colgin, L.L.: Mechanisms and functions of theta rhythms. Annu. Rev. Neurosci. **36**, 295–312 (2013)
8. Palva, S., Palva, J.M.: Functional roles of alpha-band phase synchronization in local and large-scale cortical networks. Front. Psychol. **2**, 204 (2011)

Chapter 44
Quantitative Analysis of Functional Connectivity Between Prefrontal Cortex and Striatum in Monkey

Zaizhi Wen, Jianhua Zhang, Xiaochuan Pan, and Rubin Wang

Abstract Reward prediction is essential for learning behavior and decision-making process in the brain. It is well known that the neurons in both prefrontal cortex (PFC) and striatum are involved in encoding reward information and the interplay between the PFC and striatum plays an important role in cognitive processes. However, it remains elusive how interaction between PFC and striatum is modulated in reward conditions. To investigate this issue, the local field potentials (LFPs) were simultaneously recorded in the lateral PFC and striatum of a male monkey while performing the sequential paired-association task with the asymmetric reward scheme. The nonlinear interdependence (NLI), a measure of the generalized synchrony characterizing both the coupling strength and the coupling direction, was used to quantify the strength of bidirectional functional connectivity between the PFC and striatum. The results suggest that PFC has strong bi-directional coupling with the striatum and the interaction in beta frequency band plays an important role in regulating monkey's behavior in reward prediction task.

Keywords Prefrontal cortex · Striatum · Reward prediction · Nonlinear interdependence · Functional connectivity · Synchronization measures

---

Z. Wen
School of Information Science and Engineering, East China University of Science and Technology, Shanghai, People's Republic of China

J. Zhang (✉)
School of Information Science and Engineering, East China University of Science and Technology, Shanghai, People's Republic of China

Institute of Cognitive Neurodynamics, East China University of Science and Technology, Shanghai, People's Republic of China
e-mail: zhangjh@ecust.edu.cn

X. Pan · R. Wang
Institute of Cognitive Neurodynamics, East China University of Science and Technology, Shanghai, China

J. M. Delgado-García et al. (eds.), *Advances in Cognitive Neurodynamics (VI)*,
Advances in Cognitive Neurodynamics,
https://doi.org/10.1007/978-981-10-8854-4_44

44.1 Introduction

The PFC and striatum are two important brain regions, and anatomically, they are tightly connected [1]. The anatomical connections between these two regions suggest that they may have close relations in functions. Many studies have reported that PFC and striatum are involved in many different cognitive functions like learning, reward processing, category representation, and behavior controlling [2, 3]. Although individual functions of the PFC and striatum in learning and behavior controlling processes have been demonstrated quite well [4], their interactive functions remain elusive. Some studies showed that the dysfunctional synchronization between the two regions was related to many neurological and psychiatric disorders, such as autism, depression, and schizophrenia [5]. A recent study demonstrated the synchronization of LFPs between these two areas significantly increased in beta band after monkeys had learned a classification task, which suggested that synchronized oscillations between the PFC and striatum played a vital role in category learning [6].

In this study, to investigate the role of interaction between the two areas in reward processing, LFPs were simultaneously recorded in the PFC and striatum while a monkey was performing the sequential paired-association task with the asymmetric reward schedule [7]. We first filtered the recorded LFPs to obtain the signals in three frequency bands: beta (15–30 Hz), low gamma (30–50 Hz), and high gamma (50–100 Hz). For the LFPs in each frequency band, we evaluated the synchronization between the PFC and striatum by computing the nonlinear interdependence (NLI). We found that the synchronization in small reward trials was significantly higher than that in large reward trials in beta band. As NLI can not only measure the coupling strength but also indicate the coupling direction, we found that the functional connectivity was significantly greater from the PFC to the striatum than that from the striatum to the PFC. The results suggest that PFC has strong interaction with the striatum and the interaction in beta band may play an important role in regulating monkey's behavior in reward prediction process.

44.2 Materials and Methods

44.2.1 Behavioral Task

One male Japanese monkey served as a subject in this study (Tom, 8.5 kg). A detailed description of the experimental procedure and behavioral task can be found in [7]. Briefly, the monkey was first trained to learn two associative sequences (Fig. 44.1a) in a sequential paired-association trial (SPAT) (Fig. 44.1b). After that, an asymmetric reward schedule was introduced using reward instruction trials (RITs) (Fig. 44.1c). RITs and SPATs were arranged in one block, first RITs (three trials) and then followed by SPATs. In a given block, a correct choice of A1 $\rightarrow$ B1 $\rightarrow$ C1 would enable the subject monkey to get a large reward, while A2 $\rightarrow$ B2 $\rightarrow$ C2 would be associated with the small one. The stimulus-reward

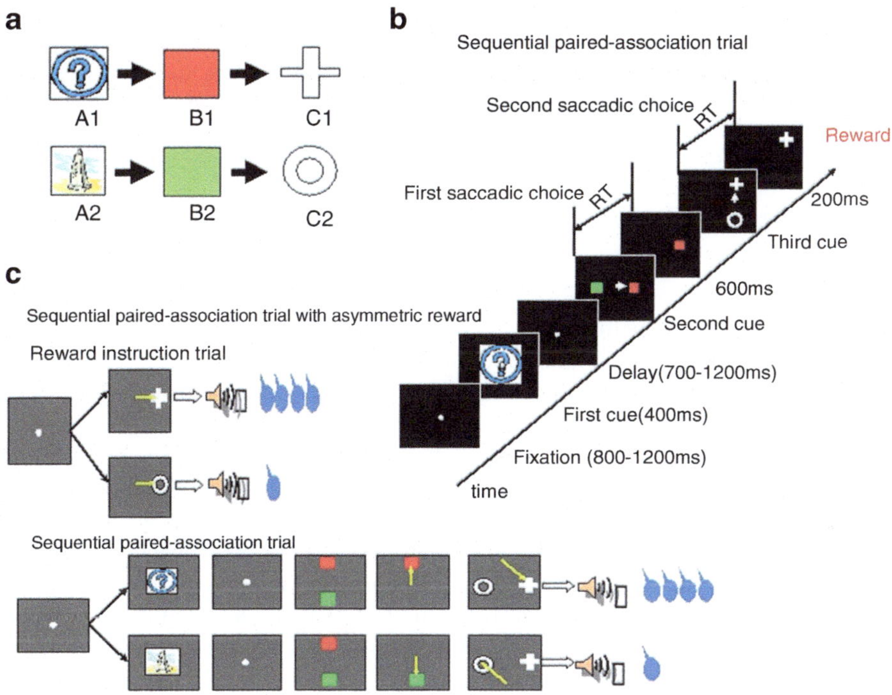

Fig. 44.1 The sequential paired-association task with an asymmetric reward scheme: (**a**) sequential paired-association tasks; (**b**) the experimental paradigm of a trial; (**c**) the asymmetric reward scheme

contingency was pseudo-randomized between blocks. A trial in which the monkey made two correct choices, whether in large or small reward condition, was considered a correct trial. The LFPs recorded in the correct trials were selected for further analysis. All surgical and experimental protocols were approved by the Animal Care and Use Committees in Tamagawa University and conducted in accordance with the National Institutes of Health's Guide for Care and Use of Laboratory Animals.

44.2.2 Data Acquisition

Extracellular recordings were conducted using linear-array multi-contact electrodes (U-probe, Plexon, USA) to obtain LFPs. In our experiment, each electrode contained eight recording contacts (impedance, 0.3–0.5 MΩ at 1 kHz) with an inter-contact spacing of 150 or 300 μm. Neuronal activity was measured against a local reference that was close to the electrode contacts (a stainless guide tube or the tube of U-probe). We performed data amplification, filtering, and acquisition with a Multichannel Acquisition Processor (Plexon, USA). The acquired signal from each contact (channel) was passed through a head-stage and then split to extract the spike and the LFP components separately. The extracted LFP signals were then filtered with a passband of 0.7–170 Hz, further amplified, digitized at 1 kHz, and saved in Plexon files. In each recording session, these two U-probe electrodes were inserted simultaneously into the PFC and striatum. Once the two electrodes reached the target positions, we did not move them any more throughout the whole session. However, for different sessions, the positions of these two electrodes were different.

44.2.3 Data Analysis Method

We analyzed the recorded LFPs off-line using custom-made MATLAB programs on a PC. As described before, we recorded the choices of the monkey in SPATs in large and small reward trials and used only the LFPs recorded in correct trials for later analysis. In total, 50 sessions of LFPs data were recorded from the monkey. For each session, the number of correct trials varies between 94 and 119, while large and small reward trials have roughly equivalent number. In this study, we concentrated on neural activity in the cue period (400 ms after the first cue onset) and the early delay period (400 ms after the first cue offset).

Synchrony is thought to play an important role in establishing functional circuitry and an essential tool used to describe neurophysiological mechanisms of communication between brain regions [8]. In this study, we adopted the synchrony measure named nonlinear interdependence (NLI) to analyze the LFPs. We first filtered the LFPs to obtain the signals in the beta (15–29 Hz) and low (30–49 Hz) and high (50–100 Hz) gamma bands, respectively. Then, for each frequency band, we adopted the same process described below.

As described before, eight channels of LFP data from the PFC and striatum were recorded simultaneously. Therefore, there were 64 pairs of channels in total (one from PFC and the other from striatum). For each pair of channels, The NLI was calculated in every big and small reward trials from every session separately. Then NLIs could be averaged across trials and sessions that had the same reward condition. However, some studies have reported that the increase of power and oscillation of signals could have an enhancement to the calculated synchrony measure value and this increase of NLI may not be caused by the inner synchronization of the two regions [6]. Considering this, for each trial, another trial in the same reward condition was randomly selected, and the NLI was calculated based on these two shuffled trials which now had no direct relation. This procedure was performed for each pair of channels, and then the NLIs in the same reward condition were averaged to obtain the $NLI_{shuffle}$. The actual NLIs used in our analysis were bias corrected by subtracting this shuffled NLI from the original value.

44.2.4 Nonlinear Interdependence Metric

Nonlinear interdependence (NLI) is measure of generalized synchrony which evaluates the interdependency according to the distance of delay vectors of two time series and is widely used in nonlinear systems [9, 10]. Unlike many other synchrony measures, such as phase-locking value and mutual information, NLI can not only compute the coupling strength but also indicate the coupling direction [11]. It was also reported that NLI is more robust to noises than most other measures [12]. These advantages make NLI particularly suitable for the analysis of neural signals. A detailed description of the algorithm for NLI can be found in [12].

44.3 LFP Data Analysis Results

We recorded LFPs simultaneously in the PFC and striatum using two U-probe electrodes while the monkey was performing the sequential paired-association task with the asymmetric reward schedule. A synchrony measure named nonlinear interdependence was used to analyze the functional connectivity between striatum and PFC in different reward conditions and different task periods. Figures 44.2, 44.3, and 44.4 show the NLIs for 64 pairs of channels in beta, low gamma, and high gamma, respectively.

As NLI is an asymmetry synchrony measure, for each frequency band, NLIs of both directions were calculated. As shown in these figures, in the cue period, only the NLIs from striatum to PFC direction in beta band differed significantly in different reward condition (ANOVA F-test: $F = 7.65, p = 0.01$). No significant

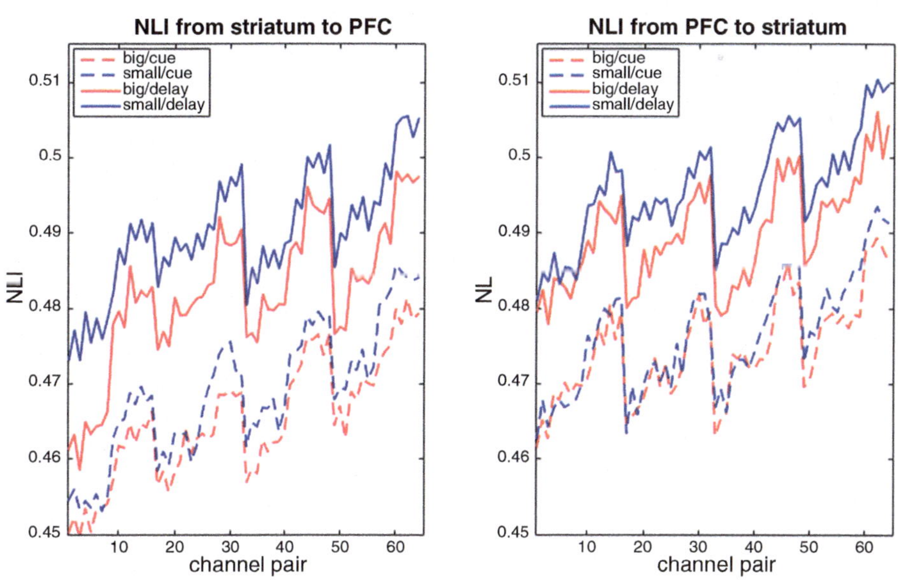

Fig. 44.2 Nonlinear interdependence (NLI) between striatum and PFC in beta frequency band (15–30 Hz)

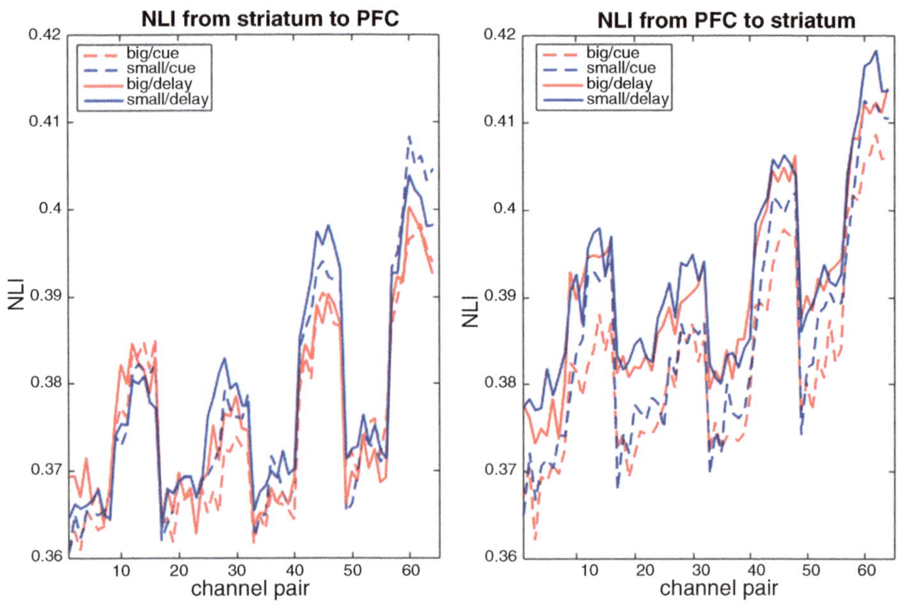

Fig. 44.3 Nonlinear interdependence (NLI) between striatum and PFC in lower gamma frequency band (30–50 Hz)

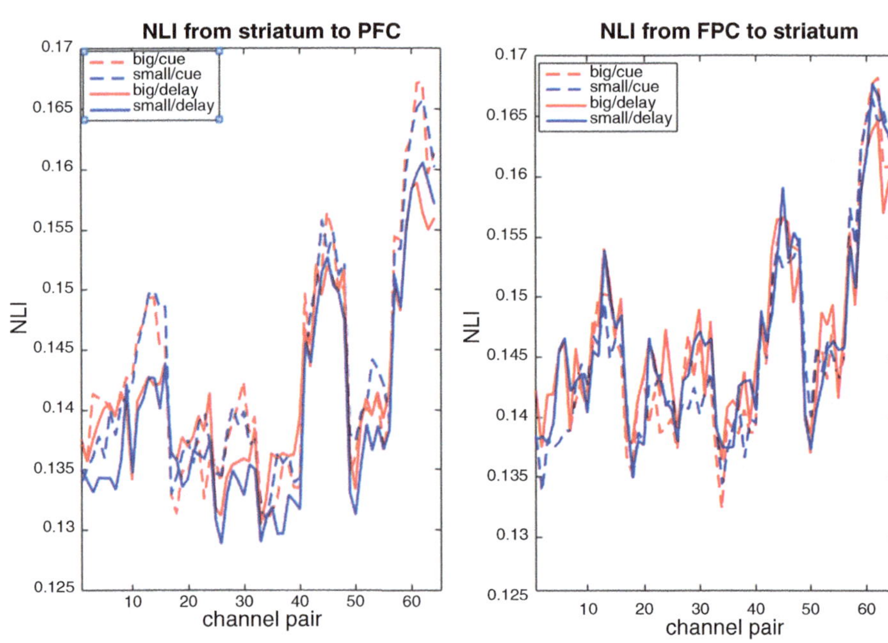

Fig. 44.4 Nonlinear interdependence (NLI) between striatum and PFC in higher gamma frequency band (50–100 Hz)

difference was found in either direction for low and high gamma band. As for the delay period, in beta band, the NLIs in small reward conditions were significantly higher than that in big reward conditions, for both striatum to the PFC (ANOVA F-test: $F = 28.02$, $p = 5.06 \times 10^{-7}$) and the opposite direction (ANOVA F-test: $F = 20.04$, $p = 1.66 \times 10^{-05}$). In low and high gamma band, significant difference was found in neither direction (ANOVA F-test in low gamma, striatum to PFC, $F = 1.17$, $p = 0.28$; PFC to striatum, $F = 0.46$, $p = 0.50$; in high gamma, striatum to PFC, $F = 1.94$, $p = 0.17$; PFC to striatum, $F = 0.15$, $p = 0.70$). Our results also showed that the NLIs from PFC to striatum were significantly higher than that from striatum to PFC, no matter in what reward condition and frequency band.

44.4 Conclusion

In this study, we recorded LFPs simultaneously in the PFC and striatum using two U-probe electrodes while the monkey was performing the sequential paired-association task with the asymmetric reward schedule. By analyzing the NLIs between striatum and PFC in different reward conditions, we found that the functional connectivity strength between these two brain regions was significantly higher in small reward conditions than that in big ones, and this difference was only found in beta band. These results implied that the interaction between the PFC and the striatum might play an important role in processing reward information and this interaction between these two brain regions was mainly related to beta band LFPs.

Moreover, our results suggested that there was significantly greater functional connectivity from the PFC to the striatum than that of opposite direction, in all the three frequency bands (beta, low, and high gamma). This result was consistent with the fact that PFC neurons have direct projections to striatal neurons, while striatal neurons do not project to LPFC neurons directly and might suggest that PFC neurons could somehow regulate the striatum neuron's activity.

Acknowledgments This work was supported in part by the National Natural Science Foundation of China under Grant No. 61075070 and Key Grant No. 11232005.

References

1. Alexander, G., DeLong, M., Strick, P.: Parallel organization of functionally segregated circuits linking basal ganglia and cortex. Annu. Rev. Neurosci. **9**, 357–381 (1986)
2. Asaad, W., Eskandard, E.: Encoding of both positive and negative reward prediction errors by neurons of the primate lateral prefrontal cortex and caudate nucleus. J. Neurosci. **31**, 17772–17787 (2011)
3. Deserno, L., Huys, Q., Boehme, R., Buchert, R., Heinze, H.: Ventral striatal dopamine reflects behavioral and neural signatures of model- based control during sequential decision making. Proc. Natl. Acad. Sci. U. S. A. **112**, 1595–1600 (2015)
4. Hikosaka, O., Isoda, M.: Switching from automatic to controlled behavior: cortico-basal ganglia mechanisms. Trends Cogn. Sci. **14**, 154–161 (2010)

5. Yoon, J.H., Minzenberg, M.J., Raouf, S., D'Esposito, M., Carter, C.S.: Impaired prefrontal-basal ganglia functional connectivity and substantia nigra hyperactivity in schizophrenia. Biol. Psychiatry. **74**, 122–129 (2013)
6. Antzoulatos, E., Miller, E.: Increases in functional connectivity between prefrontal cortex and striatum during category learning. Neuron. **83**, 216–225 (2014)
7. Pan, X., Sawa, K., Tsuda, I., Tsukada, M., Sakagami, M.: Reward prediction based on stimulus categorization in primate lateral prefrontal cortex. Nat. Neurosci. **11**, 703–712 (2008)
8. Miller, E.K., Buschman, T.J.: Cortical circuits for the control of attention. Curr. Opin. Neurobiol. **23**, 216–222 (2013)
9. Liang, Z., Ren, Y., Yan, J.: A comparison of different synchronization measures in electroencephalogram during propofol anesthesia. J. Clin. Monit. Comput. **30**, 451–466 (2016)
10. Quian-Quiroga, R., Kraskov, A., Kreuz, T., Grassberger, P.: Performance of different synchronization measures in real data: a case study on electroencephalographic signals. Phys. Rev. **65**, 1–14 (2002)
11. Chen, D., Li, X., Cui, D., Wang, L., Lu, D.: Global synchronization measurement of multivariate neural signals with massively parallel nonlinear interdependence analysis. IEEE Trans. Neural Syst. Rehabil. Eng. **22**, 33–43 (2014)
12. Pei-Chen, L., Chih-Hao, C.: Spatially nonlinear interdependence of alpha-oscillatory neural networks under chan meditation. Evid. Based Complement. Alternat. Med. 360371: 1–12 (2013)

Chapter 45
Spontaneous Theta Rhythm Predicts Insomnia Duration: A Resting-State EEG Study

Wenrui Zhao, Dong Gao, Faguo Yue, Yanting Wang, Dandan Mao, Tianqiang Liu, and Xu Lei

Abstract Increased theta power and subjective sleepiness during waking EEG had been found in many researches of sleep deprivation. However, rare studies had ever investigated the theta rhythm in awake and its cortical generators in insomnia disorder (ID). Consequently, based on the scalp EEG signal and its brain cortex distribution reconstructed by a network-based source imaging, we explored the abnormal theta power of insomniacs with different insomnia duration and its cortical generators. Results indicated that, compared to good sleepers, only ID with insomnia duration above 3 years presented sustained decreased theta power in multiple networks. Intriguingly, the theta power of frontoparietal (FPN) and deep structure network (DSN) was negatively correlated with the insomnia duration. These findings suggested that decreased waking theta power in ID may be the electrophysiological correlate of subjective sleepiness deficiency, and the theta power of FPN and DSN was good predictors for the insomnia duration.

Keywords Insomnia disorder · Insomnia duration · Network EEG source imaging · Theta

45.1 Introduction

Insomnia disorder (ID) has been one of the most prevalent and common psychophysiological disorders. Chronic insomnia induced cognitive impairments and increased risks for other psychiatric disorders [1]. Despite its considerable impacts

W. Zhao · T. Liu · X. Lei (✉)
Sleep and Neuroimaging Center, Faculty of Psychology, Southwest University, Chongqing, China
e-mail: xlei@swu.edu.cn

D. Gao · F. Yue · Y. Wang · D. Mao
Sleep Psychology Center, Daping Hospital, Third Military Medical University, Chongqing, China

J. M. Delgado-García et al. (eds.), *Advances in Cognitive Neurodynamics (VI)*,
Advances in Cognitive Neurodynamics,
https://doi.org/10.1007/978-981-10-8854-4_45

on health, the pathophysiological mechanism is still poorly understood, and previous results are controversial.

The increase in theta activity during waking after sleep deprivation was found to be an electroencephalogram markers of homeostatic sleep propensity and subjective sleepiness in both human and rat [2, 3], and this effect was most pronounced in frontal areas. However, it is questionable that whether ID also presents the same pattern. Existing studies were based on the topographic analysis of frontal electrodes; thus, an unsolved problem is its cortex localizations, which may reveal the core regions of dysfunctional sleep homeostasis in patients with insomnia. In addition, the prefrontal cortex and hippocampus were deemed to be the generators of theta oscillation [4, 5], but in insomnia patients the abnormality was consistently found in these areas [6], from which we can infer that insomniacs may also present abnormal theta activity and homeostatic dysregulation during wakefulness.

To achieve these aims, we proposed a method called resting-state cortex rhythms (RECOR) for a detailed localization of cortical sources of resting-state EEG rhythms in a common network parcellation of the human brain function [8]. The main aims of our study were to examine (1) the impact of insomnia disorder on the theta rhythm, and (2) the distribution of abnormal theta rhythm in brain networks, especially in the frontoparietal network (FPN).

45.2 Methods

Thirty-three patients with ID (23 females, age 42.7 ± 9.4 years) and 14 healthy good sleepers (HGS) (6 females, age 41.9 ± 11.7 years) participated in the study. Insomniacs were recruited from Department of Sleep Psychology Center, Daping Hospital, Third Military Medical University, and they were divided into two groups, among which 16 had insomnia duration below 3 years (B3) and 17 above 3 years (A3). All participants completed some questionnaires including Pittsburgh Sleep Quality Index (PSQI), Self-Rating Depression Scale (SDS), and Self-Rating Anxiety Scale (SAS). The diagnosis of the total 33 IDs was evaluated by experienced psychiatrist (author DG and FY) according to the International Classification of Sleep Disorders-3. A written informed consent was obtained after a detailed explanation of the study protocol. The study was approved by the Ethics Committee of Southwest University, and all procedures involved were in accordance with the sixth revision of the Declaration of Helsinki.

Eyes-closed resting-state EEG data was recorded about 5 min from the 64 scalp tin electrodes mounted in an elastic cap (Brain Products, Munich, Germany), with the sampling frequency of 500 Hz around 19:00 to 20:00 in the evening. The impedance of all electrodes was kept below 5 kΩ. The preprocessing was conducted using MATLAB scripts supported by EEGLAB (http://sccn.ucsd.edu/eeglab). The recorded EEG data with ocular, muscular, and other types of artifact were preliminarily identified and excluded. And then, continuous EEG data was filtered with band-pass between 0.1 and 45 Hz and referenced to common average.

RECOR (as provided at http://www.leixulab.net/recor.asp) was used to estimate the power of EEG rhythms in the eight large-scale brain networks [7, 8]. It included two steps to calculate the power of EEG rhythms in each brain network. Firstly, network-based source imaging (NESOI) was employed to estimate the cortical sources of EEG theta rhythm (4–8 Hz) [8]. The second step is averaging the solutions of NESOI across all vertices of the eight large-scale brain networks: visual network (VIS), somatomotor network (SOM), dorsal attention network (DAN), ventral attention network (VAN), limbic network (LIM), FPN, default mode network (DMN), and deep brain structure network (DSN).

Statistical analysis was performed by ANOVA, using RECOR solutions of the theta band as dependent variables. ANOVA had one between-participants factor of Group (B3, A3, and HGS) and one within-participants factor of Network, forming a 3×8 mixture design.

45.3 Results

Significant differences were found in PSQI (14.1 ± 2.4, 15.4 ± 2.8, 3.9 ± 1.6, respectively, $F_{[2,44]} = 95.67$, $p < 0.001$), SAS (55.5 ± 11.9, 52.8 ± 12.5, 31.3 ± 4.3, $F_{[2,44]} = 23.21$, respectively, $p < 0.001$), and SDS (59.5 ± 11.9, 55.5 ± 13.6, 34.6 ± 6.6, respectively, $F_{[2,44]} = 20.57$, $p < 0.001$) among the groups of B3, A3, and HGS.

The power of theta rhythm in HGS, B3, and A3 groups was illustrated in Fig. 45.1. Obviously, the theta power in scalp decreased gradually as the disease duration increased (Fig. 45.1a).

We performed a one-way ANOVA among the factor group (HGS, B3, and A3), and the result indicated a significant effect ($F_{[2,44]} = 3.819$; $p < 0.05$) (Fig. 45.1b). The post hoc testing found a significant lower theta power when A3 (0.914 ± 0.395) was compared with HGS (1.731 ± 1.294, $p < 0.05$) and B3 (1.421 ± 0.655, $p < 0.05$). There is no significant difference between HGS and B3 ($p = 0.43$).

Then we averaged all vertices of a given large-scale brain network; this step may minimize the effects of poor spatial sampling of scalp EEG. We performed an ANOVA between the factors Groups and Networks.

Figure 45.2 illustrated the cortex current density distribution in 3 groups of 8 networks, which evidenced a marginally significant main effect of the factor Group ($F_{[2,44]} = 2.814$; $p = 0.071$, $\eta P2 = 0.113$ with 95% confidence intervals between 0 and 0.4504) and significant main effect of Network ($F_{[1.247,54.871]} = 4.841$; $p < 0.05$, $\eta P2 = 0.099$ with 95% confidence intervals between 0 and 0.5345). Interaction effect between Group and Network was not significant.

The post hoc testing showed that, compared to HGS (2.072 ± 0.447), both A3 (1.035 ± 0.089) and B3 (1.195 ± 0.069) groups presented lower power of theta band in DAN ($t = -2.49$, $p = 0.019$ for A3 compared to HGS and $t = -2.069$, $p = 0.048$ for B3 compared to HGS). When compared to A3 in FPN of theta band (1.003 ± 0.069), both the group of HGS (1.537 ± 0.190) and B3 (1.255 ± 0.072)

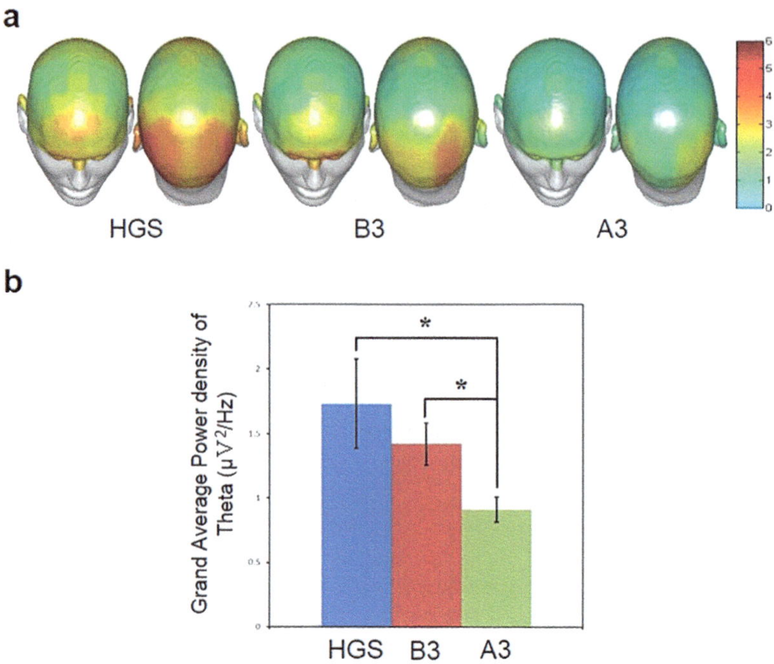

Fig. 45.1 The topography (**a**) and grand average of theta power (**b**) in HGS, B3, and A3 groups. Notice A3 group has significantly lower theta power than HGS. (*) indicate significant difference at $p < 0.05$

presented significant lower power ($t = -2.828$, $p = 0.008$ for A3 compared to HGS and $t = 2.507$, $p = 0.018$ for B3 compared to A3).

When compared to A3 in DSN of theta band (0.935 ± 0.050), both the group of HGS (1.252 ± 0.111) and B3 (1.170 ± 0.065) presented significant lower power ($t = -2.75$, $p = 0.010$ for A3 compared to HGS and $t = 2.890$, $p = 0.007$ for B3 compared to A3). Finally, only HGS (1.412 ± 0.128) and A3 (1.065 ± 0.093) groups showed significant difference in the theta power of SOM ($t = -2.236$, $p = 0.033$).

Pearson partial correlations between the theta power of eight networks and insomnia duration were performed, with age, gender, BMI, and education used as covariates. Results indicated that the insomnia duration negatively correlated with the theta power of FPN ($R = -0.507$, $p = 0.001$) and DSN ($R = -0.406$, $p = 0.007$).

45.4 Discussions

Up to now, the theta rhythm during wakefulness had not been comprehensively researched. Our study not only verified a previous study of decreased theta power during waking EEG [9] but also found that the significant decrease only existed

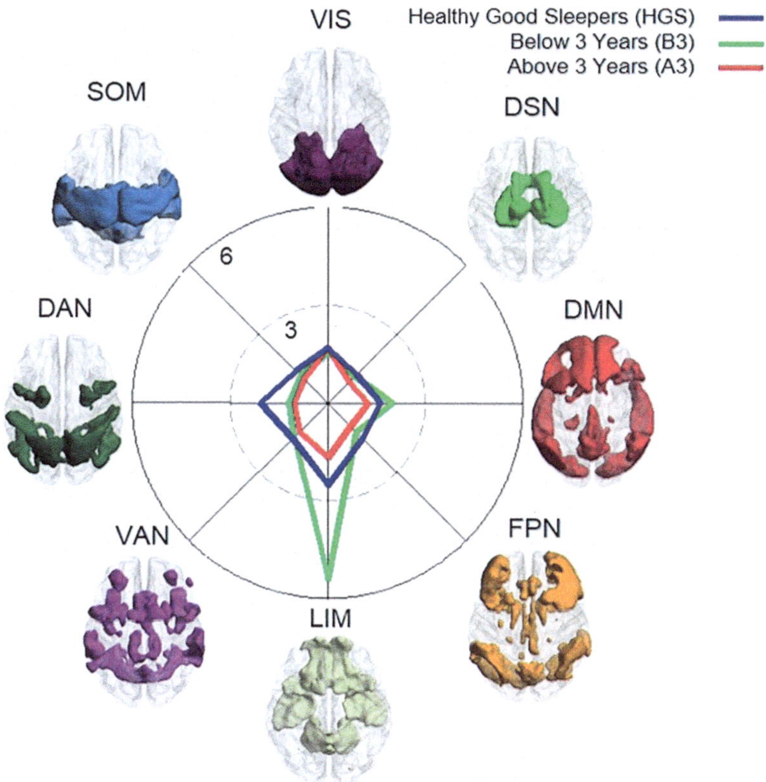

Fig. 45.2 Network normalized electroencephalographic (EEG) spectral power density distribution in eight Networks of theta band. A statistical ANOVA interaction was performed among the factors Groups (HGS, B3, and A3) and Networks (VIS, SOM, DAN, VAN, LIM, FPN, DMN, and DSN)

in the patients with longer insomnia duration. Indirect evidences from sleep deprivation demonstrated that theta activity in waking was seen as a marker of homeostatic sleep propensity [2]. Besides, the prefrontal cortex and hippocampus were deemed to be the generators of theta oscillation [4, 5], but insomnia was characterized by abnormal morphometry in the frontal cortex and hippocampus [6]. Accordingly, the decreased theta power in A3 group may be the reflection of decreased homeostatic sleep propensity or subjective sleepiness deficiency in patients with insomnia and correlate with their impaired brain structures. Obviously, insomnia duration was a key factor affecting the patients' theta power in waking and leading this homeostatic dysregulation.

What are the corresponding brain networks of the decreased theta power in ID? To answer this question, cortical sources of theta rhythm were compared among groups of B3, A3, and HGS. Our results revealed that the decrease was not widely distributed throughout the brain but concentrated on the networks of SOM, DAN,

FPN, and DSN. Additionally, we also found that the decrease of theta power in FPN and DSN negatively correlated with insomnia duration. It suggested that insomnia duration could be predicted by spontaneous theta activity in FPN and DSN.

In conclusion, by using the network-based source imaging of resting-state EEG theta rhythm, this study characterized insomniacs with different insomnia duration. Results indicated that the decreased theta power didn't exist in all insomniacs; only the chronic insomniacs with longer insomnia duration presented the most obvious decrease. The insomnia duration-related decreased theta power may be the reflection of decreased sleep propensity and abnormal brain morphometry in the frontal cortex and hippocampus. In addition, the networks of FPN, DAN, SOM, and DSN were the cortical sources of decreased theta power. However, only FPN and DSN in theta band presented negative correlations with insomnia duration, which may correlate with the atrophy of core regions in two networks and represent gradually dysfunctional cognitive control and disrupted sleep homeostasis in patients with different insomnia duration. These results motivated future researchers and clinicians to regard the insomnia duration as a highly pathogenic factor, and source localization technique of RECOR also may be an effective and convenient method to explore the neurodegenerative characteristics in insomnia disorder.

Acknowledgments This research was supported by grants from the National Nature Science Foundation of China (31571111) and Fundamental Research Funds for the Central Universities (SWU1609109).

References

1. Fortier-Brochu, E., Beaulieu-Bonneau, S., Ivers, H., Morin, C.M.: Insomnia and daytime cognitive performance: a meta-analysis. Sleep Med. Rev. **16**, 83–94 (2012)
2. Finelli, L.A., Baumann, H., Borbély, A.A., Achermann, P.: Dual electroencephalogram markers of human sleep homeostasis: correlation between theta activity in waking and slow-wave activity in sleep. Neuroscience. **101**, 523–529 (2000)
3. Vyazovskiy, V.V., Tobler, I.: Theta activity in the waking EEG is a marker of sleep propensity in the rat. Brain Res. **1050**, 64–71 (2005)
4. Buzsáki, G.: Theta oscillations in the hippocampus. Neuron. **33**, 325–340 (2002)
5. Ishii, R., Shinosaki, K., Ukai, S., et al.: Medial prefrontal cortex generates frontal midline theta rhythm. Neuroreport. **10**, 675–679 (1999)
6. Spiegelhalder, K., Regen, W., Baglioni, C., Riemann, D., Winkelman, J.W.: Neuroimaging studies in insomnia. Curr. Psychiatry Rep. **15**, 1–6 (2013)
7. Lei, X.: Electromagnetic brain imaging based on standardized resting-state networks. In: Paper Presented at: 5th International Conference on Biomedical Engineering and Informatics (BMEI), Chongqing, China, pp. 40–44 (2012)
8. Lei, X., Xu, P., Luo, C., Zhao, J., Zhou, D., Yao, D.: fMRI functional networks for EEG source imaging. Hum. Brain Mapp. **32**, 1141–1160 (2011)
9. Wołyńczyk-Gmaj, D., Szelenberger, W.: Waking EEG in primary insomnia. Acta Neurobiol. Exp. **71**, 387–392 (2011)

Chapter 46
Differences in Perceiving Narratives Through Screens or Reality

Miguel Ángel Martín-Pascual, Celia Andreu-Sánchez, José M. Delgado-García, and Agnès Gruart

Abstract Understanding a narrative requires a high level of attention. Today, we are used to perceive narratives not only in real world but also through screens. Here, we approach the visual perception of those narratives by the viewer's eyeblink rate in different situations: while watching videos with narratives, watching videos without narratives, listening to narratives with no video, watching videos with the same narrative but different editing styles, and looking the same narrative in real performance. Watching videos with narratives decreases eyeblink rate. Video editing style affects eyeblink rate, regardless of narratives. The type of stimulus, in which a narrative is viewed, screened, or performed, affects eyeblink rate. Media professionals show a significant lower eyeblink rate than non-media professionals while perceiving narratives.

Keywords Cinema perception · Attention · Vision · Professionalization · Screens · Reality

46.1 Perceiving Narratives Through Screens and Reality

Cinema appeared at the end of the nineteenth century. And, with it, narratives that had traditionally been played in stages started to be told through screens.

Cinema creators invented an audiovisual language to communicate contents to viewers. Some rules of this new language were taken from theatrical one; others were invented ad hoc [1]. Those new rules and patterns moved along with the story

---

M. Á. Martín-Pascual (✉) · C. Andreu-Sánchez
Neuro-Com Research Group, Universitat Autònoma de Barcelona, Barcelona, Spain
e-mail: Miguelangel.Martin@uab.cat

J. M. Delgado-García · A. Gruart
Division of Neuroscience, Pablo de Olavide University, Seville, Spain

J. M. Delgado-García et al. (eds.), *Advances in Cognitive Neurodynamics (VI)*,
Advances in Cognitive Neurodynamics,
https://doi.org/10.1007/978-981-10-8854-4_46

of cinema [2, 3] and built a mixture of editing styles due to narrative, duration, and economy of storytelling needs.

Today, screens surround us in everyday life. In the USA, adults spend an average of 8 h and 47 min per day watching screens [4]. We are used to distinguish between drama on screens and on reality, and no one calls police when seeing a murder in a movie [5]. But, do we perceive those narratives equally?

46.2 Eyeblinks, Attention, and Cinema Cuts

A way to approach this visual perception is by measuring eyeblinks. Visual perception is interrupted by eyeblinks continuously. They hide visual flow between 150–400 milliseconds each time [6–8]. Blinking has the physiological function [9, 10] of wetting and protecting the cornea. But it also has a psychological one [11], and it has been linked to attention [12] also while watching videos [13, 14].

In the 1990s, Walter Murch, a cinema editor and sound designer, raised an interesting question [15]: Can we connect eyeblinks with movie editing in cinematographic narratives? Murch, with three Oscar awards, had worked for several years in Hollywood, in films like The Godfather (1972), The Conversation (1974), or Apocalypse Now (1979), among others. He suspected that eyeblink might have a comprehension function in films. As an editor, he wanted to know whether there was a predictable and measurable blink that could let him know the best moment to cut a shot. The question was good enough to make some research to test it.

46.2.1 Eyeblink Synchronization and Narratives

In 2009, Nakano and colleagues [13] hypothesized that eyeblinks become synchronized while viewing video stories. They presented 3 different narrative-style stimuli to 14 subjects: videos with the same narrative (pieces of "Mr. Bean" British comedy), videos without narrative (background videos of landscapes or tropical fish), and narratives without video (audiobooks of "Harry Potter").

They obtained that mean eyeblink rate while watching videos with narrative was significantly lower than that during the rest state. This difference was not obtained in videos without narrative nor in narratives without video.

Also, they obtained that spontaneous blink rate was synchronized within and across subjects when they viewed the same narrative in a video (this was not found in the other two conditions). Thus, they concluded that following a narrative per se was not the cause of this eyeblink synchronization.

In their study, synchronous blinks occurred at points of less importance in the narrative, when the main character (Mr. Bean, played by Rowan Atkinson) disappeared of screen or at the conclusion of actions, among others. This proved that eyeblinks when watching narratives through screens are linked to attention (see Table 46.1).

Table 46.1 Eyeblink rate (number of eyeblinks per minute) according to narrative

Stimuli	Eyeblinks per minute
Videos with narrative	16.6 ± 5.4
Rest state	24.2 ± 10.9
Videos without narrative	20.0 ± 8.9
Rest state	20.8 ± 10.6
Narratives without videos	26.3 ± 12.2
Rest state	22.5 ± 7.8

Adapted from Nakano et al. [13]
Fourteen subjects participated in this study (mean ± SD)

Table 46.2 Eyeblink rate (number of eyeblinks per minute) looking at screens (different editing styles) or reality

	Eyeblinks per minute		Eyeblinks per minute
One-shot movie	13.776 ± 9.641	Screens	13.208 ± 8.897
Hollywood-style movie	13.427 ± 9.338		
MTV-style movie	12.421 ± 8.283		
Performance	14.632 ± 7.794	Reality	14.632 ± 7.794

Adapted from Andreu-Sánchez et al. [14, 16]
Forty subjects participated in this study (mean ± SD)

46.2.2 Eyeblinks in Screen and Reality

In 2017, we [16] hypothesized that looking at narratives in reality and watching them through screens would provoke a different eyeblink rate. We presented 4 stimuli to 40 subjects. Three of them were video stimuli and one was a performance. All four stimuli had the same narrative, action, character, and duration. The difference between the three video stimuli was edition: one was a one-shot movie, another was a Hollywood-style movie, and the third was an MTV-style movie. We wanted to compare perception of narratives through screens by not avoiding that contents in a movie may be represented very differently, depending on the media edition.

We obtained significant differences in spontaneous blink rate between screens and reality (see Table 46.2). Screened narrative decreased viewers' eyeblink rate, compared to performed narrative. Also, in video stimuli, the style of the edition affected eyeblink rate [14].

Thus, according to our results, the type of stimulus (live performance or screened movie) that contains a narrative, affects spontaneous blink rate of viewers.

46.2.3 Media Professionalization

On the other hand, different studies have linked professionalization to differences in cognitive patterns. Lotze et al. (2003) [17] studied the musicians' brain compared

Table 46.3 Eyeblink rate per minute (min^{-1}) of non-media professionals and media professionals looking at audiovisual editions in screens and a live performance

	Non-media professionals	Media professionals
	Eyeblinks per minute	Eyeblinks per minute
One-shot movie	18.018 ± 9.917	9.534 ± 7.386
Hollywood-style movie	17.508 ± 10.108	9.347 ± 6.462
MTV-style movie	15.9 ± 8.899	8.941 ± 6.012
Performance	18.207 ± 7.828	11.056 ± 6.042

Adapted from Andreu-Sánchez et al. (2017) [14, 16]
Forty subjects participated in this study (mean ± SD)

to amateurs. They found that professionals showed more focused activation patterns during imagined musical performances.

Zheng et al. (2012) [18] linked a decrease of eyeblink rate of surgeons with the attention needed during the development of their job. Faubert (2013) [19] analyzed athletes' skills. Maguire et al. (2000, 2006) [20, 21] found structural brain differences in taxi and bus drivers.

Keeping that in mind, we hypothesized that watching screens steadily over time, making concomitant decisions with a high level of attention as media professionals do, would provoke visual perceptive differences in spontaneous blink rate [14, 16].

We obtained significant differences between media and non-media professionals watching narratives through screens and looking them at reality (Table 46.3).

Taking into account that lower spontaneous eyeblink rate is linked to higher attention [16, 22], it may be predictable that media professionals would present a lower eyeblink rate while watching narratives through screens, since they are used to do it to make decisions as part of their jobs. The more surprising result here is that media professionals also show a significant lower eyeblink rate when looking at narratives at reality.

46.3 Conclusion

Watching videos with narratives decreases eyeblink rate, while listening to narratives without videos does not. The style of the edition in a screened content affects eyeblink rate: the same narrative with different style of edition provokes differences in eyeblink rate. MTV editing style decreases viewers' eyeblink rate. There are also differences in perceiving narratives through screens or reality. Screens decrease eyeblink rate.

Media professionals spend a lot of time watching narratives through video making decisions related to media contents. They show a significant lower eyeblink rate when watching screens, compared to non-media professionals. More interesting is the fact that media professionals also show a significant lower eyeblink rate when watching narratives performed in real world.

References

1. Burch, N.: La Lucarne de l'infini. Naissance du langage cinématographique, Nathan (1991)
2. Bordwell, D., Staiger, J., Thompson, K.: Classical Hollywood Cinema: Film Style and Mode of Production to 1960, 1st edn. Routledge, London (1985)
3. Bordwell, D.: Intensified continuity: visual style in contemporary American film. Film Q. **55**, 16–28 (2002)
4. Nielsen.: The Total Audience Report: Q1 2016 (2016)
5. Carroll, N., Seeley, W.P.: Cognitivism, psychology, and neuroscience: movies as attentional engines. In: Shimamura, A.P. (ed.) Psychocinematics. Exploring Cognition at the Movies, pp. 53–75. Oxford University Press, Oxford/New York (2013)
6. Skotte, J.H., Nøjgaard, J.K., Jørgensen, L.V., Christensen, K.B., Sjøgaard, G.: Eye blink frequency during different computer tasks quantified by electrooculography. Eur. J. Appl. Physiol. **99**(2007), 113–119
7. VanderWerf, F., Brassinga, P., Reits, D., Aramideh, M., Ongerboer de Visser, B.: Eyelid movements: behavioral studies of blinking in humans under different stimulus conditions. J. Neurophysiol. **89**, 2784–2796 (2003)
8. Stern, J., Boyer, D., Schroeder, D.: Blink rate: a possible measure of fatigue. Hum. Factors. **36**, 285–297 (1994)
9. Delgado-García, J.M., Gruart, A., Múnera, A.: Neural organization of eyelid responses. Mov. Disord., Off. J. Mov. Disord. Soc. **17**, S33–S36 (2002)
10. Bour, L., Aramideh, M., Ongerboer de Visser, B.W.: Neurophysiological aspects of eye and eyelid movements during blinking in humans. J. Neurophysiol. **83**, 166–176 (2000)
11. Fogarty, C., Stern, J.A.: Eye movements and blinks: their relationship to higher cognitive processes. Int. J. Psychophysiol. **8**, 35–42 (1989)
12. Wiseman, R., Nakano, T.: Blink and you'll miss it: the role of blinking in the perception of magic tricks. PeerJ. **4**, e1873 (2016)
13. Nakano, T., Yamamoto, Y., Kitajo, K., Takahashi, T., Kitazawa, S.: Synchronization of spontaneous eyeblinks while viewing video stories. Proceed. Biol. Sci./R. Soc. **276**, 3635–3644 (2009)
14. Andreu-Sánchez, C., Martín-Pascual, M.A., Gruart, A., Delgado-García, J.M.: Eyeblink rate watching classical Hollywood and post-classical MTV editing styles, in media and non-media professionals. Sci. Rep. **7**, 43267 (2017)
15. Murch, W.: In the Blink of an Eye: A Perspective on Film Editing. Silman-James Press, Los Angeles (1995)
16. Andreu-Sánchez, C., Martín-Pascual, M.A., Gruart, A., Delgado-García, J.M.: Looking at reality versus watching screens: media professionalization effects on the spontaneous eyeblink rate. PLoS One. **12**, 0176030 (2017)
17. Lotze, M., Scheler, G., Tan, H.-R., Braun, C., Birbaumer, N.: The musician's brain: functional imaging of amateurs and professionals during performance and imagery. NeuroImage. **20**, 1817–1829 (2003)
18. Zheng, B., Jiang, X., Tien, G., Meneghetti, A.: Workload assessment of surgeons: correlation between NASA TLX and blinks. Surg. Endosc. Other Interventional Tech. **26**, 2746–2750 (2012)
19. Faubert, J.: Professional athletes have extraordinary skills for rapidly learning complex and neutral dynamic visual scenes. Sci. Rep. **3**, 1154 (2013)
20. Maguire, E.A., Gadian, D.G., Johnsrude, I.S., Good, C.D., Ashburner, J., Frackowiak, R.S.J., Frith, C.D.: Navigation-related structural change in the hippocampi of taxi drivers. Proc. Natl. Acad. Sci. U. S. A. **97**, 4398–4403 (2000)
21. Maguire, E.A., Woollett, K., Spiers, H.J.: London taxi drivers and bus drivers: a structural MRI and neuropsychological analysis. Hippocampus. **16**, 1091–1101 (2006)
22. Colzato, L.S., Slagter, H.A., Spapé, M.M.A., Hommel, B.: Blinks of the eye predict blinks of the mind. Neuropsychologia. **46**, 3179–3183 (2008)

Chapter 47
Self-Organization with Constraints: The Significance of Invariant Manifolds

Ichiro Tsuda

Abstract Classifying self-organization phenomena into two categories, one category consists of order formation occurring at macroscopic level, which stems from cooperative and competitive interactions of elementary units of a system concerned. On the other hand, the other category implies the formation of the elementary units at microscopic or mesoscopic levels via some constraint acting at macroscopic level. The latter relates to differentiation such as cell differentiation in embryos and functional differentiation in cortical modules. We treat the latter self-organization in a framework of optimization problem.

Keywords Self-organization · Constraints · Variational principle · Functional differentiation · Chaotic itinerancy

47.1 Introduction

Recent development of cognitive neurodynamics has led us to reconsider the theories and key concepts in far-from-equilibrium systems within the framework of cognitive brain science. Among others, we have treated the theory of self-organization [1]. The theory of self-organization developed particularly from the 1960s to 1980s by the appearance of scientific heroes, Hermann Haken [2] and Ilya Prigogine [3], after scientific revolution of cybernetics [4].

Haken and Prigogine successfully made theories for far-from-equilibrium systems, extended equilibrium phase transitions, and equilibrium and linear thermodynamics, respectively. In particular, Haken's idea of synergetics based on the slaving principle was applied to many fields including neuroscience. Under the influence of synergetics, a scientific hero in neuroscience, Walter Jackson Freeman, successfully applied the idea of phase transitions to non-equilibrium state transitions appearing

I. Tsuda (✉)
Chubu University Academy of Emerging Sciences, Kasugai, Aichi, Japan
e-mail: tsuda@isc.chubu.ac.jp

J. M. Delgado-García et al. (eds.), *Advances in Cognitive Neurodynamics (VI)*,
Advances in Cognitive Neurodynamics,
https://doi.org/10.1007/978-981-10-8854-4_47

in the mesoscopic level of neural chaotic activity and extended it to adopt neural field theory [5].

Classifying self-organization phenomena into two categories, one category may consist of order formation at macroscopic scales, which stems from cooperative and competitive interactions of elements of a system concerned. Here, elementary interactions are supposed to occur at microscopic levels such as atomic or molecular levels.

On the other hand, the other category may consist of the formation of elementary units at microscopic or mesoscopic levels as subsystems of a total system, which occurs via constraints acting on the whole system. The latter relates to differentiation such as cell differentiation in embryos and functional differentiation in cortical modules. In fact, functional differentiation of the brain occurs via interactions between the brain and the environmental factors, not directly via microscopic interactions of macromolecules such as genes. Environmental factors may play a role in constraints acting on the whole brain. However, constraints include qualitative factors such as people's intention and cultural background. Pattee [6] discriminates such constraints from dynamics. In reference to Pattee's idea, we adopted variational principle to make constraints for the dynamics at both microscopic and macroscopic levels.

47.2 Variational Principle

We assume that neural dynamics follows a topological dynamical system (φ, Ω), where φ is a group action acting on phase space, i.e., state space Ω. Then, a mathematical model may be expressed by $\frac{dx}{dt} = f(x; \lambda)$, including the bifurcation parameter λ, thus representing a family of dynamical systems. Environmental factors, here expressed by G, must be introduced to represent interactions between the brain and the environment. Thus, in the extended phase space, the dynamic model is written as:

$$\frac{dx}{dt} = \mathrm{f}(x, \lambda) + G(x, t) \tag{47.1}$$

Furthermore, we introduce intentional constraints expressed as C, which are here supposed to be quantified in the form of information quantities such as transfer entropy and conditional mutual information. Then, the variational principle is introduced in the following way [1]:

$$\delta L = \delta \int_0^T \left\{ \mathtt{C} + \mu \left(\frac{dx}{dt} - f(x, \lambda) - G(x, t) \right) \right\} dt = 0, \tag{47.2}$$

In Eq. (47.2), μ is a Lagrange multiplier.

In self-organization of neural systems, various types of synaptic learning algorithm may play a role in internal constraints. It is typically seen in Malsburg's studies of self-organization of orientation-sensitive cells in primary visual cortex [7] and also in Kohonen's studies of self-organizing map [8]. Amari developed a mathematical theory for topographic mapping by introducing a generalized Hebb rule, extending a Hebb rule to incorporate inhibitory synapses in the learning rule [9].

47.3 Significance of the Generation of Invariant Manifold

In the use of variational principle expressed in eq. (47.2), the Lagrange multiplier is a function of state variables, their time derivatives, and may also be a function of time itself, and thus the equation of motion of Lagrange multiplier is derived. Here, one question arises: what is an initial condition of Lagrange multiplier? There is, however, no reason to determine the initial condition.

The ad hoc determination of the initial condition often leads to unrealistic dynamical trajectories such as trajectories diverging to infinity. However, recent studies on the state space dimension of neural activity under sensory stimuli show the existence of invariant manifold with relatively low dimension, which consists of various evoked activity [10]. Thus, if the neural activity is restricted to such a low-dimensional invariant manifold, then the instability via addition of the equation of motion of the Lagrange multiplier may be restricted to sustain the topology of invariant manifold (see Fig. 47.1).

This idea of topological invariance of invariant manifolds, which can occur through restricted instability via constraint, leads us to sustainable reorganization of

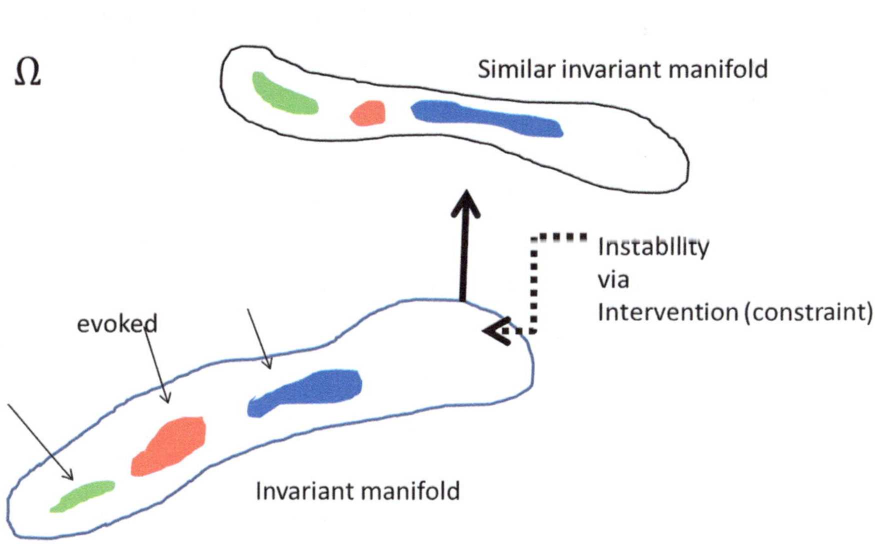

Fig. 47.1 Topological change by restricted instability via constraints

neural networks under the influence of environmental factors, hence leading neural mechanism of functional differentiation.

On the other hand, the introduction of constraints C will not bring about fatal instabilities of invariant manifold. This is because dynamical systems embedded in neural network systems may well be changed and stabilized toward evolutionary stable states in its optimization process, associated with globally stabilized complex transitions between cognitive states such as chaotic itinerancy.

Acknowledgments This work was partially supported by Grant-in-Aid for Scientific Research on Innovative Areas (Non-linear Neuro-oscillology: Towards Integrative Understanding of Human Nature, KAKENHI grant number 15H05878) from the Ministry of Education, Culture, Sports, Science and Technology, Japan.

References

1. Tsuda, I., Yamaguti, Y., Watanabe, H.: Self-organization with constraints—a mathematical model for functional differentiation. Entropy. **18**(74) (2016) doi:https://doi.org/10.3390/e18030074
2. Haken, H.: Advanced Synergetics. Springer, Berlin (1983)
3. Nicolis, G., Prigogine, I.: Self-Organization in Nonequilibrium Systems. Wiley, New York (1977)
4. Von Claus Pias, H. (ed.): Cybernetics|Kybernetik. The Macy—Conferences 1946–1953. Bd.1 Transactions/Protokolle. Diaphanes, Zurich/Berlin (2003)
5. Freeman, W.J., Kozma, R.: Scale-free cortical planar networks. In: Handbook of Large-Scale Random Networks, pp. 1–48. Springer, Heidelberg (2015)
6. Pattee, H.H.: The complementarity principle in biological and social structures. J. Social Biol. Struct. **1**, 191–200 (1978)
7. Von der Malsburg, C.: Self-organization of orientation sensitive cells in the striate cortex. Kybernetik. **14**, 85–100 (1973)
8. Kohonen, T.: Self-organized formation of topologically correct feature maps. Biol. Cybern. **43**, 59–69 (1982)
9. Amari, S.: Topographic organization of nerve fields. Bull. Math. Biol. **42**, 339–364 (1980)
10. Luczak, A., Bartho, P., Harris, K.D.: Spontaneous events outline the realm of possible sensory responses in neocortical populations. Neuron. **62**, 413–425 (2009)

Chapter 48
On the Nature of Coordination in Nature

Emmanuelle Tognoli, Mengsen Zhang, and J. A. Scott Kelso

Abstract Aiming to identify general principles governing collective behavior at multiple levels, we visit complex systems whose dynamic patterns traverse neural, behavioral, and social levels. Rather than approaching such systems from their distinct scientific perspectives, e.g., neuroscience, psychology, or sociology, we unite them in the study of their coordination dynamics. A study of multiple people coordinating their behavior, dubbed "the human fireflies experiment," reveals spatiotemporal metastability. Another study of real fireflies, often taken as the poster child for strong synchronization, also reveals a telltale spatiotemporal mixture of integration and segregation, as had an earlier investigation into the coordination of neural ensembles. Empirical data is contextualized with a theoretical model of coordination dynamics and confirms its prediction that weak coupling and broken symmetry play key roles. We conclude that nature, in all its diversity and uninterested in subsuming itself to the simpler organizing phenomena favored by scientists, such as synchronization, in fact revels in spatiotemporal metastability.

Keywords Broken symmetry · Multiscale · Multi-agent · Spatiotemporal · Metastability · Extended HKB · Weak coupling · Complexity

---

E. Tognoli (✉) · M. Zhang
The Human Brain and Behavior Laboratory, Center for Complex Systems and Brain Sciences, Florida Atlantic University, Boca Raton, FL, USA
e-mail: tognoli@ccs.fau.edu

J. A. S. Kelso
The Human Brain and Behavior Laboratory, Center for Complex Systems and Brain Sciences, Florida Atlantic University, Boca Raton, FL, USA

Intelligent Systems Research Centre, Ulster University, Derry Londonderry, Northern Ireland, UK

J. M. Delgado-García et al. (eds.), *Advances in Cognitive Neurodynamics (VI)*, Advances in Cognitive Neurodynamics,
https://doi.org/10.1007/978-981-10-8854-4_48

48.1 Introduction

Living systems carry a complexity the understanding of which represents an enormous challenge to experimental and theoretical science. The aspiration persists though that eventually those systems will find their fundamental laws, the "Newtonian breakthrough" of complexity (e.g., [1–6]). Complexity science has recognized that interactions play out between elements at the same level of description but importantly also across levels [7–12], both within and across system boundaries. Accordingly, to reach the goal of finding laws of complex living systems, paradigms are required that look beyond domain-specific characteristics, so that all levels may fall under the same investigational scope.

Dynamical approaches have the ability to unite multiple levels of description in a single language, because they focus on *flows*: mathematical formalisms are not restricted to domain-specific phenomenologies, nomologies, and incommensurable quantifications. The dynamical perspective also gives due respect to adaptation [13–15, 49] and exposes processes [9, 16–20]. Useful empirical efforts include the measurement of state variables continuously [21, 22] and ideally at multiple levels of description [11, 23–27]. A striking property of living systems seems to be the emergence of functionality [16, 28–30]. Because functions emerge from interactions [30], a specific dynamical approach was developed to emphasize how the dynamics of coordinative states emerge from the parts' coupling and symmetry [9, 31]. The work presented hereafter – examining collective dynamics of a variety of systems – is inscribed within this theoretical framework of coordination dynamics [9, 32]. Examples are examined to underline the common phenomenology across all levels, especially the concept of metastability [33]. Current research expands on earlier neurocognitive and neurobehavioral work [34–36] but shifts the focus from time to space-time (see also [30]) in order to more fully grasp the complex interplay of multiple parts. The pervasive phenomenology of metastability in nature and its interpretation via mathematical/computational models are cues toward fundamental laws of coordination in living systems.

48.2 Coordination Dynamics of Fireflies

When the coordinated patterns of male firefly flashes in trees from Thailand was documented (e.g., [37, 38]), the phenomenon became a poster child for synchronization [39]. Here we take synchronization, not loosely but within its formal definition that components lock their phase to each other (e.g., [9]). Mathematically, that requires attractor(s) [9, 17, 35]. Figure 48.1 shows a spatiotemporal analysis of fireflies, *Photinus carolinus*, that were video-recorded in the Smoky Mountains, USA. A Raster (Fig. 48.1a) and flash density plots (Fig. 48.1b) reveal their collective flashing at a strikingly regular period of about 1 Hz, whose emergence is the outcome of many fireflies' behavior. But the details of fireflies' spatiotemporal dynamics are more nuanced than synchrony.

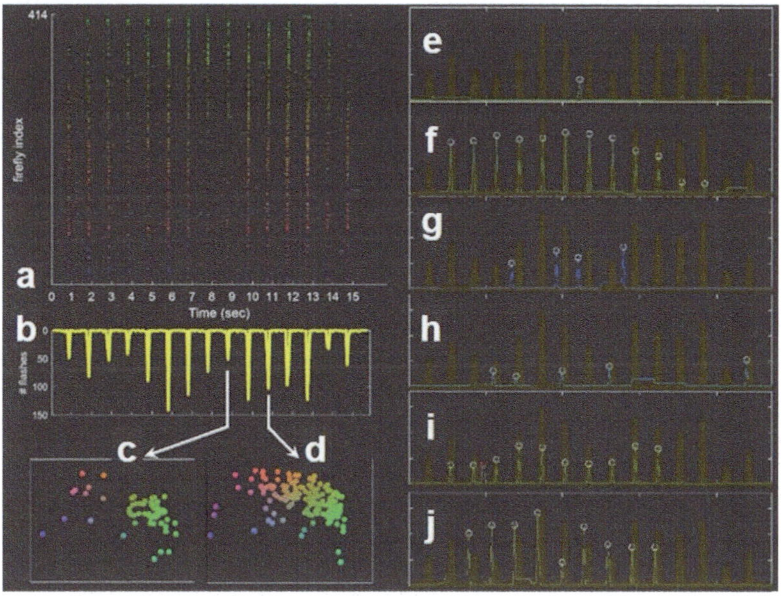

Fig. 48.1 Coordination dynamics of firefly flashes carries the hallmark of metastability. A raster (**a**), flash probability density (**b**), and spatial maps of flash events (**c**–**d**) demonstrate a spatiotemporal dynamics that has more complexity than (attractor-based) synchrony. Examples of individual behaviors (**e**–**j**) recognize that the coupling between individual and collective behavior is weak and presents key attributes previously seen in metastability, including bistable tendencies, dwell-escape dynamics, and broken symmetry. See details in text

Ensembles of fireflies vary in size over time, from small (Fig. 48.1c), with e.g., 1/8 of the population partaking in an event, to large (Fig. 48.1d) with almost a third participating, as revealed by spatial maps and analysis of flashing events. Individuals firefly behavior (e.g., Fig. 48.1e–j, bright lines representing luminance of individuals over time – for reference, the thick brown line indicates flashing density of the population as in Fig. 48.1b) varies from single contribution (Fig. 48.1e) to sustained (Fig. 48.1f, i–j) and sporadic behavior (Fig. 48.1g–h). At the temporal scale of this sample observation, phase coordination patterns tend to in phase (Fig. 48.1f), but also to a lesser extent antiphase (Fig. 48.1e.g.,), and out of phase coordination (Fig. 48.1h).

Dwell-escape dynamics (Fig. 48.1i, notice that most flashes align to the collective behavior, and one extra flash occurs, marked red, putatively releasing the intrinsic tendency to flash faster than the group) and quasi-unlocked coordination (Fig. 48.1j, notice individual flashing is slower than the group, and initial phase-lead gets smaller, as if group and individual just follow their own pace without influencing each other) are also observed. These behaviors are compatible with an underlying spatiotemporal dynamics that is metastable (see also [30]).

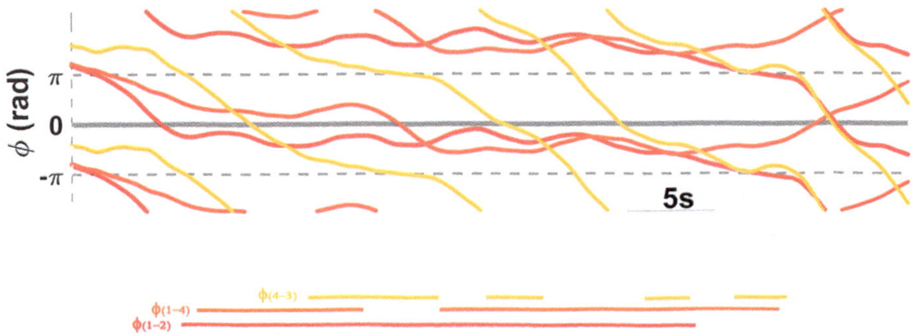

Fig. 48.2 Coordination dynamics of multiple people engaging in a sensorimotor coordination task reveals the dwell-escape coordination dynamics characteristic of spatiotemporal metastability. The relative phase of four participants (three dyads) is shown on top. Markers on the bottom indicate the location of dwells in the relative phase

48.3 Spatiotemporally Coordinated Human Behavior

To explore the possible generality of spatiotemporal metastability, we studied sensorimotor coordination among groups of eight people who sat at booths around an octagonal table. People faced an array of LEDs to see their own and others' behavior, using a touchpad to flash one of the LEDs (a human "firefly" experiment, Fig. 48.2, [40]). Human behavior was less periodic than fireflies' but otherwise revealed similar phenomenology, that is, bistability, attracting tendencies, and metastability [41]. In an exemplary trial, pairwise relative phases between four people that coordinated with each other showed dwells that persisted for longer (red) or shorter durations (yellow) and, accordingly, ensembles that included more or fewer participants over time (see also Fig. 48.1 in [25]). The dwells (horizontal segments of the relative phase) were interspersed with escape (wrapping), a hallmark of metastable coordination dynamics.

Quantitative analysis of 120 participants (15 groups, not shown) revealed some attracting tendencies, mainly near inphase, and in a selective analysis of the strongest instances of coordination, some bistable tendencies near inphase and antiphase that echoed the essential bistable tendencies observed in the brain and behavior [30, 36, 42–44]. These data suggest that just like fireflies, multi-agent human sensorimotor coordination exhibits spatiotemporal metastability.

48.4 Discussion

Examples of social behavior – from fireflies to humans – share common features with the spatiotemporal coordination dynamics of the brain, from microscale to macroscale [30]. In particular, the existence of metastability and tendencies for synchronization within as well as across frequency bands is ubiquitous [17, 30,

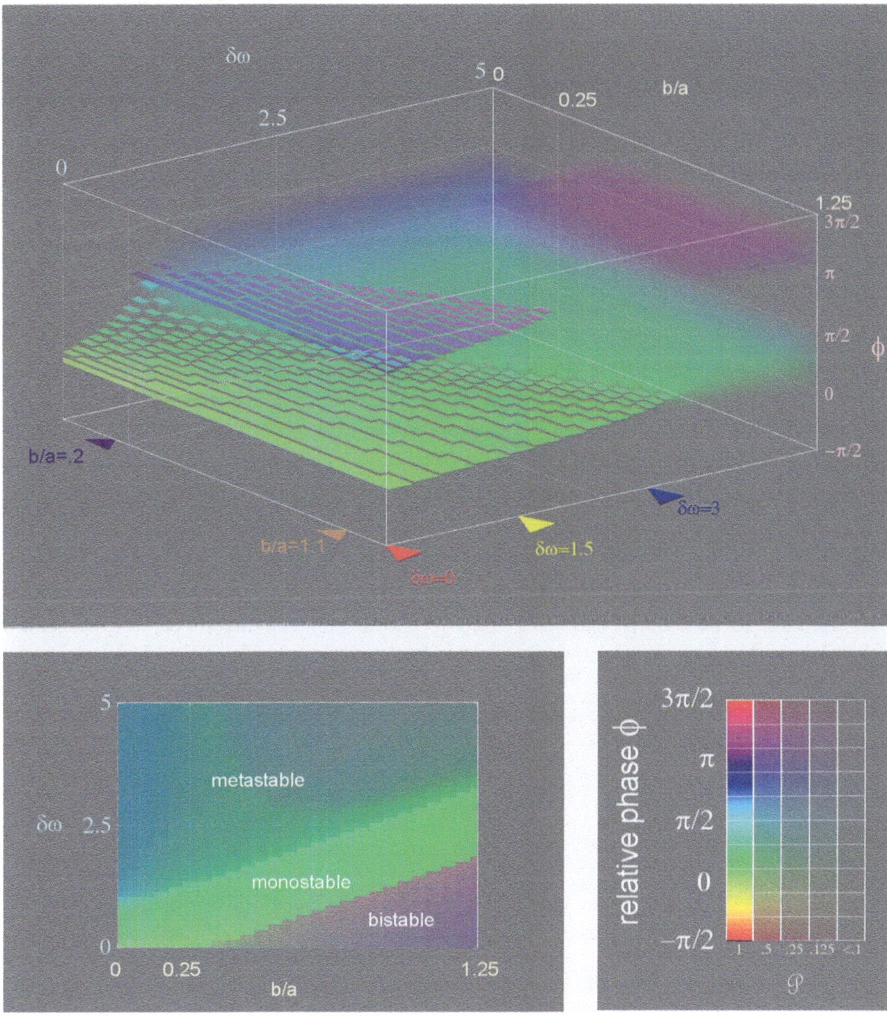

Fig. 48.3 Side view (upper panel) and top view (lower left) of a parameter space for the extended HKB model from [47], whose relative phase probability density has been encoded according to the legend on the right. Metastability emerges for weak coupling (b/a → 0) and broken symmetry (δω large)

45, 46]. Models of coordination dynamics such as the extended HKB model [47] posit two key factors that drive the dynamics of metastability (Fig. 48.3): one parameter, δω, specifies a difference in the components' intrinsic frequencies (how diverse are the self-sustained oscillators when left to themselves) and the other, b/a, expresses the coupling strength binding the components together (how intense their mutual "pull" is on each other's dynamics). Both factors play into the opportune disappearance of attractors: for example, when coupling is weak and components are diverse, metastability emerges.

Starting from empirical investigations of synchrony [16, 30, 34, 36], an easy phenomenon to query [17], we have now accrued evidence that metastability is common, if not pervasive (see also [30, 36, 45, 48]). It is likely that we only notice metastable dynamics when it possesses a striking symmetry. As a consequence much of the less orderly collective behavior typical of complex systems is misclassified. On the other hand, it is also possible that epochs appearing as synchronized are decontextualized from their broader dwell-escape dynamics due to finite windows of observation.

Once recognized, pervasive spatiotemporal metastability should not be all that surprising. Nature has many parts interacting at multiple levels, each with distinct properties often weakly coupled. According to coordination dynamics, such conditions are both necessary and sufficient for the emergence of metastability. We have suggested elsewhere that delays in appreciating the full scope of metastability may be due to methodological biases that tend to sweep dwell-escape dynamics under the rug of quasi-synchronization [17]. A common phenomenology governing the collective behavior of neural, behavioral, and social systems at multiple levels is one small step toward the formulation of laws of complex living systems. It is hoped that such yet-to-be-discovered laws open up useful generalizations for cognitive and social neurodynamics.

Acknowledgments This work was supported by NIMH Grant MH080838.

References

1. Kelso, J.A.S., Haken, H.: New laws to be expected in the organism. Synergetics of brain and behavior. In: Murphy, M., O'Neill, L. (eds.) What Is Life, pp. 137–160. Cambridge University Press, Cambridge (1995)
2. Perez Velazquez, J.L.: Finding simplicity in complexity: general principles of biological and nonbiological organization. J. Biol. Phys. **35**, 209–221 (2009)
3. Anderson, P.W.: More is different. Science. **177**, 393–396 (1972)
4. Von Bertalanffy, L.: An outline of general system theory. Br. J. Philos. Sci. **1**, 134–165 (1950)
5. Atlan, H.: Entre le cristal et la fumée: Essai sur l'Organisation du Vivant. Éditions du Seuil, Paris (1979)
6. Yates, F.E.: Order and complexity in dynamical systems: homeodynamics as a generalized mechanics for biology. Math. Comput. Model. **19**, 49–74 (1994)
7. Campbell, D.T.: Downward causation in hierarchically organised biological systems. In: Ayala, F.J., Dobzhansky, T. (eds.) Studies in the Philosophy of Biology, pp. 179–186. Macmillan Education, London (1974)
8. Pattee, H.H.: Complementarity vs. reduction as explanation of biological complexity. Am. J. Physiol.-Reg. I. **236**, R241–R246 (1979)
9. Kelso, J.A.S.: Dynamic Patterns: the Self-Organization of Brain and Behavior. MIT Press, Cambridge, MA (1995)
10. Ellis, G.F.: On the nature of causation in complex systems. Trans. Roy. Soc. S. Afr. **63**, 69–84 (2008)

11. Kelso, J.A.S., Dumas, G., Tognoli, E.: Outline of a general theory of behavior and brain coordination. Neural Netw. **37**, 120–131 (2013)
12. Kozlowski, S.W., Chao, G.T., Grand, J.A., Braun, M.T., Kuljanin, G.: Advancing multilevel research design: capturing the dynamics of emergence. Organ. Res. Methods. **16**, 581–615 (2013)
13. Holland, J.H.: Adaptation in Natural and Artificial Systems: an Introductory Analysis with Applications to Biology, Control, and Artificial Intelligence. MIT Press, Cambridge, MA (1992)
14. Kauffman, S.A.: The Origins of Order: Self-Organization and Selection in Evolution. Oxford University Press, New York (1993)
15. Freeman, W.J., Kozma, R., Werbos, P.J.: Biocomplexity: adaptive behavior in complex stochastical dynamical systems. Biosystems. **59**, 109–123 (2001)
16. Bressler, S.L., Tognoli, E.: Operational principles of neurocognitive networks. Int. J. Psychophysiol. **60**, 139–148 (2006)
17. Tognoli, E., Kelso, J.A.S.: Enlarging the scope: grasping brain complexity. Front. Syst. Neurosci. **8**, 122 (2014)
18. Rabinovich, M.I., Simmons, A.N., Varona, P.: Dynamical bridge between brain and mind. Trends Cogn. Sci. **19**, 453–461 (2015)
19. Russo, E., Treves, A.: Cortical free-association dynamics: distinct phases of a latching network. Phys. Rev. E. **85**, 051920 (2012)
20. Tsuda, I.: Chaotic itinerancy and its roles in cognitive neurodynamics. Curr. Opin. Neurobiol. **31**, 67–71 (2015)
21. Tognoli, E., de Guzman, G.C., Kelso, J.A.S.: Interacting humans and the dynamics of their social brains. In: Wang, R., Gu, F. (eds.) Advances in Cognitive Neurodynamics (II), pp. 139–143. Springer, Dordrecht (2011)
22. Dumas, G., de Guzman, G.C., Tognoli, E., Kelso, J.A.S.: The human dynamic clamp as a paradigm for social interaction. Proc. Natl. Acad. Sci. U. S. A. **111**, E3726–E3734 (2014)
23. Bassett, D.S., Gazzaniga, M.S.: Understanding complexity in the human brain. Trends Cogn. Sci. **15**, 200–209 (2011)
24. Bar-Yam, Y.: Multiscale complexity/entropy. Adv. Complex Syst. **7**, 47–63 (2004)
25. Tognoli, E., Kelso, J.A.S.: On the brain's dynamical complexity: coupling and causal influences across spatiotemporal scales. In: Yamaguchi, Y. (ed.) Advances in Cognitive Neurodynamics (III), pp. 259–265. Springer, Dordrecht (2013)
26. Devor, A., Bandettini, P.A., Boas, D.A., Bower, J.M., Buxton, R.B., Cohen, L.B., Dale, A.M., Einevoll, G.T., Fox, P.T., Franceschini, M.A., Friston, K.J.: The challenge of connecting the dots in the BRAIN. Neuron. **80**, 270–274 (2013)
27. Kelso, J.A.S., Schöner, G., Scholz, J.P., Haken, H.: Phase locked modes, phase transitions and component oscillators in coordinated biological motion. Phys. Scr. **35**, 79–87 (1987)
28. Koch, C., Laurent, G.: Complexity and the nervous system. Science. **284**, 96–98 (1999)
29. Sporns, O.: Network analysis, complexity, and brain function. Complexity. **8**, 56–60 (2002)
30. Tognoli, E., Kelso, J.A.S.: The metastable brain. Neuron. **81**, 35–48 (2014)
31. Haken, H.: Principles of Brain Functioning: a Synergetic Approach to Brain Activity, Behavior and Cognition, vol. 67. Springer, Berlin (2013)
32. Kelso, J.A.S.: Coordination dynamics. In: Meyers, R.A. (ed.) Encyclopedia of Complexity and System Science, pp. 1537–1564. Springer, Heidelberg (2009)
33. Kelso, J.A.S.: Multistability and metastability: understanding dynamic coordination in the brain. Philos. Trans. R. Soc. B. **367**, 906–918 (2012)
34. Bressler, S.L., Kelso, J.A.S.: Cortical coordination dynamics and cognition. Trends Cogn. Sci. **5**, 26–36 (2001)
35. Kelso, J.A.S., Tognoli, E.: Toward a complementary neuroscience: metastable coordination dynamics of the brain. In: Perlovsky, L.I., Kozma, R. (eds.) Neurodynamics of Cognition and Consciousness, pp. 39–59. Springer, Berlin/Heidelberg (2007)
36. Tognoli, E., Kelso, J.A.S.: Brain coordination dynamics: true and false faces of phase synchrony and metastability. Prog. Neurobiol. **87**, 31–40 (2009)

37. Smith, H.M.: Synchronous flashing of fireflies. Science. **82**, 151–152 (1935)
38. Buck, J., Buck, E.: Mechanism of rhythmic synchronous flashing of fireflies. Science. **159**, 1319–1327 (1968)
39. Ermentrout, B.: An adaptive model for synchrony in the firefly pteroptyx malaccae. J. Math. Biol. **29**, 571–585 (1991)
40. Zhang, M., Kelso, J.A.S., Tognoli, E.: Critical diversity: divided or united states of social coordination. PLoS One. (2018) https://doi.org/10.1371/journal.pone.0193843
41. Kelso, J.A.S., de Guzman, G.C., Holroyd, T.: The self organized phase attractive dynamics of coordination. In: Babloyantz, A. (ed.) Self Organization, Emerging Properties and Learning, Series B, vol. 260, pp. 41–62. Plenum, New York (1991)
42. Kelso, J.A.S.: Coordination dynamics of human brain and behavior. Springer Proc. Phys. **69**, 223–234 (1992)
43. Tognoli, E., Lagarde, J., de Guzman, G.C., Kelso, J.A.S.: The phi complex as a neuromarker of human social coordination. Proc. Natl. Acad. Sci. U. S. A. **104**, 8190–8195 (2007)
44. Kelso, J.A.S., de Guzman, G.C., Reveley, C., Tognoli, E.: Virtual partner interaction (VPI): exploring novel behaviors via coordination dynamics. PLoS One. **4**, e5749 (2009)
45. Tognoli, E., Kelso, J.A.S.: Spectral dissociation of lateralized pairs of brain rhythms. arXiv preprint (2013) arXiv:1310.7662
46. Bhowmik, D., Shanahan, M.: Metastability and inter-band frequency modulation in networks of oscillating spiking neuron populations. PLoS One. **8**, e62234 (2013)
47. Kelso, J.A.S., Del Colle, J.D., Schöner, G.: Action-perception as a pattern formation process. In: Jeannerod, M. (ed.) Attention and Performance 13: Motor Representation and Control, pp. 139–169. Lawrence Erlbaum Associates, Inc, Hillsdale (1990)
48. Bressler, S.L., Kelso, J.A.S.: Coordination dynamics in cognitive neuroscience. Front. Neurosci. **10**, 397 (2016)
49. Nordham, C.A., Tognoli, E., Fuchs, A., Kelso, J.S.: How interpersonal coordination affects individual behavior (and vice versa): experimental analysis and adaptive HKB model of social memory. Ecological Psychology, (just-accepted) (2018) https://doi.org/10.1080/10407413.2018.1438196

Chapter 49
Beyond Prediction: Self-Organization of Meaning with the World As a Constraint

Jan Lauwereyns

Abstract In recent years predictive coding has gained considerable popularity as a powerful theoretical approach in neuroscience. However, one recurring problem with predictive-coding models is that in order to minimize prediction errors, the most efficient strategy would be to avoid unpredictable situations as much as possible. I propose to use the concept of "intrinsic reward" as a key addition to predictive-coding models, where the ultimate goal of prediction is not to minimize error but to maximize reward. Predictive coding with intrinsic rewards would explain not only how the typical utilitarian behaviors work but also how seemingly spurious activities such as reprocessing can be tied to the mechanisms of predictive coding. Essentially, the reprocessing would lead to the self-organization of a richer, intrinsically rewarding experience of meaning, with real-world information as a constraint. This "intensive approach" to information processing would tend toward the expansion of meaning and predictive power.

Keywords Predictive coding · Intrinsic reward · Intensive approach · Reprocessing

49.1 Introduction

The basic concept of "predictive coding" [1, 2] is that the brain endeavors to minimize prediction errors: it tries to anticipate the current situation so as to reduce the loss of energy required to deal with the unexpected. Yet, as pointed out several decades ago in a critique of behaviorist models [3], curiosity is a very typical and basic characteristic of many animals, particularly also humans: we tend to seek a certain level of new stimulation even if there is no guaranteed material or immediate benefit, sometimes even if there is a considerable risk of adverse outcomes. More-

J. Lauwereyns (✉)
Faculty of Arts and Science, Kyushu University, Fukuoka, Japan
e-mail: jl@artsci.kyushu-u.ac.jp

J. M. Delgado-García et al. (eds.), *Advances in Cognitive Neurodynamics (VI)*, Advances in Cognitive Neurodynamics,
https://doi.org/10.1007/978-981-10-8854-4_49

over, much brain activity is devoted to a type of coding in which the relevant information is fully known and requires no analysis in terms of prediction (e.g., savoring the taste of wine, or ruminating on a negative experience). Humans often spend large amounts of time in such "postdictive" information processing. How can these phenomena be reconciled with the framework of predictive coding? The present chapter, written as a tribute to the creative mind of the great neuroscientist and occasional poet Walter J. Freeman, offers an essayistic exploration of the question.

49.2 Shared Consciousness

Predictive coding gives us the framework, the theory to spar with. To be fair, we should acknowledge some of the things that it explains very well. There are at least two big stories in neuroscience that look as if they were crafted exclusively for the theory of predictive coding. Of course, the truth is probably the other way around. With the emergence of the two big stories, there was the need, the urgency, to think about how they (and other observations) could be integrated into something unifying. Theorists had to come up with predictive coding to accommodate the two big stories.

Both stories really originated at the grassroots. Story Number One was about dopamine; the neurotransmitter of which everyone knew it had something to do with pleasure, or reward. In the 1980s researchers started relating the activity of neurons that release dopamine to the control of action, when monkeys tried to obtain a food or liquid reward. In the 1990s those same and some other researchers discovered it was a bit more specific than that. When a monkey suddenly got a reward, the dopamine neurons "fired" (they exhibited a short burst of action potentials, or nerve impulses—the electrical action that effectively released the dopamine). But when the monkey knew a reward was going to come, the dopamine neurons did not do anything special at the time the reward was delivered (they continued occasionally producing an action potential, at their normal rate, their base rate). Instead, the dopamine neurons fired at the time when the monkey found out that a reward was going to come, that is, when it received a "cue," some sensory signal, say, a beep or a flash of light that indicated a reward was impending.

There was more. When a reward was predicted by a sensory cue, but then omitted (when the experimenter cheated and did not actually give the reward), the dopamine neurons completely stopped firing at the time the reward was supposed to come but did not, as if the neurons wallowed in angry silence at the disappointment—noticeably more silent than the base rate. The pattern of data was best understood as a form of information processing with respect to reward. Not signaling anything physical about rewards (not the taste, for instance), but more abstract information about rewards. It was a form of updating on predictions. If everything was as expected, the dopamine neurons remained neutral, occasionally producing an action potential (about seven or eight per second). For good news ("positive error"), they fired a quick salvo. For bad news ("negative error"), they briefly shut up completely.

Later we realized that this kind of updating, or news reporting, actually applied to many more types of neural coding, in various perception, memory, and decision-making paradigms. It looked as if our societies of neurons are not representing all the information in the surround, not drawing elaborate copies of the world, but working from summaries and predictions, assuming everything will be as it is or should be, only to focus on updating, reserving most energy for specific bits that do not fit, or require a reassessment. Indeed, this is predictive coding.

Story Number Two was about Broca's area, so-called area 44, connected with the production of language, the organizing of words into sentences, down in the left frontal lobe of humans. Anatomists had compared this area to the ventral premotor area in monkeys, a part of the brain that would be involved in motor planning—how the monkeys decided to use which part of the body in what way in order to get to where they wanted. While trying to figure out the neural coding for this motor planning, researchers stumbled across a strange class of neurons. Some of these neurons fired when a monkey was about to pick up a nut. This was fine, in accordance with what we knew. But the same neurons also fired when the monkey was restrained, could not reach out, and instead witnessed the experimenter picking up the nut. This did not make sense at first. The monkey was not planning any movement, but the neurons were firing as if it did. On second thought, it did make sense. The coding was abstract, referring to the action of "picking up a small object," regardless of who was the agent, the monkey's ego ("I") or the experimenter's id ("you"). There was a commonality, a form of sharing. A similarity. The researchers, and many inside and outside the community of neuroscientists, got very excited. We spoke of "mirror neurons" and fantasized about how this bit of neural tissue formed a precursor to our ability to perceive and understand action. Perhaps it gave us a glimpse of a protolanguage, one in which actions (verbs) were distinguished from objects and agents (later to evolve into nouns and pronouns). It was not a stretch of the imagination to connect this idea to something more inspiring. Was this like taking other people's perspective? To see what they see, possibly even to feel what they feel. Empathy. Sharing. There we have it, that inevitable word of the day. We like and we share.

Predictive coding can deal very well with the notion of sharing. Our theory of someone else's mind works from the assumption that what goes for me goes for you. It is a powerful assumption, both very specific and reasonably accurate, plus a great shortcut to the inaccurate (which now requires only a bit of editing, to correct the first draft; and editing is so much easier than writing from scratch).

The sharing is not neutral. We like and we share. More often than not, the sharing implies a positive value. Some gain a return, a reward, an expanding circle, an enrichment, a cultivation, the growth of a group, an identity, the insertion of something on our timeline, the writing of a little note in our autobiography. But is "the same" really the same? What gets repeated in a repetition? Physically speaking, this current configuration of atoms here and now can never be the same as that next configuration there and then. The sameness is merely a simile, a metaphor, a carrying over, across an abstract bridge, of one information structure to another.

49.3 The Nature of the Universe

The key observation about sharing consciousness is that it takes the cognitive processing out of the head, beyond the territory of a single mind—and then back into the head, or actually multiple heads, the heads of all the ones we're sharing consciousness with. Neural Darwinism applies not just to a society of neurons inside one brain, but also to societies of brains (to quote a book title by Walter J. Freeman). Societies of societies of neurons: with dynamics within and between brains. Somehow my consciousness may be modulated by your consciousness. What you say, what you fancy, and what you hate is likely to influence my thinking and the thinking by others who are listening too. I can agree or disagree, or decide to think more, but this is already a matter of negotiation, conversation, speaking in or out of turn, and replying, more or less politely, writing back and forth, yes, activities propagated by language (human language, with its infinities, thanks to the symbols and recursive structures).

The sharing is about spreading—not dividing but multiplying. Spreading the word, spreading knowledge. The negotiation often focuses on novelty, aiming to enlarge or revise the collection of things known about the world, about the nature of the universe. How do we reconcile this with the notion of the brain as the seat of a soul desperately set on minimizing prediction errors? We do not, would be my short answer. Ultimately I think prediction serves primarily as a tool, not to minimize prediction error but to maximize reward. The predictive information processing can and must be analyzed from the perspective of affect. There can be infinitely many truths or correct predictions, but there are only so many that matter to us. Value comes into it, from the beginning. With reward on the foreground, it is a lot more straightforward to connect the information processing to our bodily needs and wants, the homeostasis, which implies that not all correct predictions are equal. Some are preferable over others. More importantly, with the concept of reward, we can deal much better with a fundamental paradox about pleasure and prediction error. Often the pleasure resides precisely in the positive prediction error, in the unexpected, the happy surprise. Shall we call bizarre the theory that tells us with a straight face we are naturally trying to minimize fun? The poet of *The Flowers of Evil* wrote closer to the truth when he warned us in his preface: the worst of all demons is boredom.

Again, value comes into it, for prediction errors as well. Some prediction errors are rewarding. Some are not. We need a theory that explains how we manage to focus on rewarding prediction errors, how we become aware of the desired range of error, the optimal type of error. Predictive coding will certainly always be included as an important computational trick. There is no denying that it economizes on energy expenditure in the brain. But our consciousness relishes the new as if life depends on it, and perhaps life does really depend on it. Of course we do not relish everything new; only the relevant stuff that moves within reach of our eagerly grasping intellects.

We go beyond prediction, from the determined to the free, the unpredictable. It could be that a hierarchical model with several layers of predictive coding and meta-coding provides the answer, as long as it gives a leading role to reward. Perhaps we have a hunch about where to look for nice surprises, maybe we already have a vague idea about which kind of new is the right kind of new. This would be a prediction-based search for a rewarding type of error. And might there be a boredom predictor at work, as a meta-coder, to shift our attention to another area of predictive processing when things begin to draw a yawn? The concept of prediction remains our best point of departure. It challenges us to think more sharply.

The contours of all things in the field of thought are alive with interactive, mutually formative dynamics. The boundaries are active. The negotiation invites the twin ghosts of self-organization and the observer's paradox. When you or I see a certain something, we actually do not, because it has already changed. It *was* already changed: by our gaze. Our gaze has shaped it into something else, and in the process, the something else has shaped you or me into someone else. This suffices to drive us crazy? Not quite. The changes tend to be slight and cumulative, in fact so slight that we rarely realize the dynamics. We look at our faces in the mirror and do not discern the daily evidence of aging until we must suddenly come to terms with that unbelievable portrait from 10 years ago. Our senses work in a different timescale, with different types of data. The changes pertain not directly to physical variations in the world, but to the *implications* we gather from the physical variations that we sense. We use those implications in the stories we write, the thoughts we think, about things as they are. I would like to quote here from "Poetry of Brains," a talk given by Walter Freeman in August 2005—particularly the preview of ideas as transcribed for a YouTube video (search for "Walter Freeman Poetry of Brains"). I subscribe to these ideas (not perfectly but well enough), and I could not relay them any faster or more efficiently with other words:

1. *Brains are illogical.*
2. *All we know is an illusion created by our brains. Sensing (seeing, hearing, smelling, tasting,* etc.*) is impossible. The idea of neural representation, stimulus/response, input/output,* etc. *is false.*
3. *Our illusion of reality is updated as we move our bodies, provoking responses from reality that impact our senses, which then pass/fails* (sic) *the current illusion, causing the brain to keep it, update it, or create a new one = learn.*
4. *To share experiences is impossible, but if we move, test, and act together, we can force our brains to create similar illusions of reality that feel like shared experiences and help us to create trust.*

The negotiation entails the creation and updating of something conceptual, something we cannot honestly call a "representation" because it does not in fact represent anything. There is no one-to-one mapping to any physical information. No exact copy. "Illusion" sounds perhaps a little too strong (with its connotation of deception); Freeman probably used that word as a spicy stimulus for his live audience. Importantly, the emphasis is on the creating and updating, the active boundaries of a thing inside the head that does not quite exist in the same way anywhere else. This conceptual thing, our knowledge, our imagination, functions as a compilation of predictions about what the world will give back to us as we move around in it. This is the predictive-coding side of it. But it does more; it makes sense.

It gives meaning to the world. It tells us a story about the nature of the universe that we can marvel at, just for its own sake. There is something intrinsically rewarding about it. In the meantime, to be sure, this imagination reflects not only, not even mainly, a visual faculty. I prefer to think of it as a domain of writing rather than drawing. Writing, not literally in a natural language, but figuratively, as a composition with meaningful elements (possibly, but not necessarily, words). Through compositions, through weavings, we establish all kinds of connections, which help to characterize the objects and subjects, the body and the mind, the self and the world. This form of connecting can be quite real, if not neatly linear. The text shapes the activity of the active boundaries. It guides our interactions, our negotiations: where and how we move in the world, and what we try to share with whom. The better we write, the tighter we connect. Being an optimist, I will add: in most cases the better we write, the happier we connect.

49.4 The Language of Inquiry

Writing linearizes the nonlinear. In that sense already, writing edits reality. It is a selective process that implies valuation, a choice of words that describe some things, but not others. One might try to interpret the meaning-giving operation as an exaptation (a shift in function) from basic forms of predictive coding that were oriented toward obtaining specific "objects," from food to mating partners. We could think, for instance, that predictive coding emerged as a set of information-processing mechanisms to maximize reward, where reward primarily reflected a desirable based on tangibles in our environment, connecting to bodily needs and wants. The needs and wants constrain the range of desirables: water when thirsty, a mate when horny, a sweater when cold. The information processing, then, would have come about as something aimed at extrinsic rewards—extrinsic in the sense that the reward finds its target in the outside world. This line of thought quickly gets fuzzy when we try to apply it to our present-day situation. Our most important utilitarian rewards today—money, prestige—are not easily or accurately described as tangible. Much revolves around their constructions inside our heads, with abstract ideas that are connected to the outside world in complex, nonlinear ways. Still, rewards relating to power and wealth can properly be understood as extrinsic, with predictive coding serving as a cognitive tool to obtain targets in the outside world.

That shiny little abstract thing, the poem, works differently. It has famously been claimed to make nothing happen, or to be written solely for its own sake. The pleasure does not come from outside. The reward is intrinsic. Happiness and sweet pain ride alongside the thoughts generated by the poem, in reading or writing the poem. If this is predictive coding, it must be headed for a crash, being a cognitive tool to obtain more predictive coding: thoughts spiraling out, spinning out of control, in infinite regress, an intrinsic attractor lost within itself. There is an enlightening nothingness that underscores the pleasure of a poem, or of a poem of poems (a novel). We immerse ourselves in the virtual imagery, to the point that we sometimes describe it as a feeling of being lost in it, wholly absorbed by the fiction, completely captivated by a toy model of the world. Negatively speaking, this might be condemned as an escape. Positively speaking, this might be applauded as … an escape. From the mundane and the contingent to the extraterrestrial and transcendent. From reality to nothingness, something virtual, fictional, something that has no urgent business with the material world. And yet, this nothingness is enlightening. It allows us to make sense of things, or some portion of the things on our mind. We get to know a bit more about the world, and this extra bit of knowledge, though of no practical usage, offers a real sense of enrichment. We gain a bit of truth or beauty, even if it is kicked into action by pain or ugliness (sometimes referred to as the ecstatic or sublime by technicians desensitized to the simpler, older vocabulary). It makes the world infinitesimally more—but *more*—meaningful. It doesn't give us the meaning of life, but it adds to its meanings.

The enlightening nothingness brings us to the religious domain, where the curious word "trust" reappears. Trust, or what remains beyond doubt, is all we have in the face of reality. Beauty and truth are predicates we apply to the things we perceive. Given that perception cannot represent reality, but only produce some kind of fiction or abstraction through rather unreliable and irrational methods, we are naturally forced to take a leap of faith whenever we accept something to deserve a special predicate. More often than not the leap of faith is what actually makes the thing true or beautiful. Hence the infamous placebo effects, the empowering self-fulfilling prophecies, and the tragic, inescapable oracles.

The leap of faith, to be sure, should not be confused with the domain of deistic religions. The poem—intrinsic reward, endless mirror—works religiously, at least by one (albeit obscure) etymological reading of *religio*. The poem reconnects things that became unbound in the being-multiple of reality. It creates a focal object that gives meaning to a range of past and future experiences. One cognitive psychologist suggested that attention integrates multiple sensory features into a single object of thought, while one moral philosopher likened the giving of attention to the saying of a prayer. Both ideas are valid, I think. Attention, as a form of prayer, integrates features into an object file, a semantic unit that brings together a set of disparate bits of information. The poem shapes our attention. It does this esemplastically, to use a Romantic neologism, transforming the many things into a single entity.

This poetic function gives us discriminative power: the ability to organize information into object files, where the first model, the precedent or paradigm, serves as a template by which we can compare whatever comes next. This undoubtedly also implies some utilitarian benefits, but it is not just a matter of cognitive economy. (In its utilitarian form, we could recognize the poetic function in the textbook information processes studied by cognitive psychologists: language, memory …) Ultimately, the poetic function cares only about truth and beauty, about making things accessible to something beyond the immediate needs of the body. It defies death. It is turned toward meanings that transcend the limits of here and now. Also in this more conventional notion of religion, the poetic function aims at something larger than life. Paradoxically, it does so by digging deeply into experience, through the most intense living and writing. It does so the way Walter J. Freeman did, and the way his writings still do.

References

1. Hohwy, J.: The Predictive Mind. Oxford University Press, Oxford New York (2013)
2. Clark, A.: Surfing Uncertainty: prediction, Action, and the Embodied Mind. Oxford University Press, New York (2016)
3. Berlyne, D.E.: Curiosity and exploration. Science. **153**, 25–33 (1966)

Chapter 50
Bias Versus Sensitivity in Cognitive Processing: A Critical, but Often Overlooked, Issue for Data Analysis

Jan Lauwereyns

Abstract One of the goals of cognitive neuroscience is to characterize the information-processing mechanisms by which individuals respond to sensory or mnemonic data in a variety of contexts. Many theoretical approaches, based on rates of responding and on reaction time data, have indicated two parallel, independent, but not mutually exclusive dimensions by which the information processing can be influenced: on the one hand, the quality of information, conceptualized as a signal-to-noise ratio, or the ability to accurately distinguish a target from among distractors, often expressed as a sensitivity measure; on the other hand, the response tendency, referring to the a priori likelihood of a certain response, often expressed as a bias measure. Although these two dimensions can be readily distinguished in behavioral measures, they are often overlooked in the data analysis of contemporary neurophysiological studies. Through a critical reading of several recent high-profile studies, I demonstrate the pitfalls of this oversight.

Keywords Bias · Sensitivity · Selective attention · Signal-to-noise ratio

50.1 Introduction

In the study of decision-making and signal detection, it has been well established that we can consider two fundamentally different types of information processing, one driven by prediction (which I will call "bias") and one driven by incoming information (which I will call "sensitivity") [1, 2]. Although the concepts seem straightforward and noncontroversial, their proper application turns out to be surprisingly slippery and problematic. In this paper I offer a critical review of recent research, with two illustrative case studies, to highlight some of the issues.

J. Lauwereyns (✉)
Faculty of Arts and Science, Kyushu University, Fukuoka, Japan
e-mail: jl@artsci.kyushu-u.ac.jp

J. M. Delgado-García et al. (eds.), *Advances in Cognitive Neurodynamics (VI)*,
Advances in Cognitive Neurodynamics,
https://doi.org/10.1007/978-981-10-8854-4_50

50.2 Case Study One: Engel et al., *Science*, 2016

Consider two judges, one in an on state and one in an off state. The on state would be a racist, white supremacist state, whereas the off state would be a neutral one, where most citizens are committed to the concept of diversity. Data from the statistics department tells us that the judge in the on state has a 78% correct conviction rate for black criminals. The judge in the off state has a 65% correct conviction rate. Shall we conclude that the judge in the racist, white supremacist state has a superior ability to detect crime by blacks?

"Not so fast," I hear most of us responding (especially those of us committed to real news). What about the false conviction rates?

It should be too obvious to have to spell out, but without the false conviction rates, we really cannot assess the performance of the judge in a meaningful way.

Does the judge in the on state perform better than the judge in the off state, or is she simply more likely to convict, regardless of the evidence before her? It could be that the judge in the on state just says "guilty" in 78% of all cases, without even looking at the police reports. She would be convicting a lot of innocent people.

In the context of my example, with its racial overtones, we probably do know better. If we spell out the obvious, somewhat pedantically, then we do so primarily to put the spotlight on potential unfairness.

In the context of neuroscience, however, similar issues tend to move below the radar, unnoticed. In fact, I deliberately construed the example with the on and off states and the two judges after data from a recent article in *Science* on "selective modulation of cortical state during spatial attention" [3]. The article is based on neurophysiological data from two monkeys performing a visual discrimination task.

The authors analyzed the on-off dynamics of spiking across cortical layers in visual area V4, while the monkeys had to detect a change in a cued target to either perform an anti-saccadic eye movement when there was a change or maintain fixation on a spot at the center of the screen when there was no change. There was a 50% probability of a change in the cued target.

The data did show an interesting modulation in the on-off dynamics of spiking, but the interpretation was hampered by the problematic analysis. After observing the modulation in the on-off dynamics, the authors aimed to assess whether this modulation correlated with the behavioral performance. They noted: "When the cued orientation change occurred in the RFs [receptive fields] of recorded neurons, the probability of detecting that change was significantly greater when it occurred during an On-phase than during an Off-phase (median detection probability 64.8% during Off-phase, 78.3% during On-phase, difference 13.5%, $p < 10^{-3}$, Wilcoxon signed rank test" (p. 1143).

Looking at the supplementary materials, it becomes clear that this analysis of performance was based solely on trials in which there was a change in the cued target. They offered no data on the relationship between on-off dynamics and the likelihood of a false alarm, that is, responding with an eye movement in the absence of a visual change. As a result, the data are not interpretable at the most basic

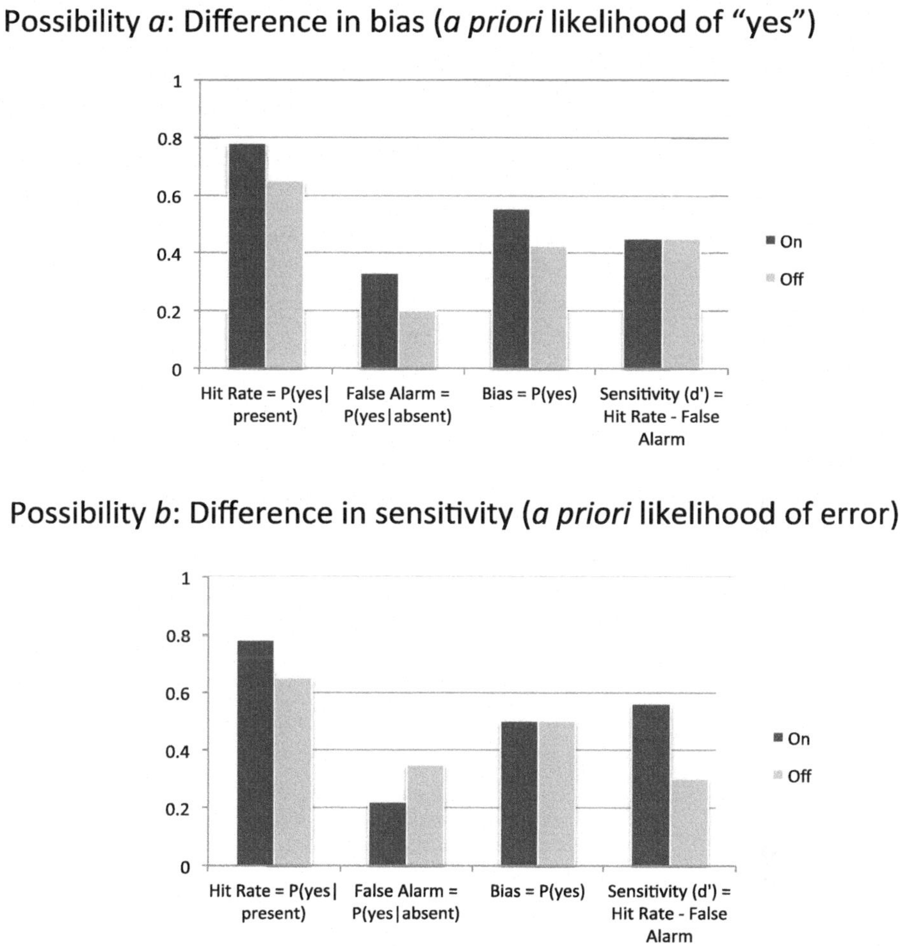

Fig. 50.1 Illustration of different possibilities, using the data reported about the hit rates in the on versus off phase in the study by Engel et al. (2016) [3]

level of performance. It could be that, in the on phase, there is simply a higher a priori likelihood of a behavioral response indicating detection (i.e., the probability of a "yes" response), regardless of whether there was an actual change. Conversely, it is possible that, in the on phase, there is an improved ability to distinguish a change from no change. Figure 50.1 illustrates the different possibilities. Crucially, it is necessary to compare the performance in cases when there was no change (particularly, the likelihood of false detection). Without this information, we cannot conclude that performance was improved in the on phase.

Indeed, the entire paper by Engel et al. (2016) [3] offers a classic example of misleading data, by which effects of selective attention are presented as if they reflect improved information processing (i.e., heightened sensitivity), whereas they could actually be due to bias (i.e., increased likelihood of responding). Throughout their paper, the authors consistently presented an impoverished dichotomy of possibilities when considering the on-off dynamics with respect to cortical state.

The on-off dynamics would reflect "arousal or selective attention" (p. 1140), and attention "involves the selective recruitment of local neuronal populations encoding behaviorally relevant stimuli at one retinotopic location and the simultaneous suppression of populations encoding irrelevant stimuli in other retinotopic locations" (pp. 1140–1141). Added to this statement are references to well-known articles by the Maunsell and Desimone labs, which have consistently shown effects of improved sensitivity under conditions of selective attention. However, between general arousal and specific improvement of sensitivity, we can posit a different mechanism of specific anticipation – *bias*.

In this respect, it is also important to note that the authors focus their analysis predominately on the interval between cue and target, that is, *before* the crucial moment of potential visual change and during an epoch in the trial when the monkey is actively anticipating the next event. Such an anticipation period is exactly when bias mechanisms may be most relevant.

Intriguingly, the authors distinguished between two types of attention conditions, one labeled "covert attention" and the other "overt attention." In the covert attention condition, the monkey was cued to attend a visual target in the receptive field of the recorded neuron (implying that, for an anti-saccadic response, the monkey would have to make an eye movement away from the receptive field). In the overt attention condition, the monkey was cued so that, for an anti-saccadic response, it would have to make an eye movement toward the receptive field. This latter labeling seems inappropriate, because in actuality, throughout the trial, the monkey's overt attention had to be centered on the fixation spot at the center of the screen.

The actual eye movement at the end of the trial should be better understood as an operant response, a conclusive act, not a matter of further, ongoing information processing (as would be implied by "attention").

In fact, both attention conditions must be understood as covert, but one would be aimed at processing visual information (to discriminate a change in the visual target), whereas the other would be aimed at preparing a motor response (to make an anti-saccadic eye movement). Perhaps the "overt attention" condition would be better labeled as the "covert motor attention" condition (as opposed to the "covert visual attention" condition).

Interestingly, the authors noted that both attention conditions produced notable on-off dynamics, more so than control conditions (in which neither the visual discrimination target nor the target for the anti-saccadic eye movement was in the receptive field of the recorded neuron).

The authors further noted that there was modulation of both the on and off phases in the covert visual attention condition but only modulation of the on phases in the alternative attention condition (the covert motor attention).

This raises the possibility that there were two parallel mechanisms at work: one mechanism (e.g., a bias mechanism) that leads to boosting in the on phases, without affecting the off phases, and could be related to a heightened likelihood of responding, most clearly seen to operate in the covert motor attention condition; and another mechanism (e.g., a sensitivity mechanism) that operates through on as well as off phases to improve the signal-to-noise of information processing.

In this respect, it would be important to analyze the relationship between the on-off dynamics and signal-detection measures, in both attention conditions. Presently, the authors showed only a general measure of detection probability based on change trials and this only in a comparison between the covert visual attention condition and the control condition.

Instead, we would learn much more about the underlying mechanisms if the authors showed d-prime and criterion measures (based on change and no-change trials) for all conditions. It may well be that one mechanism, especially operating in the covert motor attention condition, affects the criterion in signal detection, whereas another mechanism, especially operating in the covert visual attention condition, affects the d-prime in signal detection.

Currently, the analysis in the Engel et al. (2016) [3] paper does disservice to the important data collected by the authors. As a result, the paper presents premature, misleading statements about selective modulation of cortical state, as if it reflects heightened sensitivity to visual changes, whereas in reality the data could be due to bias, with a change in the likelihood of responding. Alternatively, there might be multiple mechanisms at work, which could effectively, and usefully, be separated, merely by improving the analysis.

50.3 Case Study Two: Fiser et al., *Nature Neuroscience*, 2016

In other research, the issues of bias and sensitivity are overlooked in the analysis of the information properties of neuronal activity, when there is effectively no opportunity to estimate a correlation with behavior. As an example, I focus on an important recent study on "experience-dependent spatial expectations in mouse visual cortex [4]." In the abstract of this research paper by Fiser et al. (2016) [4], it is boldly asserted that "the activity of neurons in layer 2/3 of mouse primary visual cortex (V1) becomes increasingly informative of spatial location" (p. 1658). In this study, mice repeatedly ran through a corridor on the way to a liquid reward. In the corridor, there were five visual landmarks. The study focuses on the changes in neuronal activity in response to these landmarks. While it was not possible to connect the changes in neuronal activity to behavioral performance in this paper, there were evident influences from experience in the neurodynamics. Unfortunately, however, the authors missed the opportunity to characterize the activity fully with respect to types of information processing involved.

As comprehensively argued elsewhere [1], the neuronal activity with respect to sensory information in different contexts (e.g., tasks or level of expectancy) can be distinguished in terms of the temporal profile as well as in terms of the signal-to-noise ratio. Particularly, the concept of bias matches well with prospective mechanisms (e.g., changes to the baseline activity before the appearance of relevant

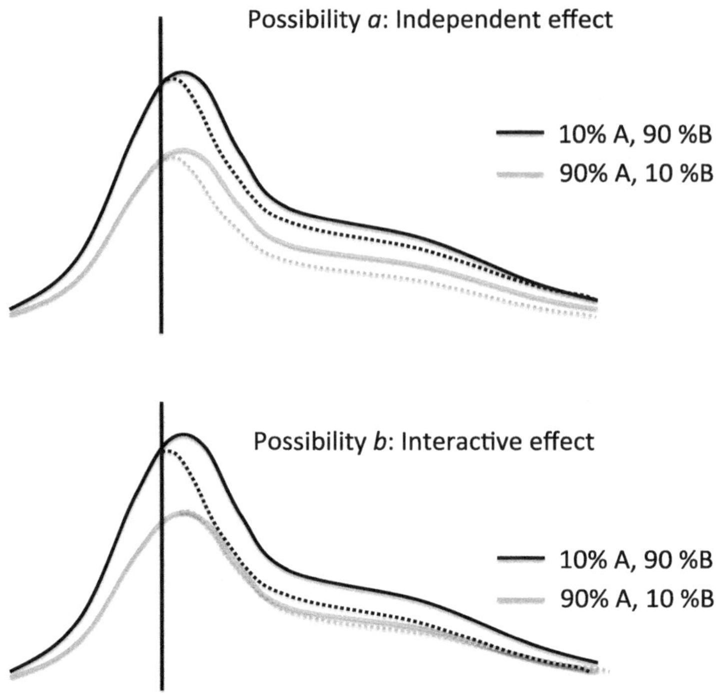

Fig. 50.2 Illustration of different possibilities, using the data reported on the response to visual landmarks by predictive neurons in the study by Fiser et al. (2016) [4]

information) that tend to be independent from the following sensory processing, leading to additive or linear effects in the neuronal activity. For instance, if a neuron is tuned to one particular sensory feature (firing more for stimulus A than for stimulus B), then this difference in firing may hold irrespective of the changes in the baseline (see Fig. 50.2, possibility *a*). Such a bias effect in neuronal activity is equivalent to setting an a priori likelihood of responding (or probability of responding one way rather than another, irrespective of the actual incoming signal).

Conversely, the concept of sensitivity is compatible with synergistic mechanisms (i.e., changes to the stimulus-related activity) that tend to interact with sensory processing, producing multiplicative or nonlinear effects in the neuronal activity. For instance, if a neuron is tuned to one particular sensory feature (firing more for stimulus A than for stimulus B), then this difference in firing may be seen predominately in one task context rather than another (see Fig. 50.2, possibility *b*). Such a sensitivity effect in neuronal activity is equivalent to a change in the quality of information processing (typically, a heightened sensitivity).

To the credit of the authors, an effort was made to distinguish predictive neurons from visual neurons by considering the temporal profile of the neuronal activity. Unfortunately, this was not followed up with an analysis of the information properties of these neurons. Particularly, the authors checked only the visual responses to the neurons' preferred features (analogous to checking only the hit rates, or correct yes responses), without considering the visual responses to the neurons' non-preferred features (analogous to ignoring the false alarms, or the incorrect yes responses). Consequently, we cannot check whether the neuronal activity showed a selective improvement of information processing or a boost of activity in advance of stimulus presentation without an improvement in the signal-to-noise ratio. Yet, the authors go on to proclaiming prematurely that there was an improvement of information processing.

Again, as in the paper by Engel et al. (2016) [3], the analysis by Fiser et al. (2016) [4] fails to make appropriate use of the important data collected by the authors. As a result, the paper presents misleading statements about the effects of expectation in visual neurons, as if it reflects heightened sensitivity to visual information, whereas in reality the data could be due to bias, with a change in the likelihood of responding. Alternatively, there might be multiple mechanisms at work, which could effectively, and usefully, be separated, merely by improving the analysis.

References

1. Lauwereyns, J.: The Anatomy of Bias: how Neural Circuits Weigh the Options. The MIT Press, Cambridge, MA (2010)
2. Lauwereyns, J.: Brain and the Gaze: On the Active Boundaries of Vision. The MIT Press, Cambridge, MA (2012)
3. Engel, T.A., Steinmetz, N.A., Gieselmann, M.A., Thiele, A., Moore, T., Boahen, K.: Selective modulation of cortical state during spatial attention. Science. **354**, 1140–1144 (2016)
4. Fiser, A., Mahringer, D., Oyibo, H.K., Petersen, A.V., Leinweber, M., Keller, G.B.: Experience-dependent spatial expectations in mouse visual cortex. Nat. Neurosci. **19**, 1658–1664 (2016)

Chapter 51
Mindful Education and the Kyoto School: Contemplative Pedagogy, Enactivism, and the Philosophy of Nothingness

Anton Luis Sevilla

Abstract This presentation is a dialogue between eastern philosophy (the Kyoto School of Philosophy) and western cognitive science (particularly psychology and neuroscience) on the topic of education. As a discussion point, I take up contemplative pedagogy/mindful education, which is a movement that attempts to integrate contemplative practices into the process of education. I will examine four points of connection between this movement and the ideas of the Kyoto School. The four points concern the idea of mindfulness itself, the subject-object relationship, well-being, and creativity. To represent the Kyoto School, I will focus on Nishida Kitarô's idea of pure experience and Nishitani Keiji's idea of the standpoint of emptiness. Additionally, I will try to bring these ideas into dialogue with the enactivist view of Francisco Varela (who directly cites Nishitani's philosophy in *The Embodied Mind*).

Keywords Japanese philosophy · Emptiness · Pure experience · Philosophy of education

51.1 Introduction

How can we connect the discourses of mindful education and the philosophy of emptiness of the Kyoto School of Philosophy in a way that mutually benefits research in both fields?

The names "mindful education," "contemplative pedagogy," and "contemplative education" all refer to educational movement that aims to introduce mindfulness practices (practices to develop nonjudgmental, open, and creative attention) to education and studies the methods, objectives, and effects of these practices. There are many streams of this movement—those that trace to Jon Kabat-Zinn's

A. L. Sevilla (✉)
Faculty of Arts and Science and Graduate School of Education, Kyushu University, Fukuoka, Japan

J. M. Delgado-García et al. (eds.), *Advances in Cognitive Neurodynamics (VI)*, Advances in Cognitive Neurodynamics,
https://doi.org/10.1007/978-981-10-8854-4_51

mindfulness-based stress reduction (MBSR), those that trace to Ellen J. Langer's mindful learning, those that are tied to various schools of holistic education, and those tied to organized religions like Buddhist education. While keeping this plurality in mind, I will focus on the first two streams.

As research on mindful education grows, it is beginning to encounter philosophical issues concerning its view of the learner, education, its relationship with religion, ethics, politics, etc. Because of this, it is helpful to articulate a philosophical framework by which this movement can be assessed (see attempts by Orr (2014) [1] and Brown and Cordon (2009) [2]). This paper attempts to contribute to this endeavor by trying to ground mindful education in the ideas of the Kyoto School of Philosophy.

The Kyoto School of Philosophy is a school of modern Japanese philosophy that, from its inception, integrated western philosophy with eastern meditative practices and ideas. It was founded by Nishida Kitarô (1870–1945), whose first book, *An Inquiry into the Good* (1911), contains a view of experience and learning that, like MBSR, is grounded in the experience of Buddhist meditation. Furthermore, his leading disciple Nishitani Keiji (1900–1990) developed a view of "the standpoint of emptiness" that can be connected to both the MBSR-based and Langer-based definitions of mindfulness. Additionally, I will attempt to *provisionally* connect these ideas to the enactivism of Francisco Varela, in hopes of strengthening the connection between the neuroscientific, psychological, and philosophical discourses.

This grounding serves a second purpose. There have been growing attempts in both Japanese and English to develop the philosophy of education of the Kyoto School (see Sevilla (2016) [3] for a summary). There are many western ideas that can link up with those of the Kyoto School—Herbart, Bollnow, Dewey—and one very promising possibility is the connection with mindful education.

To explore this two-fold purpose, I shall focus on four points of continuity between mindful education and the Kyoto School, before proceeding to note some differences and areas of continued research.

51.2 Mindful Education and Emptiness

I suggest that there are four main points of contact between mindful education and the Kyoto School. First, there are similarities between the idea of mindfulness itself and the standpoints offered by Nishida and Nishitani. Second, there are similarities in the way they view the learning that occurs between subject and object. Third, they make the connection between learning, emotion, and overall well-being. Fourth, these views allow for an escape from routine and the possibility of creative problem solving due to an appreciation of contradictoriness.

51.2.1 Mindfulness and Nothingness

While the definitive meaning of "mindfulness" is up for debate, [4] we can begin with a simple definition by Jon Kabat-Zinn: "Mindfulness is the awareness that arises by paying attention on purpose, in the present moment, and non-judgmentally to things as they are" [5]. This is cultivated through various practices (or meditations), the most common of which is "mindfulness of breath," where one focuses on the feeling of breathing (in the belly, chest, or nose), without controlling the breath in any way [6]. This meditation can also be done on the feelings in the body, emotions, etc. The result of such meditation is a sense of being in the present moment and being one with experience (rather than separated from it via constant judgment) without being attached to it (a refusal of the experiential truth of change).

The very idea of "mindfulness" is heavily influenced by Buddhism, particularly the *vipassanā* practices of Theravada Buddhism. In the Kyoto School, similar meditative practices in Zen Buddhism influenced Nishida Kitarô's ideal of "pure experience."

Nishida Kitarô's idea of pure experience refers to experience that is prior to the separation between subject and object, prior to the abstraction of thought from feeling and from willing. In this "experience," it is not an "I" experiencing an "object," but self and world emerging from pure experience. Like mindfulness, pure experience sheds the judgments by which an individual tries to control an external reality and returns to a more foundational relationship between the I and the world, where there is a sense of unity without attachment.

Heisig (2001) [7] notes that despite similarities to William James' idea of pure experience, Nishida goes beyond James by insisting that this pure experience is not merely the foundation of subjective consciousness but is the foundation of objective reality itself. This brings Nishida close to philosopher and neuroscientist Francisco Varela (1946–2001), whose idea of "enactivism" was primarily a critique of the idea that there is a mind and there is a world and the former represents the latter through cognition [8]. He writes, "We propose as a name the term *enactive* to emphasize the growing conviction that cognition is not the representation of a pregiven world by a pregiven mind but is rather the enactment of a world and a mind on the basis of a history of the variety of actions that a being in the world performs" (9). He includes within this view thinkers like Walter J. Freeman, Richard Rorty, Jean Piaget, and Jerome Bruner.

This notion of "no-mind" (or no pregiven mind) and the metaphysical unity of subject and object might seem to take us beyond mindfulness, for mindfulness does not have these metaphysical presumptions. But when we examine mindful *education*, we find that things might be much closer to the ontologies that have been discussed.

51.2.2 Unity of Subject and Object

In mindful education, mindfulness practices are integrated into learning processes in various ways. For example, students might meditate at the beginning of a period, before any lessons take place. Or students might watch a film on social injustice and then meditate on their feelings after, before discussing what they watched. Another possibility is that students may read texts or data graphs with a meditative (rather than analytic) mindset, before analyzing these.

The aims of these practices are varied, but Barbezat and Bush (11) see four main objectives: building focus and attention, developing a contemplative and introspective approach to course material, cultivating compassion, and improving self-understanding, creativity, and insight. Allow me to briefly explain these. First, the repeated meditative practice of paying attention (to one's breath or feelings) improves the ability of students to sustain focus on the sensory and cognitive material presented in class. This is clearly connected to improved learning in all its forms. Second, students learn about a lesson *in connection* with their own experience, reducing the sense of alienation from subject matter, and possibly increasing their sense of involvement. Third, meditative practices have an interpersonal component (I leave this to a further paper). And fourth, these all come together to allow unique, personal approaches to the lesson, allowing for the creation of new knowledge.

The capacity of mindful education to achieve these objectives has been studied via empirical research. This research demonstrates the effects of these practices on awareness, emotions, prosocial attitudes, and even changes in the brain (prefrontal cortex and right anterior insula) (see Barbezat and Bush (2014) [6], 24–38; and Kaszniak (2014) [9]).

At the core of how this "works" is a shift in the relationship between subject and object. When one begins a class with meditation, or deliberately adopts an open, nonjudgmental attitude when examining a lesson or data set, this alters how students perceive the lesson. Instead of seeing a lesson as something external to the self, to be solved and mastered by the self, the lesson is seen as something that is inseparable from the self. For example, when reading a poem, the different emotions and thoughts that rise up while reading are part of the self-unfolding of the poem. The same can hold for data from an economic survey. Because of this connection between experience and the lesson, the lesson seems much more relevant to the learner. As the student learns about the lesson, the student is also learning *through* the self and *about* the self. And conversely, as the student learns about the self, the student is learning through the lesson and about the lesson. This allows for creativity and insight into the subject matter—in a manner very different from an "analytic" or "objective" approach.

Zajonc's (2009) [10] idea of "contemplative inquiry" is one attempt to theorize this mode of learning. By first opening oneself to what one is learning, breaking one's reactive ways of interpreting what one experiences, and taking a perspective of open (rather than focused) awareness, one is able to approach the object via an "epistemology of love"—where, rather than knowledge as a control of the subject over the object, one has a delicate respect, an intimate, participatory, and vulnerable *union with the object.*

While this clearly leads to a better self-understanding and engagement, does this lead to a better understanding of the *subject matter*? In other words, does this way of knowing better correspond to the ontological nature of the world? I argue that by integrating mindfulness into education, this movement is *suggesting* that mindfulness is not merely a positive psychological state but is *truer* to reality and how it is disclosed. This brings it much closer to Nishida's ontology than merely clinical versions of mindfulness. Nishida saw pure experience as the foundation not just of consciousness but of reality itself. Thus, a return to pure experience allows not only for self-understanding but as a most intimate understanding of reality.

Nishitani (1991) [11] points out in *Nishida Kitarô* that this approach was a novel response to the battle of modernity: Do we know reality as it is by ideas/reason internal to the knower? Or do we know it through scientific observation of what is *outside*? (74–79). This philosophical problem is mirrored in philosophy of education. As Gutek (1997) [12] writes, the former view suggests a platonic, humanities-based education, which, while able to have a strong sense of connection between learner and learned, tends to fall into dogmatism with the lack of a clear external reality. The latter view suggests a more Aristotelian, scientific education, which while critical, tends to alienate the subject from the object.

Nishida refuses both these options. In contrast to the metaphysical/religious approach, Nishida's "pure experience" refuses all dogmatic presuppositions and takes its standpoint in experience. This is akin to the scientific approach, but it goes deeper than the dualistic approach of science to reality, wherein the observer is separate from the observed. Nishida (3–4) writes, "When one directly experiences one's own state of consciousness, there is not yet a subject or an object, and knowing and its object are completely unified."

This gives learning a very different character. In a move very similar to Zajonc, Nishida (174) writes, "To love a flower is to unite with the flower, to love the moon is to unite with the moon... In this way, the [sic] knowledge and love are the same mental activity; to know a thing we must love it, and to love a thing we must know it."

The similarity of these two ways of thinking suggests the radical ontological implications of mindful education as well as the practical possibilities of pure experience.

51.2.3 Well-Being and the Unity of the Self

The previous point has deep implications. But it is possible that the widespread acclaim of the mindfulness movement has more practical reasons: mindfulness practices have been shown to be quite effective in a number of clinical settings at addressing problems like depression, anxiety, addiction, etc. [see Barbezat and

Bush (2014) [6], 27–29). Mindful education thus promises to improve emotional well-being in schools, reducing incidence of violence and self-harm.

Why does mindfulness improve well-being? There are two seemingly contradictory answers to this: first, improved awareness of one's inner thoughts and feelings allows one to better understand oneself and respond to one's own needs. Educationally, connecting learning to the fullness of experience reduces the sense of alienation (see Barbezat and Bush (2014) [6] and Langer (1997) [13]). Second, as Williams et al. suggest, while mindfulness practices increase one's awareness of one's feelings, that awareness is not a uniting with feelings but rather a dis-identification from them. I see my feelings and I know that they come and they go, while awareness remains. Thus, I am not unduly perturbed by my feelings.

This seeming contradiction connects to Nishida and Nishitani. Seen from Nishida, in the ordinary learning experience, subject and object are dualistically separated; intellect, affect, and volition are fragmented, and the self has no unity. This presents an ethical problem (of alienation). Nishida writes, "The good refers to that which satisfies the internal demands of the self. Because the greatest demands of the self—that is, the demands of personality—are the fundamental unifying power of consciousness, to satisfy these demands and thereby actualize personality is for us the absolute good" (132). The return to pure experience, becoming one with the object of learning, does not merely result in a deeper form of learning that is open to scientific rigor and insight. It results in a return to the unity of consciousness, healing the brokenness of the self, and realizing what Nishida calls our "True Self."

Nishitani has a slightly different approach. In *Religion and Nothingness* (1961), he refers to ordinary experience as "the standpoint of consciousness," which is characterized by suffering because it forces things into set standpoints (through cognition, affect, and volition). By going through crisis (the standpoint of nihility), it arrives at "the standpoint of emptiness," wherein things are seen as both A and not-A—that is, cognitive, affective, and volitional grasping are relativized and freed from their compulsion [14]. Interestingly, Varela et al. (1993) [8] summarizes Nishitani's views as well, see 241–245.

Nishida's views are similar to the view in mindfulness of becoming more "one" with one's embodiment and feelings, whereas Nishitani's views are similar to the view of mindfulness as dis-identifying from internal phenomena through a broader, nonjudgmental awareness. However, if one considers Nishitani's own discussions of his mentor's ideas (see Nishitani (1991), 116, 118), one sees that the two thinkers are simply articulating two ways of approaching the same standpoint, suggesting a unity between the two seemingly separate aspects of mindfulness. Dis-identification is merely a way of undoing a twisted way of unity (compulsiveness with feelings)—the end product is a unity in difference where one feels openly but is not attached to feelings.

51.2.4 Creative Mindfulness

The connection between Nishida and Nishitani brings us to a possible connection between the Kabat-Zinn view of mindfulness (which we have been discussing at length) and a more "active" view of mindfulness from Ellen Langer. Langer critiques "learned mindlessness," arguing that a main problem in education is its tendency to instill damaging cognitive habits—presuming that there is only one answer to questions, presuming that tasks always come with the needed information to solve them, and presuming that there is only one way to approach a problem. Thus, mindfulness contributes by promoting "the creation of new categories, openness to novelty, awareness that there is more than one perspective ... alertness to distinctions, and orientation in the present" (see Reber (2014) [15] for a summary).

This view is often seen as distinct from Kabat-Zinn's approach. It is less focused on clinical improvements, less focused on passive reception of experience, less focused on meditation, more actively creative. But with Nishitani's view above, we can see one possibility of how these two approaches can link up more clearly.

First, in Nishitani's view, one takes a more active approach to "negating" ordinary dualistic experience. Rather than merely "returning to pure experience," the foundations of the representational ego must be shaken through exposure to what representation cannot fathom. In Rinzai Zen Buddhism, this is usually carried out through koans (e.g., Hakuin's "What is the sound of one hand clapping?"). These drive the ego from the standpoint of consciousness to the standpoint of nihility, where its entire being becomes a question.

Pedagogically, this is similar to the practical suggestions of Langer—to make statements conditional, offer opposing evidence, and literally muddle one's teaching. While this may increase cognitive load on the short term (see Reber), in the long term this improves flexibility of thinking. This view is not opposed to the more meditative approach of Kabat-Zinn. But as we see in Rinzai Zen Buddhism, koan practice (active negation) and meditation (passive return to oneness) work together toward the same goal of awakening to unity with reality.

Second, Nishitani clarifies the relationship between nothingness (non-knowing, passivity) and being (knowing, active creation of categories and distinctions). The standpoint of emptiness is not a denial of knowing. But by momentarily denying knowing, one is able to attain a deeper standpoint that includes both knowing and not-knowing. Nishitani (124) writes, "On the field of emptiness, however, the selfness of a thing cannot be expressed simply in terms of its 'being one thing or another'. It is rather disclosed precisely as something that cannot be so expressed.... Should we be forced to put it into words all the same, we can only express it in terms of a paradox, such as: 'It is *not* this thing or that, therefore it *is* this thing or that. / Being is only being if it is one with emptiness." To put it simply, emptiness (non-knowing) includes being (knowing) without being reducible to its pre-existing, one-sided manifestations, allowing emptiness to be the ground for the active creation of new ideas.

To put it in Nishida's terms, pure experience is not merely prior to all judgments; it is also the womb from which all distinctions are born (see Nishida, ch. 11). This view can strongly tie together the meditative mindfulness of Kabat-Zinn and the creative, koan-like mindfulness of Langer.

51.3 Conclusion

As we have seen above, there are many productive points of contact between mindful education and the Kyoto School of Philosophy's ideas. Mindfulness is similar to the ideas of pure experience and the standpoint of emptiness. Applying this standpoint to education and learning suggests more than a mere therapeutic method but an *ontology* where subject and object are one—similar to Nishida, Nishitani, and Varela. Furthermore, mindful education and the Kyoto School build well-being by both unifying and dis-identifying the self. And finally, Langer's and Kabat-Zinn's mindfulness can be connected via Nishitani's view of how the self transforms in its relationship to learning and the truth.

However, needless to say, there are differences between these ideas. Mindfulness seems to focus very much on the normative practice rather than the foundations of knowing, which is why the ontological premises of mindful education are merely suggestions in the background of practice. Furthermore, mindful education does not suggest a total transformation of the person akin to Buddhist enlightenment—which is strongly suggested, especially by Nishitani. The focus is more on gradual changes in the present and near future. Finally, mindfulness focuses on an *intentional* awareness of experience, but pure experience is closer to "no-mind," where the awareness and nonjudgment of experience is no longer intentional but completely internalized.

There is much room for continued research in this direction. To what extent does mindful education require this radical ontology? How does it relate with deeper deconstructions of the concept of subjectivity/self? To what extent might it connect to a mindfulness of "no-mind"?

Furthermore, these questions expand to a broader range of philosophical issues: Is mindfulness religiously neutral, and can it be used in public education? Mindfulness also has a host of claims for building compassion and prosocial behavior—are these intrinsically connected to mindfulness, or not? Barbezat and Bush suggest this, and so does Varela—with direct reference to Nishitani's view of how emptiness ties to compassion. Furthermore, how does this compassion relate to culturally bound forms of ethics? The Kyoto School of Philosophy dealt extensively with these questions, not only through the work of Nishida and Nishitani but also (and perhaps more so) in the work of Tanabe Hajime, Hisamatsu Shin'ichi, and Watsuji Tetsurô. Perhaps a continued dialogue could continue to be mutually beneficial to both mindfulness theory and the philosophy of the Kyoto School.

Acknowledgments This paper was written with the aid of funding from Japan Society for the Promotion of Science (JSPS) Grants-in-Aid for Scientific Research (Project Number: 17K13988).

References

1. Orr, D.: A philosophical framework for contemplative education. In: Gunnlaugson, O., et al. (eds.) Contemplative Learning and Inquiry across Disciplines, pp. 31–50. State University of New York Press, Albany (2014)
2. Brown, K.W., Cordon, S.: Toward a phenomenology of mindfulness: subjective experience and emotional correlates. In: Didonna, F. (ed.) Clinical Handbook of Mindfulness, pp. 59–81. Springer, New York (2009)
3. Sevilla, A.L.: Education and empty relationality: thoughts on education and the Kyoto School of Philosophy. J. Philos. Educ. **50**, 639–654 (2016)
4. McCown, D.: East meets west in the pedagogy of the mindfulness-based interventions. In: Ie, A., Ngnoumen, C., Langer, E. (eds.) The Wiley Blackwell Handbook of Mindfulness, pp. 1085–1104. Wiley, Chichester (2014)
5. Williams, M., et al.: The Mindful Way Through Depression: Freeing Yourself from Chronic Unhappiness. The Guilford Press, New York (2007)
6. Barbezat, D., Bush, M.: Contemplative Practices in Higher Education: Powerful Methods to Transform Teaching and Learning. Jossey-Bass, San Francisco (2014)
7. Heisig, J.: Philosophers of Nothingness: An Essay on the Kyoto School. University of Hawaii Press, Honolulu (2001)
8. Varela, F., Thompson, E., Rosch, E.: The Embodied Mind: Cognitive Science and Human Experience. MIT Press, Cambridge, MA (1993)
9. Kaszniak, A.: Contemplative pedagogy: perspectives from cognitive and affective science. In: Gunnlaugson, O., et al. (eds.) Contemplative Learning and Inquiry across Disciplines, pp. 197–211. State University of New York Press, Albany (2014)
10. Zajonc, A.: Meditation as Contemplative Inquiry: When Knowing Becomes Love. Lindisfarne Books, Great Barrington (2009)
11. Nishitani, K.: Nishida Kitarô. University of California Press, Berkeley (1991)
12. Gutek, G.: Philosophical and Ideological Perspectives on Education. Allyn & Bacon, Needham Heights, MA (1997)
13. Langer, E.J.: The Power of Mindful Learning. Perseus Books, Cambridge, MA (1997)
14. Nishitani, K.: Religion and Nothingness. University of California Press, Berkeley (1983)
15. Reber, R.: Mindfulness in education. In: Ie, A., Ngnoumen, C., Langer, E. (eds.) The Wiley Blackwell Handbook of Mindfulness, pp. 1054–1070. Wiley, Chichester (2014)

Printed by Printforce, the Netherlands